Human Populations,

Genetic Variation,
and
Evolution

Human Populations, Genetic Variation, and Evolution

Laura Newell Morris

UNIVERSITY OF WASHINGTON
REGIONAL PRIMATE RESEARCH CENTER

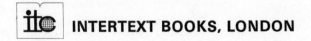 INTERTEXT BOOKS, LONDON

Published by
International Textbook Company Limited
158 Buckingham Palace Road, London, SW1 W9TR

© International Textbook Company Limited 1972

First published by Chandler Publishing
Company, an Intext publisher, USA.

This edition published 1972
ISBN 0 7002 0173 4

To Arval,

Source and Critic

Contents

Preface

The central topic of this book of readings is the study of the evolution of human populations from the point of view of genetics. The book results from my experience in teaching a course entitled "Human Population Genetics" in the Department of Anthropology, University of Washington. Because no text existed which could have been used by physical anthropologists as opposed to mathematically oriented population geneticists, the class assignments were drawn from various pertinent articles which were scattered in the physical anthropology, genetics, and medical literature. This book is a compilation of the types of readings that proved most useful in the course.

The primary criteria for the selection of articles were relevance and readability. In each section I tried to include at least one paper by a person considered to be a recognized authority on the subject in question; other articles were selected because they illustrate a specific concept well. Each set of selections is preceded by an "overview" designed to guide the reader into the materials. A certain amount of background information is provided, with special attention given to the theoretical or methodological problems encountered in human-population research. The overviews are lengthier than introductions usually found in volumes of collected readings, and to the sophisticated reader they may appear superfluous. However, I have written them in the knowledge gained from teaching that it is always helpful, and often necessary, to provide students with a map of the terrain before expecting them to find their way through the complexities of data and conclusions. But the readings themselves are the core of this book. They represent the thoughts and research of the people eminent in the field, and no amount of abstracting or rephrasing can ever do justice to the original work. Only by careful reading and analysis of the included articles will the student gain a proper understanding and appreciation of the field of human-population genetics. It should be emphasized, however, that both the overviews and the included papers are meant to address a reader with previous knowledge of the elements of genetics and statistics. The brief glossary provides some basic terminology, but students not possessing some background may find this book difficult to use.

My deepest gratitude goes to all my former students who have suffered through the growing pains of a new course in the process of development. They have provided immeasurable inspiration and help. My thanks go to Derek F. Roberts, who initiated me into the field of human-population genetics, and who has provided innumerable helpful suggestions for this book. Further, I wish to thank the authors of the articles and the editors and publishers of the journals who have given permission to reproduce the materials included here. I am indebted and grateful for the atmosphere conducive to scholarly research and writing provided by Drs. Theodore C. Ruch and Orville A. Smith, Regional Primate Research Center, University of Washington. Finally, thanks to my typist, Mrs. Ann Gibbs.

Laura Newell Morris
Seattle, Washington

Human Populations,
Genetic Variation,
and
Evolution

CHAPTER 1

Introduction
to Population Genetics

Overview

"Population genetics," it has been said, "assumes the existence of mechanisms for heredity and variation and inquires into the ways in which the genetic makeup of the population is altered or is held in equilibrium by the multiple influences of selection, migration and breeding structure" (Crow, 1961).

The assumption that there are "mechanisms for heredity and variation" is the premise that is basic to the general science of *genetics*. It is difficult, however, to talk meaningfully about genetics per se because there has been such a rapid proliferation of knowledge and subdiscipline specialization. Nevertheless, even though they are expanding at an increasing rate, the subdisciplines that are included within the general field of genetics continue to share a common body of theory and many methods of collecting and analyzing data. These common factors give coherence to the general field of genetics and to its various subdivisions. One such subdivision is the field of *human genetics*, which itself is further subdivided into the areas of medical genetics, biochemical genetics, cytogenetics, somatic cell genetics, immunogenetics, formal (mathematical) genetics, and, the subject of this book, *population genetics*. In practice, however, these areas of human genetics are not clearly demarcated, but are closely interrelated and interdependent. For example, the study of the distribution and evolution of the abnormal hemoglobins in human populations has witnessed the union of medical genetics, biochemical genetics, formal genetics, and population genetics.

1

The basic premise of all the subdivisions of genetics—that there are "mechanisms for heredity and variation"—is essential to the study of evolution, for evolution necessarily may proceed only in the presence of heredity which provides for genetic continuity between generations. Therefore, the hereditary mechanisms which involve gene behavior during individual reproduction are basic to the operation of evolution, but evolution is a holistically oriented process, involving as it does gene behavior within a total population. *Evolution* is defined, therefore, as *the change that occurs through time in the genetic composition of a population.*

The study of evolution in human populations is focused on the so-called four *forces of evolution* which operate to bring about genetic changes. The four forces are *mutation, natural selection, genetic drift,*[1] and *gene flow.* In summary, inquiry "into the ways in which the genetic makeup of the population is altered or is held in equilibrium" by these multiple forces constitutes the study of human population genetics.

The Mendelian Laws

The mechanisms for heredity and variation were first set forth by the Augustinian monk Gregor Mendel in 1865. In a paper presented at a meeting of the Brünn Society for the Study of Natural Science and published the following year in the *Proceedings* of the Society, Mendel reported the results of his hybridization experiments with the common garden pea. This paper may well be considered the original treatise on the theory of the *gene*, as the basic unit of heredity was later named. Here for the first time the particulate, or unblending, nature of heredity was demonstrated. Mendel based his conclusions on seven years of careful observations and statistical analyses of the behavior of seven pairs of "differentiating characters" through successive generations of peas. From this work came the so-called "Mendelian Laws" of the *independent segregation* and *recombination* of dominant and recessive characters. These Mendelian Laws constitute the cornerstone of the modern science of genetics, but a cornerstone which was not publicly laid until thirty-five years after Mendel's original experiments. During those years his work went totally unnoticed by the scientific community, and not until 1900 did three botanists independently "rediscover" Mendel's paper and verify his conclusions in their own experiments (de Vries, 1900; Correns, 1900; Von Tschermak, 1900). Recognition of the mechanisms of heredity led to a rapid growth of knowledge, and within the next two years Mendel's Laws were extended to many species of plants and animals, including man. Thus began the development of a new science, which in 1906 was given the name *genetics* by William Bateson (Stern, 1968).

---

[1]Genetic drift would perhaps be better labeled as a "process" rather than a "force" of evolution in that it is random in its operation and cannot be strictly said to exert force per se. However, for convenience it is included here with the other so-called "forces of evolution."

The neglect of Mendel's work by the scientific community is of particular significance because Charles Darwin, a contemporary of Mendel, remained ignorant of the latter's important conclusions about the mechanisms of heredity. In 1859, seven years before the publication of Mendel's paper, Darwin had proposed his famous theory of natural selection and evolution in the *Origin of Species*. Here he presented impressive data to argue for the evolutionary role of natural selection as it operates on the variability present in all species. Yet, as the geneticist T. Dobzhansky (1965) has suggested, Darwin's theory of natural selection was a "colossus with feet of clay." For Darwin had postulated species variation as the fountainhead of evolution, but he was unable to explain the mechanisms accounting for either the origin or the maintenance of the variation that he had observed in all organisms. As did many other scientists of that day, Darwin advocated a blending theory of inheritance, which stated that any offspring is a blend or mixture of the characteristics of both parents. Thus Darwin was caught in a contradiction between two of his ideas: species variation and the blending theory of inheritance. For, if the blending theory of inheritance were true, then generations of intermating with blending would have produced homogeneous species within which all variation would have disappeared. In brief, maintenance of variation would not be possible. Darwin was aware of this contradiction in his work but was unable to resolve it. One then wonders what impact Mendel's work would have had on Darwin and his fellow scientists if they had been aware of it. Certainly the Mendelian Laws would have resolved one problem inherent in Darwin's ideas by explaining how progeny may differ from their parents, and how variants in a species may retain their distinctiveness generation after generation.

Conceivably, the scientific atmosphere of the times was not ready for Mendelism. In 1865 chromosomes had not yet been discovered, and without an understanding of chromosomes Mendel was unable to demonstrate either the underlying factors that are responsible for the external hereditary "characters" he had observed, or the nature of the biological mechanism that is involved in segregation and recombination. However, during the thirty-five-year "non-Mendelian period" important research was done, especially in the field of cytology, which provided scientists with basic knowledge about the cell and its behavior during meiosis and mitosis. This information was essential to the subsequent development of genetics, and in 1900 the scientific world was ripe for a restatement of Mendelian genetics.

The Hardy-Weinberg Law

The Mendelian Laws of independent segregation and recombination also provided the basis for the first mathematical description of populations. Actually, Mendel himself had approached the problem of inheritance, not only by considering the heredity of individuals and their offspring, but by statistically analyzing whole populations. The whole-population approach to heredity was also recognized and carried foward by the American pioneer in Mendelian genetics W. E. Castle (1903). He demonstrated that the Mendelian ratio of 1:2:1 would remain constant in the randomly mating generations following the mating of the heterozygous F_2 individuals from the original cross $AA \times aa$. Yet, like Mendel's work before him, Castle's paper, with all its implications for population genetics, remained unnoticed. Instead, the English mathematician Godfrey Hardy, working independently, formulated the law that was to provide the mathematical cornerstone of population genetics. The story of Hardy's entrance into the field of genetics is an interesting one.

In 1908 the English geneticist Reginald Punnett gave a lecture before The Royal Society of Medicine in London entitled "Mendelism in relation to disease." In his lecture he referred to brachydactyly, or "short-fingeredness," an anomaly of the hand which is inherited as a dominant Mendelian trait. During the discussion of Punnett's paper, a statistician raised a question: Why are dominants not always the most frequent types in any population, when theoretically they should be present in the Mendelian proportion of three dominants to one recessive? The question was a puzzling one, and at the time Punnett was unable to provide a satisfactory answer. He later posed the problem, as a purely mathematical one, to his friend and colleague Godfrey Hardy (Stern, 1965). Hardy's unpretentious paper, published in *Science* in 1908 and reproduced later in this chapter, presents the solution to Punnett's problem. In it Hardy demonstrated mathematically that under certain conditions the numerical distribution of the three genotypes for any trait (pure dominants, heterozygotes, and pure recessives) for a pair of Mendelian characters will remain stable from generation to generation in a population, whatever their initial proportions.

The mathematical theorem which describes this stable distribution of genotypes in a population was known for over twenty-five years as "Hardy's Law." But in 1943, the geneticist Curt Stern pointed out in the brief note to *Science* included here, that a German obstetrician, Wilhelm Weinberg, had independently set forth the equivalent formula in a lecture delivered three months before Hardy's publication. The names of both men are now attached to the population theorem known as *the Hardy-Weinberg Law* or *the Hardy-Weinberg formula*.

Both Hardy and Weinberg dealt with the ideal distribution of *genotypes* in a population. Hardy used the proportions $p : 2q : r$ to designate the three genotypes; Weinberg used the symbols m and n to represent the original pure male and female "representatives" of the genotypes AA and BB, respectively. After randomly crossing these pure genotypes, Weinberg obtained the now familiar formula, which came to be called the Hardy-Weinberg formula:

$$m^2\,AA + 2mn\,AB + n^2\,BB.$$

This theorem is considered a cornerstone of population genetics because it mathematically describes the behavior of genetic traits through time within a specific unit—the *population* (see Chapter II for a discussion of the population as a unit of study).

Actually, the population assumed under the Hardy-Weinberg Law is a unique population. It does not change genetically; that is, it cannot and does not evolve. It is a so-called *ideal population*, which means that within it certain "ideal" conditions must necessarily be fulfilled. The ideal population is a mathematical abstraction, because no real population ever fulfils all of the necessary conditions; that is, the population must be large, the sexes must be equally distributed, mating must be random (*panmictic*), and all parents must be equally fertile. But even these conditions, as originally postulated by Hardy and Weinberg, are not sufficient to constitute a fully ideal population. One further stipulation must be added: The population must be free from the four forces of evolution; that is, mutation, natural selection, genetic drift, and gene flow.

In part, Hardy recognized the unbalancing effects that some of the forces of evolution could exert on a real population, although he did not specifically label the forces as such. By stipulating the necessary conditions for the ideal population, Hardy indirectly excluded his population from being subject to certain forces of evolution. For example, by postulating a size of "fairly large numbers," he implicitly excluded the ideal population from being subject to the sampling effect resulting in *genetic drift*, a process which operates primarily in small populations (see Chapter V). As a mathematician, Hardy was aware of the inverse statistical relationship which exists between the size of the sample drawn and the amount of its deviation from the expected theoretical proportions. In a large population the statistical deviation which occurs in the sample comprised of each new generation is small, but as a result of even this small statistical deviation a new "stable" distribution is established each generation. Yet the population maintains stability through time because, as Hardy pointed out, each new distribution differs "but slightly from the original distribution." With respect to the conditions of sampling, an actual large population and Hardy's ideal population approximate each other. By contrast, in a small population the statistical deviation of each generation may be large enough to effect a significantly different new distribution. In this manner, the genetic composition of a small population may change through time. Although the concept of genetic drift was later fully elaborated by the geneticist Sewall Wright (see Chapter V), Hardy's comments on the operation of statistical sampling effects in a population must be regarded as the first treatment of the concept.

Hardy also pointed out that if a character influenced fertility, the hypothetical population equilibrium would be "greatly complicated." In this way he indirectly recognized and excluded the force of natural selection as it operates through the agent of *differential fertility*. But, nowhere did he

mention, or exclude, the other equally important agent of natural selection, namely *differential mortality* (see Chapter IV). Therefore, to Hardy's
original stipulation that all parents must be equally fertile must be added
the further condition for an ideal population—the survival of all offspring
must be random.

It is obvious that the Hardy-Weinberg ideal population can never be
found in the real world of human populations. Thus the question arises, Of
what possible use is this formula to the population geneticist who has to
work in populations which are not infinitely large and panmictic, and which
are subject to the forces of evolution so that they have evolved in the past
and continue to change genetically? First, the formula provides a standard
against which genetic change in a population may be measured and predicted. In this manner it serves as a basic theorem which can be expanded
and elaborated by other mathematical models that deal with change in
populations. This approach characterizes the work of the formal geneticists R. A. Fisher, Sewall Wright, and J. B. S. Haldane, who, beginning in
the 1920's, elaborated the theoretical mathematical foundations of modern
population genetics. They have enabled geneticists to better deal with the
genetic parameters of actual populations, and it is now possible to mathematically "correct" the Hardy-Weinberg formula for some of the various
factors which remove actual human populations from the ideal.

Second, the Hardy-Weinberg formula may be applied to large populations to provide an estimate of gene frequencies at a single point in time.
The modern population geneticist who is primarily interested in quantitatively describing, comparing, and measuring changes in the genetic composition of populations uses the Hardy-Weinberg formula in this manner.
In order to accomplish these goals, and rather than considering a population solely in terms of the distribution of individual genotypes, as Hardy and
Weinberg did, he analyzes a population from the perspective of the distribution of genes in its *gene pool*. The article included here by William C.
Boyd is based on this type of analysis.

The *gene pool* is that conceptually abstract reservoir of genetic information which contains all the genes carried by a group of interbreeding individuals within their sex cells (gametes). Today, the Hardy-Weinberg Law
is phrased in terms of *alleles* and their *frequencies* in the gene pool of a
population, rather than in terms of genotypes. Weinberg's terms m and n
now refer to alleles in a gene pool rather than to genotypes in a population
as they did originally; in practice the terms p and q are commonly used,
usually referring to the dominant and recessive allele respectively.
Theoretically, the gene pool of a population can be fully described in terms
of the percentage frequencies of the alleles present at each of the total
number of loci carried by man. However, given our lack of knowledge about
the total numbers and types of genetic loci in any individual, a gene pool is
characterized on the basis of very few loci, usually less than twenty.

TABLE I-1. RANDOM-MATING FREQUENCIES

		Male gametes	
		p	q
Female gametes	p	p^2	pq
	q	pq	q^2

The Hardy-Weinberg Law deals with the simplest genetic case, that of a single locus carrying only two alleles, p and q. The manner in which genetic stability is maintained under a two-allele system is best understood if the gene pool is visualized as divided into two component sexual units: one unit containing all the male gametes (spermatozoa), carrying the alleles p and q; the other unit containing all the female gametes (ova) in equal numbers. The relative proportions of p and q are identical between the two sex units. If all the male and female gametes mate randomly, the offspring will be distributed as shown in Table I-1. Whether or not the gene pool is initially in equilibrium, after one generation of random mating, genetic equilibrium at a single locus is established and then perpetuated at the same gene frequencies through subsequent generations (see Table I-2). For purposes of clarity the three possible genotypes are represented by D, H, and R respectively in Table I-2, and

$$D + H + R \rightarrow p^2 + 2pq + q^2$$

The original Hardy-Weinberg formula was applied only to a locus carrying two alleles, but in human-population gene pools many genetic loci carry more than two alleles. The mathematician F. Bernstein, in his popu-

TABLE I-2. ESTABLISHMENT OF EQUILIBRIUM UNDER RANDOM MATING

Type of mating	Frequency of mating	Offspring*		
		D	H	R
$D \times D$	D^2	D^2	—	—
$D \times H$	$2DH$	DH	DH	—
$H \times H$	H^2	$\frac{1}{4}H^2$	$\frac{1}{2}H^2$	$\frac{1}{4}H^2$
$D \times R$	$2DR$	—	$2DR$	—
$H \times R$	$2HR$	—	HR	HR
$R \times R$	R^2	—	—	R^2
Total	1.00	$(D + \frac{1}{2}H)^2$ or p^2	$2(D + \frac{1}{2}H)(\frac{1}{2}H + R)$ or $2pq$	$(\frac{1}{2}H + R)^2$ or q^2

*The expected distribution of offspring from each type of mating is derived from Mendelian probability ratios. Thus, from the mating of two heterozygotes, $pq \times pq$, the offspring will appear theoretically in the Mendelian ratio of 1 : 2 : 1.

Source: Modified from Li, 1963.

TABLE I–3. DISTRIBUTION OF BLOOD GROUPS
UNDER BERNSTEIN'S FORMULA*

Genotype	I^{AA}, I^{AO}	I^{BB}, I^{BO}	I^{AB}	I^{OO}
Blood groups	A	B	AB	O
Proportion	$p^2 + 2pr$	$q^2 + 2qr$	$2pq$	r^2

*Consider the problem of estimating gene frequencies from a random sample of unrelated individuals. The symbols A, B, AB, and O denote both the blood groups and their corresponding proportions in the sample; the symbols p, q, and r denote the gene-pool frequencies of the blood-group A, B, and O respectively. The table shows that

$$A + O = (p + r)^2, \quad B + O = (q + r)^2, \quad O = r^2;$$

hence,

$$p = 1 - \sqrt{(B + O)}, \quad q = 1 - \sqrt{(A + O)}, \quad r = \sqrt{O}.$$

Source: Modified from Li, 1963.

lation-genetic analysis of the frequencies of the blood groups A, B, O, and AB (Bernstein, 1924; 1925), was the first to apply an expanded version of the Hardy-Weinberg formula to a genetic system involving several alleles. On the basis of the blood-group work done previously (Landsteiner, 1901; Hirschfeld and Hirschfeld, 1919), the hypothesis had been set forth that the blood-group phenotypes were determined by two independent Mendelian pairs, one pair consisting of the alleles A and non-A, and the other pair of alleles B and non-B. Applying the Hardy-Weinberg formula, Bernstein compared the blood-group frequencies expected under a two-allele system with the frequencies actually observed. He found consistently significant differences, and concluded that the two-allele hypothesis was incorrect. He then proposed a triple-allele system at a single locus and applied an expanded Hardy-Weinberg formula to the data (Table I–3). In this way he obtained excellent agreement between the observed and expected frequencies. Bernstein's formula is used today for the calculation of gene frequencies at loci involving more than two alleles.

Genetic Methods and Human Populations

Bernstein's work, in conjunction with the Hirschfelds' discovery that the frequencies of the ABO phenotypes differ in various populations, marked the entry of physical anthropology into genetics. Physical anthropologists had long been interested in the description, comparison, and genetic classification (taxonomy) of the groups of mankind. For these purposes they had used the measurement and description of various external physical characteristics of the body such as height, weight, cephalic index, hair form and color, eye color, and the like. It is true that populations could be described on the basis of such traits, but assessing a group's exact genetic relationship was exceedingly difficult. First, methodological problems, such as the errors involved in taking body measurements or the degree of subjectivity involved in classifying hair form, introduce a certain amount of bias into the data of the populations to be compared. Second, because the characters in question are under the control of innumerable genes (*polygenic characters*), and for the most part are extremely sensitive to environmental influences (height and weight differentials between individuals, for example, may be more reflective of nutritional than genetic differences), and because many genotypes cannot be distinguished from each other phenotypically, the statistical methods of studying these characters are very complex.

For a proper genetic analysis and comparison of populations, criteria were needed that could be objectively assessed, that were relatively unaffected by environmental factors, and that ideally were nonadaptive. To the early workers the blood groups seemed to satisfy these criteria completely, and the quantitative genetic analysis, description, and comparison of human populations began. The Hirschfelds' original study of blood-group distribution in racial groups (1918) was followed by the discovery of the MN blood groups in 1927 (Landsteiner and Levine, 1927), and, in 1939 to 1941, that of the Rh blood-group system and its relation to disease (Levine and Stetson, 1939; Landsteiner and Wiener, 1941). The impetus which this blood-group research gave to population genetics is shown by the rapid progress made in the study of the genetic composition of human populations in the 1940's and 1950's. Yet, admittedly, in the earlier anthropological studies progress was often accompanied by a tendency to overrate the exclusive use of the blood groups for taxonomic purposes. This simplistic approach on the part of a few workers led some of their more morphologically oriented colleagues to question the general value of genetic methods in taxonomic studies. In the article included here, the serologist William Boyd discusses some of the objections that were, and continue to be leveled against the use of the blood groups, because, as he points out, some physical anthropologists "are still doubtful about the value of blood grouping in their science." Boyd maintains, however, that genetic methods can make important contributions to physical anthropology, and he describes four populations where application of the methods has provided supplementary or new information about their genetic affinities.

Although Boyd's position may continue to have its adversaries, it seems increasingly evident that genetics can and does provide a valuable research tool to the physical anthropologist interested in human-population evolution and taxonomy. As laboratory methods improve, more genetic loci in man become susceptible to testing and description. In particular, the techniques of biochemical genetics have provided the investigator with an impressive arsenal of new alleles for the quantitative analysis of populations. Advances in this field date back to the 1940's when new methods in enzyme and amino-acid analysis revived investigation into the so-called "inborn errors of metabolism," originally described by the physician

A. E. Garrod in the early 1900's (Garrod, 1909). This work has led to the discovery of numerous biochemical genetic variants in human populations. The biochemical technique of *starch gel electrophoresis* devised by Smithies (1955) has been especially useful for examining large population samples of blood for various genetic systems (see Harris, Chapter IV). Starch gel is a supporting medium which permits the separation of proteins on the basis of their differences in electrical charge. By the use of specific buffers and stains, a number of proteins and their variants which are present in the blood can be detected. The blood proteins which have been most extensively investigated in human populations are *hemoglobin*; various *red-cell enzymes* such as glucose-6-phosphate dehydrogenase (G-6-PD); and the *serum proteins*, including *haptoglobin* which binds hemoglobin, *transferrin* which transports iron in the serum, and *albumin* which participates in osmotic-pressure regulation of the blood.

It should be kept in mind that the number of alleles available to the population geneticist for analysis of a gene pool is entirely determined by the methods presently used to identify them. This number in no way reflects the number of different alleles which are actually contained in human gene pools, and which number in the thousands; estimates vary from 10,000 to 100,000. In addition, ease in collection and analysis has dictated to a large extent which genetic characters of all those known are the most commonly used. For this reason the most extensively studied genetic traits are those identifiable from easily collected tissues such as blood, saliva, and urine, and from easily administered tests such as those for color-blindness and PTC taste sensitivity.

Hundreds of human populations throughout the world have been studied for these genetic traits. On the basis of these studies, the statement can be made that most human gene pools differ *quantitatively* in the proportions of their allele frequencies rather than *qualitatively* in kind. However, there are a few alleles which are common in some populations and absent in others. One example of such a qualitative difference is the absence of the common blood-group alleles I^A and I^B from many South American Indian tribes whose gene pools are 100 percent allele I^O. Furthermore, the quantitative differences between gene pools may be considerable. For example, the FY^a allele of the Duffy blood group appears in Caucasian populations at a frequency of about 0.43, but in African populations it is probably not over 0.02 (Reed, Chapter VI). If an allele such as Fy^a is unique in either its presence or its high frequencies to a population or a group of related populations it is called a *marker gene*, in that it serves as a genetic "label," and is of particular usefulness in taxonomic and admixture studies.

In summary, heredity as described by the Mendelian and Hardy-Weinberg Laws is a conservative process whereby individuals and populations reproduce replicas of themselves. But heredity is counterbalanced by evolutionary forces which interfere with exact replication and change gene frequencies between generations. Sewall Wright (1949) has distinguished three primary modes of immediate change in gene frequency (q), according to their degree of determinacy when acting on the gene pool of a population:

1. **Change from Systematic Pressure.** Three forces of evolution exert systematic pressure on the gene pool to change the frequency of q. The direction and magnitude of the change in q (Δq) is said to be *determinate in principle*. The three forces capable of exerting systematic pressure are: (a) *recurrent mutation*, (b) *intrapopulation selection*, and (c) *recurrent immigration* and *cross-breeding* (gene flow). Mathematical prediction of genetic change under systematic pressure assumes that there are no other processes operating on the gene pool which are indeterminate in nature. Eventually, in the absence of a change in environmental conditions, the systematic pressures effect a state of equilibrium, rather than evolution, in a gene pool.

2. **Change from Fluctuations**. Fluctuations in the systematic pressures or, from accidents of sampling (genetic drift) also produce changes in the frequency of q. Unlike the predictable changes due to systematic pressures, fluctuations cause *deviations* in q (σq) which are *indeterminate in direction*; that is, it is not predictable whether q will increase or decrease in the gene pool. However, the deviations in q are *determinate in variance* ($\sigma^2 \delta q$), and a probability curve ($\rho(q)$) describes the distribution of these deviations in single populations over an extended period of time.

3. **Change from Unique Events.** Five events can produce changes in a gene pool which are wholly *indeterminate in principle*. Their impact on the gene pool is therefore totally unpredictable, and each of the five events must be considered in terms of its own unique effects. These events are (a) a mutation that is favorable from its beginning, (b) a unique selective incident, (c) unique hybridization, (d) swamping by mass immigration, (e) a unique reduction in numbers.

Some or all of these three modes of change have been operative to varying degrees in the populations which are described in the following sections of this book. Yet, as is repeatedly emphasized in the following chapters, the investigation of genetic evolution in human populations is a difficult, if not at times an impossible, task. Although new biochemical techniques are providing the population geneticist with more raw data, the analysis of the data in evolutionary terms is limited for several reasons: the lack of information spanning several generations in human populations, the complexity of the genotype-environment relationship, the interaction of multiple forces of evolution simultaneously acting on one gene pool, the necessity of large representative samples for statistical analysis, and so on. Yet some of these problems are becoming more amenable to analysis with the widespread use of modern computers. The application of computer techniques to population and genetic data, like those done by Cavalli-Sforza on the Italian population he was studying (see Chapter V), has opened up new methods of research to the population geneticist and other people working with human populations. They can now analyze data, link records, construct mathematical models, and simulate situations too complex to be handled by less advanced means. As the geneticists James Neel and William Schull (1968) predict:

Possibly one of the more far-reaching contributions which digital computers can make to the problem before us as well as to all of biology will come through their use as a means to simulate processes too complex to be tractable to more conventional methods of analysis . . . . There is a growing awareness of the possibilities of these machines in this context, however, and one now finds demographers, anthropologists, geneticists, and communications scientists, to mention but a few of the disciplines interested in the biology of populations, actively engaged in the construction for the computer of models more nearly approximating reality than those for which analytical solutions exist. These constructs are quite possibly still overly naïve, but they have nonetheless occasioned a more orderly "thinking out" of the sequence of events involved in the persistence and spread of mutant genes in a population, and the relationship of population structure to these latter events. Patently, the opportunity which the computer affords to watch the passage of a population from one state to another can provide insight into the dynamics of a genetic system which no end of steady state analysis will ever elicit.

The fossil record tells us that the gene pools of primate populations have changed through time as man has evolved from a nonhuman primate form through the various types of fossil man to become modern *Homo Sapiens* of today. We may predict that as long as the forces of evolution continue to operate as effectively in the present and future as they have in the past, man will continue to change—gradually, perhaps imperceptibly, but definitively. Hopefully, population genetics will provide us with an increased understanding of these forces and our own evolution.

The Mathematical Foundation of Human Genetics: The Hardy-Weinberg Law

See the following readings: "The Hardy-Weinberg Law," by Curt Stern, p. 15; and "Mendelian Proportions in a Mixed Population," by Godfrey H. Hardy, p. 18.

Applications of the Hardy-Weinberg Law

See the following reading: "Four Achievements of the Genetical Method in Physical Anthropology," by William C. Boyd, p. 20.

REFERENCES TO THE LITERATURE CITED IN THE OVERVIEW

1. BERNSTEIN, F., 1924. Ergebnisse einer biostatistischen zussammenfassenden Betrachtung über die erblichen Blutstrukturen des Menschen. *Klin. Wochenshr.*, 33: 1495–1497.
2. _____, 1925. Zussammen fassende Betrachtungen über die erblichen Blutstrukturen des Menschen. *Ztschr. Abstq. u. Vererbgsl.*, 37: 237–270.
3. CASTLE, W. E., 1903. The laws of heredity of Galton and Mendel and some laws governing race improvement by selection. *Proc. Amer. Acad. Arts and Sci.*, 39: 223–242.
4. CORRENS, C., 1900. G. Mendels Regel über das Verhalten der Nachkommenschaft der Rassenbastade. *Deutsch. Bot. Ges. Ber.*, 18: 158–168.
5. CROW, J. F., 1961. Population genetics. *Amer. J. Hum. Genet.*, 13:137–149.
6. DE VRIES, H., 1900. Sur la loi de disjonction des hybrides. *Contes Rendus Acad. Sci. Paris*, 130: 845–847.
7. DOBZHANSKY, T., 1965. Mendelism, Darwinism, and evolutionism. *Proc. Amer. Phil. Soc.*, 109: 205–215.
8. GARROD, A. E., 1909. *Inborn Errors of Metabolism*. London, Oxford University Press.
9. HIRSCHFELD, L. AND H. HIRSCHFELD, 1919. Serological differences between the blood of different races. The result of researches on the Macedonian front, pp. 32–43. In Boyer, S. H. (ed.) 1963. *Papers on Human Genetics*. Englewood Cliffs, N.J., Prentice Hall.
10. LANDSTEINER, K., 1901. On agglutination phenomena of normal human blood, pp. 27–31. In S. H. Boyer, (ed.), 1963, *Papers on Human Genetics*. Englewood Cliffs, N.J., Prentice Hall.
11. _____ AND P. LEVINE, 1927. Further observations on individual differences of human blood. *Proc. Soc. Exp. Biol.*, 24: 941–942.
12. _____ AND A. S. WIENER, 1941. Studies on agglutinogen (Rh) in human blood reacting with anti-Rhesus sera and with human isoantibodies. *J. Exp. Med.*, 74: 309–320.
13. LEVINE, P. AND R. E. STETSON, 1939. An unusual case of intragroup agglutination. *J. Am. Med. Assn.*, 113: 126–127.
14. LI, C. C., 1963. *Population Genetics*. Chicago, University of Chicago Press.
15. MENDEL, G., 1865. Experiments in plant hybridization, pp. 1–20. In J. A. Peters (ed.), 1962, *Classic Papers in Genetics*. Englewood Cliffs, N.J., Prentice Hall.
16. NEEL, J. V. AND W. J. SCHULL, 1968. On some trends in understanding the genetics of man. *Perspec. Biol. and Med.*, 11: 565–602.
17. SMITHIES, O., 1955. Zone electrophoresis in starch gel: group variations in the serum proteins of normal human adults. *Biochem. J.*, 61: 629–641.

18. STERN, C., 1965. Mendel and human genetics. *Proc. Amer. Phil. Soc.*, 109: 216–226.

19. _____, 1968. *Genetic Mosaics and Other Essays.* San Francisco, W. H. Freeman.

20. VON TSCHERMAK, E., 1900. Über künstliche Kreuzung bei Pusum sativum. *Deutsch. Bot. Geo. Ber.*, 18: 232–239.

21. WRIGHT, S., 1949. Population structure in evolution. *Proc. Amer. Phil. Soc.*, 93: 471–478.

The Hardy-Weinberg Law

CURT STERN

One of the basic relations in the genetics of populations is expressed by the statement that in a very large random-mating population in which two alleles A and A' occur in the frequencies p and q ($= 1 - p$) the three types AA, AA' and A'A' are expected to remain in equilibrium from generation to generation at frequencies of p^2, 2pq and q^2, in the absence of mutation or selection. This theorem, of which a special case was discovered by Pearson (1904), is known in its general formulation as Hardy's law, or Hardy's formula (*e.g.*, Sinnott and Dunn, 1939, Sturtevant and Beadle, 1939, and Dobzhansky, 1941). It is the purpose of this note to point out that the important population formula was independently and simultaneously recognized by the Stuttgart physician, Wilhelm Weinberg (1862–1937). On January 13, 1908, Weinberg gave a lecture before the "Verein für vaterländische Naturkunde in Württemberg" under the title "Über den Nachweis der Vererbung beim Menschen." In the course of a keen exposition of both the difficulties to be met by students of human heredity and of statistical approaches which should help to overcome these difficulties, he derived the equilibrium law. The full lecture was printed in the Jahreshefte of the Verein, Volume 64: 368–382 (1908), and appeared sometime before the fall of 1908 as judged by the stamped entry on the title page of the volume which I have consulted: "Academy of Natural Sciences of Philadelphia, Sept. 28, 1908." Hardy's note in Science is signed April 5, 1908, and is published in the July 10, 1908, number.

The following is a translation of the relevant section of Weinberg's communication making corrections for three minor typographical errors:

Quite different is the situation when one considers Mendelian inheritance under the influence of panmixis. I start from the general premise that there are originally present *m* each of pure male and female representatives of type A and correspondingly *n* of each pure representatives of type B. If these cross at random one obtains, by applying the symbolism of the binomial theorem, the following composition of the filial generation:

$$\frac{(mAA + nBB)^2}{(M + n)^2} = \frac{m^2}{(m + n)^2} AA + \frac{2mn}{(m + n)^2} AB + \frac{n^2}{(m + n)^2} BB$$

or if $m + n = 1$

$$m^2 AA + 2mnAB + n^2 BB.$$

Reproduced by permission of the publisher and Curt Stern from Science, *97:137–138 (February 5, 1943).*

Curt Stern, *a zoologist and geneticist, was born in Hamburg, Germany in 1902. He received his Ph.D. in 1923 from the University of Berlin and his D.Sc. in 1958 from McGill University. He taught at the University of Berlin until 1933 when he migrated to the United States to assume a teaching position in the Department of Zoology, University of Rochester. In 1947 he moved to the University of California at Berkeley where he is presently Professor of Zoology and Genetics. He has done extensive research on the genetics of Drosophila, in 1931 proving the theory of crossing over, and in 1936 demonstrating crossing over in somatic as well as germ cells. Dr. Stern has served as president of the American Society of Zoologists, the Genetics Society of America, and The American Society of Human Genetics. In recognition of his contributions to the field of genetics he was awarded the Kimber Genetics Medal by the American Academy of Sciences in 1963. He conducted human-population field work in Vellore, India in 1963. Author of numerous scientific publications, his books include* Principles of Human Genetics *(San Francisco, 1960);* Genetic Mosaics and Other Essays *(Cambridge, Mass., 1968).*

If now the male and female members of the first generation are crossed at random among themselves one obtains the following frequencies of the various cross combinations:

$m^2 \cdot m^2 \cdot (AA \times AA) + m^4 AA$
$4m^2m \, n \, (AA \times AB) = 2m^3n \, AA + 2m^3n \, AB$
$2m^2n^2 \, (AA \times BB) = 2 \, m^2n^2 \, AB$
$4 \, (mn)^2 \, (AB \times AB) = m^2n^2 \, AA + 2m^2n^2 \, AB \, + m^2n^2 \, BB$
$4m \, n \, n^2 \, (AB \times BB) = 2 \, m \, n^3 \, AB + 2m \, n^3 \, BB$
$n^2n^2 \, (BB \times BB) = n^4 \, BB$

or the relative frequencies

$$AA : m^2 \, (m + n)^2$$
$$AB : 2m \, (m + n)^2 n$$
$$BB \; : \; (m + n)^2 n^2$$

and the composition of the second filial generation is again

$$m^2AA + 2m \, n \, AB + n^2BB.$$

Thus we obtain under the influence of panmixis in each generation the same proportion of pure and hybrid types. . . .

While Weinberg's paper, like Mendel's, appeared in an obscure journal, its failure to be recognized can not be ascribed to this fact alone. His later contributions dealing with extensions of the statistical treatment of the genetics of populations are found in the "regular" journals. These papers have received some attention (*e.g.*, Sewall Wright, 1930) and in them Weinberg refers to his 1908 pioneer work. However, both Weinberg and Hardy were ahead of contemporary thought and similar problems were not generally considered for at least eight years. At that time perhaps Hardy's name and the prominent place of his publication both helped to leave Weinberg's contribution neglected.

Hardy as a mathematician did not follow up his discovery by any further consideration of its genetic implications. Weinberg in 1909 re-

Wilhelm Weinberg (1862–1937), an obstetrician in Stuttgart, Germany, devoted the time not taken by patients to problems of human genetics. He was the first to provide a full answer to the question, How do the numerical consequences of Mendelian inheritance behave under the influence of random mating? In 1908, the mathematician G. H. Hardy also demonstrated the operation of Mendel's law of segregation under random mating, but his contribution marked his first and last effort in the field of genetics. Weinberg, on the other hand, continued to work extensively on genetic problems, often anticipating work done many years later. He devised methods for dealing with the incompleteness inherent in the records of human families selected because of an observed recessive trait. He was also the first to realize that the phenotypic variance observed in relatives could be divided into genetic and environmental components. Weinberg was truly an outstanding early worker in the field of genetics, but his work went unrecognized and unused by his contemporaries, and even today the insight he provided into the inheritance of quantitative traits has not been fully developed.

formulated his theorem in terms valid for multiple alleles—at a time when no case of multiple alleles had been discovered in man nor in plants and even Cuénot's demonstration of multiple alleles in the mouse had remained unnoticed. He also for the first time investigated polyhybrid populations and recognized their essentially different method of attaining equilibrium. Considering these facts it seems a matter of justice to attach the names of both the discoverers to the population formula.

Mendelian Proportions in a Mixed Population

GODFREY H. HARDY

To the Editor of Science: I am reluctant to intrude in a discussion concerning matters of which I have no expert knowledge, and I should have expected the very simple point which I wish to make to have been familiar to biologists. However, some remarks of Mr. Udny Yule, to which Mr. R. C. Punnett has called my attention, suggest that it may still be worth making.

In the *Proceedings of the Royal Society of Medicine* (Vol. I., p. 165) Mr. Yule is reported to have suggested, as a criticism of the Mendelian position, that if brachydactyly is dominant "in the course of time one would expect, in the absence of counteracting factors, to get three brachydactylous persons to one normal."

It is not difficult to prove, however, that such an expectation would be quite groundless. Suppose that Aa is a pair of Mendelian characters, A being dominant, and that in any given generation the numbers of pure dominants (AA), heterozygotes (Aa), and pure recessives (aa) are as $p : 2q : r$. Finally, suppose that the numbers are fairly large, so that the mating may be regarded as random, that the sexes are evenly distributed among the three varieties, and that all are equally fertile. A little mathematics of the multiplication-table type is enough to show that in the next generation the numbers will be as

$$(p + q)^2 : 2 (p + q) (q + r) : (q + r)^2,$$

or as $p_1 : 2q_1 : r_1$, say.

The interesting question is—in what circumstances will this distribution be the same as that in the generation before? It is easy to see that the condition for this is $q^2 = pr$. And since $q_1^2 = p_1 r_1$, whatever the values of p, q and r may be, the distribution will in any case continue unchanged after the second generation.

Suppose, to take a definite instance, that A is brachydactyly, and that we start from a population of pure brachydactylous and pure normal persons, say in the ratio of $1 : 10,000$. Then $p = 1$, $q = 0$, $r = 10,000$ and $p_1 = 1, q_1 = 10,000$, $r_1 = 100,000,000$. If brachydactyly is dominant, the proportion of brachydactylous persons in the second generation is $20,001 : 100,020,001$, or practically $2 : 10,000$, twice that in the first generation; and this proportion will afterwards have no tendency whatever to increase. If, on the other hand, brachydactyly were recessive, the pro-

Reproduced by permission of the publisher from Science, 28:49–50 (July 10, 1908).

Godfrey H. Hardy *(1877–1947), an English mathematician, was the first to postulate the mathematical keystone of population genetics, the Hardy-Weinberg Law. At the time, he was a fellow at Trinity College, Cambridge. In 1919 he moved to the University of Oxford where he was professor of geometry until 1931. He then returned to Cambridge where he served as professor of pure mathematics until his retirement in 1942. He was the author of numerous books in the field of mathematics.*

portion in the second generation would be 1:100,020,001, or practically 1:100,000,000, and this proportion would afterwards have no tendency to decrease.

In a word, there is not the slightest foundation for the idea that a dominant character should show a tendency to spread over a whole population, or that a recessive should tend to die out.

I ought perhaps to add a few words on the effect of the small deviations from the theoretical proportions which will, of course, occur in every generation. Such a distribution as $p_1:2q_1:r_1$, which satisfies the condition $q_1^2 = p_1r_1$, we may call a *stable* distribution. In acutal fact we shall obtain in the second generation not $p_1 : 2q_1 : r_1$, but a slightly different distribution $p':2q_1':r_1'$, which is not "stable." This should, according to theory, give us in the third generation a "stable" distribution $p_2 : 2q_2 : r_2$, also differing slightly from $p_1 : 2q_1 : r_1$; and so on. The sense in which the distribution $p_1 : 2q_1 : r_1$ is "stable" is this, that if we allow for the effect of casual deviations in any subsequent generation, we should, according to theory, obtain at the next generation a new "stable" distribution differing but slightly from the original distribution.

I have, of course, considered only the very simplest hypotheses possible. Hypotheses other than that of purely random mating will give different results, and, of course, if, as appears to be the case sometimes, the character is not independent of that of sex, or has an influence on fertility, the whole question may be greatly complicated. But such complications seem to be irrelevant to the simple issue raised by Mr. Yule's remarks.

P.S. I understand from Mr. Punnett that he has submitted the substance of what I have said above to Mr. Yule, and that the latter would accept it as a satisfactory answer to the difficulty that he raised. The "stability" of the particular ratio 1 : 2 : 1 is recognized by Professor Karl Pearson [*Phil. Trans. Roy. Soc.* (A), vol. 203, p. 60].

Four Achievements of the Genetical Method in Physical Anthropology[1]

WILLIAM C. BOYD

The use of genetical methods in physical anthropology, which really began with the Hirszfelds' classical paper (1919:675) is no longer new and is now widely accepted. Genetical methods are referred to, in an approving tone, in many books and articles published during the last few years. The late Prof. Hooton (1946) devoted 18 pages out of 788 to genetical methods, and Montagu (1960:327) devotes 93 out of 771 to the subject, giving it a very up-to-date treatment, and stating, ". . . if we are to trace the relationships of the varieties of man to one another, it is necessary that we rely on criteria which possess a more permanent character than the shifting sands of head shape. . . . Such characters are available in the blood groups, in the M-N in the Rh-Hr blood types and in the hemoglobin and haptoglobin types of man."

In addition to the characteristics mentioned by Montagu, we now know of a number of other inherited traits of man that either are or will doubtless become useful. One might well ask, are there any longer any skeptics? The answer to this question seems to be: yes, there are skeptics, and some of them are rather vocal. Worse than this, examination of some of the books and papers written by authors who seem to welcome the genetical method suggests that they, in reaching their conclusions, actually make very little use of the data and modes of reasoning that such methods provide.

Layrisse and Wilbert (1960) state that in their experience about four out of ten physical anthropologists are still doubtful about the value of blood grouping in their science. Probably most of us have had one or more of the older physical anthropologists inquire, in conversation, "Now frankly, what have blood groups ever proved that we didn't know already?"

L. Oschinsky (1959:1) has not hesitated to carry the war into his enemy's country. He says, "Nowhere has Professor Boyd posed the question as to whether or not the characteristics he is choosing are taxonomically relevant. . . . Unfortunately for Professor Boyd, it is the polygenic features

[1]The work reported in this paper was aided by use of a desk calculator purchased with grants from the American Academy of Arts and Sciences and the Wenner-Gren Foundation for Anthropological Research, New York.

Reproduced by permission of the American Anthropological Association and William C. Boyd from the American Anthropologist, 65:243–252 (1963).

William C. Boyd, an immunologist, was born in Missouri in 1903. He received his B.A. in 1925 and his M.A. in 1926 from Harvard University, and his Ph.D. in 1930 from Boston University. At present he is Professor Emeritus, Department of Biochemistry, Boston University. He has served as president of the American Association of Human Genetics and the American Association of Immunologists. He is presently on the editorial board of the American Journal of Physical Anthropology. His major research has been in immunochemistry, in which field he has published over two hundred scientific papers. He has retained a long-standing interest in the use of blood groups in racial studies and has done serological field work among populations in the Middle East, Pakistan, and the Navajo Indians of the United States. His books on blood grouping and genetics include Blood Grouping Technic, with Fritz Schiff (New York, 1942); Fundamentals of Immunology (New York, 4th ed. 1966); Genetics and the Races of Man, with Isaac Asimov (New York, 1955).

such as skin colour, hair texture, nose shape, lip thickness, which have the greatest taxonomic value. *And why is it necessary to understand the mechanism of inheritance* if one is concerned with the question of distinguishing between the various racial groups which, Professor Boyd states, is one of the chief aims of physical anthropology?" (my italics)

A few years ago, at a seminar at Columbia University, a distinguished American physical anthropologist delivered a detailed attack on the blood groups as anthropological criteria, partly on the grounds that they may respond too readily to selection pressures. More recently, two Polish anthropologists, Bielicki (1962:3) and Wierciński (1962:2) have also attacked the use of genetical methods, and readvocated the use of conventional morphological criteria.

Perhaps it is true that I never posed or attempted to answer the question brought up by Oschinsky. Let me attempt to answer it now. Believing that races are the result of the adaptation of different populations to the environments in which they live, I believe that any feature in which two populations differ has taxonomic value. I am not convinced that the characters enumerated by Oschinsky have greater taxonomic value than have, for example, the blood groups, and I have pointed out elsewhere (Boyd 1956:993) certain advantages that genetically analyzed characters such as the blood groups possess.

Back of many of these recent attacks on genetical methods there is an argument that has come to find increasing favor with those of the old school who continue to doubt the usefulness of genetic methods. This argument seems to be based on Birdsell's demonstration (1951:259; 1952: 355) that genetic drift would be more likely to bring about fixation or extinction of extreme types when these types are determined by genes acting at one or a few loci than it would when a considerable number of loci are involved. Some writers seem to have concluded from this that the same is necessarily true of the action of natural selection. Statements implying this are even found in certain textbooks.

Actually, whether selection will act more slowly on multifactorial characters than on those controlled by a single locus is a difficult and comprehensive question that has not been sufficiently investigated, even from the theoretical angle. A preliminary mathematical study of the problem shows that the answer depends on the particular genetic model we choose.

For the sake of simplicity, let us restrict our multifactorial case to the situation where two loci are involved. If the two genotypes are determined by two complementary factors, the rate of elimination of the unfavorable genotype *would* be slower than in the unifactorial case (C.C. Li, unpublished calculations).

On the other hand, if we assume that to get the unfavorable trait the cooperation of two loci is required, and that the trait does not appear in an individual unless some particular one of the alleles at that locus is represented in double dose, then the rate of removal of this undesirable trait from the population is *exactly the same* as the rate of removal in the unifactorial case.

But . . . Dr. Li has also shown (unpublished calculations) that if the gene effects are not all additive and three phenotypes result, then if the selective coefficients against two of the phenotypes are the same as in the unifactorial case, the unfavorable multifactorial trait is removed *faster* then the unfavorable unifactorial trait.

Since we do not know which of these models (if any) corresponds to the actual situation with respect to any actual human multifactorial trait, it would seem premature to base a decision on a hasty and probably emotion-packed extrapolation from Birdsell's demonstration in relation to genetic drift.

If we look for experimental evidence, it is mostly conspicuous by its absence, but Hiernaux (1962:29) does state that he has observed a shift in the morphology of the Bantu towards that of the Pygmies, in the tribes who have recently migrated into the equatorial forest of the Eastern Congo, without any similar shift in blood group frequencies.

In the light of all this, there does not seem to be at the present time any good reason to reject genetically analyzed traits and rely solely on morphological and other multifactorial characters. In fact, as we shall see below, there is other evidence that suggests that in some places at least selection has altered the incidence of frizzy hair, dark skin, etc. much faster than it has altered the blood group frequencies.

It is to be feared that our comparatively recent discovery that the blood groups are not selectively neutral (although the selective forces demonstrated thus far are mostly very weak) has reawakened the hope formerly held that the physical characteristics previously used for racial classification will turn out to be less rapidly altered by selection and therefore of more value in reconstructing the past. But if it was naive of us to reason that since we could not think of any way in which selection could attack the blood group frequencies it did *not* affect them, it would be doubly naive of us to argue now that since we have not found the nature of the selective pressures that affect the frequency of the genes controlling hair form, skin color, lip thickness, etc., there are no such pressures. Fleure (1945:580) showed fifteen years ago that this puzzle is not "beyond all conjecture," and suggested that pigmentation, at least, is connected with the intensity of ultraviolet radiation from the sun and sky.

In so far as genetic drift (the Sewall Wright effect) is concerned, I believe Birdsell's demonstration is valid. But how big a role has drift played in the formation of our present human races? Admittedly, many authors have made extensive use of the concept of drift in their speculations on the past history of man, and I have been one of the worst offenders. It is now my conviction that I and others employed this notion too freely; drift may actually have operated only in certain special cases. All authorities now agree that the most important agency of race formation is natural selection acting in isolation on random mutations. R. A. Fisher, who usually had his feet on the ground, was never convinced that drift was an important agency.

However this may be, it is now clear that the advocates of the use of genetical methods did not win the victory they thought they had, but got instead an uneasy truce that is now being disturbed by occasional potshots from certain unsubdued *franc-tireurs*. This being so, we ought perhaps to ask ourselves if the question, quoted above as coming up frequently in conversation, has a good answer, or indeed any answer at all. *Have* genetic methods made any new contributions to physical anthropology?

I believe we can answer this question in the affirmative, and in support of this would like to cite four examples. These concern the Gypsies, the American Negro, the Lapps and the Papuan pygmies.

A. The Gypsies

Confirmation of the Indian origin of the Gypsies was the earliest achievement of the genetical method, but it did not lead to any radical reversal of opinion, for it merely served to strengthen an opinion that was already generally accepted. The method was used, even by its originators, mainly as a test of the possibility of using serological data when reasoning about the past history of a human group.

The Gypsies were known to claim an Indian origin for themselves, and linguistic studies by George Borrow and others had established the fact that their language was a debased and diluted form of one of the mod-

ern derivatives of Sanscrit. But though the Gypsies, like the Basques and
Jews, maintain that they never mix with the surrounding peoples (or maintain that if they do, any resulting progeny are not considered Gypsies),
some Gypsy groups, particularly in Britain, are so like the people
they live among that some authorities doubt their Indian origin. I once
took blood from several Gypsies in a Welsh hospital, supposing them to
be Welsh, without noticing anything different about them. R. B. Dixon,
in his controversial *Racial History of Man,* said of the Gypsies, "It
seems probable that *if of northwest Indian origin,* they may, before leaving there, have had some Alpine mixture. . ." (my italics).

In 1921 Verzar and Weszeczky (1921:33) removed whatever doubt
remained by determining the blood groups of some Hungarian Gypsies.
The results showed that the blood groups of the Gypsies still agreed quite
well with those of the Indian soldiers tested by the Hirszfelds at Salonika,
and differed significantly from those of the Hungarians (Table 1). This result was obtained with the aid of the ABO blood groups alone, as the
twenty or so additional systems now available to us had not yet been
discovered.

Note that in Table 1 the blood group frequencies of the Hungarian
Gypsies, though strongly supporting the idea of an Indian origin, differ
from the Indian frequencies slightly, and in the direction to be expected
(more A, less B) if the Gypsies had mixed somewhat with other stocks,
such as the surrounding Hungarians. We could even calculate from these
figures (see Section B), if we wished to assume that the ancestors of the
Gypsies had the blood group frequencies of Hirszfeld's Indian soldiers,
what proportion of Hungarian genes Verzár's Gypsies had acquired.
Which brings me to the next achievement of the genetical method.

B. Quantitative Treatment of Race Mixture

When two populations having different frequencies of a gene mix, the
frequencies of the gene in the resulting mixed population can be predicted
if the proportions in which the two populations mix are known. It is simply
a question of taking a weighted average of the sets of two gene frequen-

TABLE 1. BLOOD GROUP FREQUENCIES OF GYPSIES, INDIANS AND HUNGARIANS

Population	Number Tested	Per Cent in Group			
		O	A	B	AB
Gypsies (Hungary)	385	34.3 ± 2.4	21.0 ± 2.1	39.0 ± 2.5	5.7 ± 1.4
Indians*	1000	31.3 ± 1.5	19.0 ± 1.2	41.2 ± 1.6	8.5 ± 0.9
Difference		3.0 ± 2.8	2.0 ± 2.4	2.2 ± 2.9	2.8 ± 1.7
Hungarians	1500	31.1 ± 1.2	38.0 ± 1.3	18.7 ± 1.0	12.2 ± 0.9
Difference between Gypsies and Hungarians		3.2 ± 2.7	17.0 ± 2.4	20.3 ± 2.7	6.5 ± 1.6

(The numbers following the various percentages are standard deviations.)

*From Hirszfeld's original paper (1919:675)

cies. Bernstein (1931) showed thirty years ago how to make such calculations and mentioned that they could be reversed to estimate from the gene frequencies of a mixed population the proportions in which two ancestral populations must have mixed to produce it. This prodecure is exactly like the well known procedure called "indirect analysis" in quantitative chemical analysis. Various authors have made use of the method (see Boyd 1950). Stevens (1952:12) has shown how to apply the method of maximum likelihood to the problem. His procedure is, however, too lengthy to summarize here.

An excellent example of the use of genetical methods to calculate the degree of mixture a given population represents is provided by the paper by Glass and Li (1953:1) In a careful study making use of the frequencies of seven genes (R_0, R_1, T, r, B, A, and R_2) Glass and Li reached the conclusion that the amount of White mixture in the North American Negro is 30.565 per cent. It is hardly necessary to remark that such a result could never have been obtained by a study of genetically unanalyzed traits such as skin color or morphology. In dealing with such problems genetical methods make a unique contribution to anthropology.

C. The Lapps

The exact position of the Lapps among the peoples of Europe has long been a controversial question. For a long time they were classified with the Mongoloids. Haddon (1925) included them in his "Urgian" (evidently a misprint for Ugrian) group with the Tungus, Cheremis, etc. Deniker (1900) classified the Lapps as "Hyperborean", apparently a branch of his Mongolian group. C. S. Coon states, " . . . the original ancestral Lapps represented a stage in the evolution of both the Upper Paleolithic Europeans and the mongoloids" (1939:305).

These and other authors recognized that the Lapps were a highly specialized human group, and genetical methods certainly confirm them in this. Examination of nearly any of the isogene maps in Mourant's book (Mourant 1954) shows that the Lapps have characteristically different frequencies of nearly all of the blood group genes (see Fig. 1). On nearly all of the maps a little area, like a "high" or "low" on a weather map, can be recognized, of which one can say, "There are the Lapps." And also, just as is the case with another isolate, the Basques, the gradual gradation of the Lapp gene frequencies into those of the populations around them shows that isolation has been imperfect, either during the formation of the Lapp race, or afterwards, or both.

But the earlier writers were wrong in thinking there was anything mongoloid about the Lapps. The Lapps have very low frequencies of the blood group B gene, whereas mongoloids have high B. The Lapps have a high frequency of the A_2 gene (the highest in the world) (Allison et al. 1956:87); this variant of A is absent in the mongoloids. The M frequen-

Isogenes for Gene A

Isogenes for Gene B

Isogenes for Gene O

Isogenes for Rh Gene C

Isogenes for Rh Gene D

Isogenes for Gene M

Fig. 1. [Redrawn from Mourant (1954)]

cies of the Lapps are low, whereas those of the mongoloids are "normal" or high.

All the genetic evidence agrees in indicating that the Lapps evolved into their very distinctive race *in situ,* and none of the evidence suggests that they are even partly of mongoloid origin. The Lapps are Europeans. Actually, some physical anthropologists were coming to this conclusion on the basis of morphology, but it seems likely that without genetical evidence to reinforce these conclusions there would have remained many a doubt.

D. The Oceanic Pygmies

In several inaccessible areas of the islands that extend from the southwest of Asia there are found groups of natives characterized by their short stature, close-curled or frizzy hair, highly pigmented skin, and other "negroid" features. Such individuals, who are generally referred to as Negritos (Spanish, "little Negroes"), have been found in the Philippines, the Andaman Islands, the Malay peninsula, southern Siam and New Guinea. There has always been a strong temptation for anthropologists to consider these Negritos as somehow derived from Africa. When some were found to be pygmies, the temptation became irresistible. Stirling (1943) stated that they all belonged to the same basic stock. Montagu (1960:439) hedges a little, "Africa and southeastern Asia and Oceania are quite a long way from each other. The Negroids of today are, generally, poor seafarers; were their ancestors so, too? It is difficult to say; Melanesians often make voyages by sea in small outrigger craft of 50 miles or more. There is no reason to suppose that their ancestors could not do likewise." Howells (1959:329) is more emphatic, " . . . it is as plain as the nose on your face that the Negritos are intimately related to fully developed Negroes—a specialized kind of man—in skin, hair form, nose shape, and so on. The Negritos are really all similar and must have a common origin. And Negritos and Negroes cannot have appeared on separate continents; they too must have had a common origin. How do we get the Negritos into the Pacific from Africa?"

At least one writer has gone to the trouble of describing a hypothetical migration route of the Negritos from Africa to the Pacific, with way-station stops in Arabia and India.

It is apparent that such writers are making certain assumptions, though they may not state them explicitly: (a) They assume that the local environment has virtually no influence in determining the physical type of the human race that evolves in it. (b) They assume that if two human stocks look alike superficially, they must be related by common descent, at least partially. (c) They assume that similar adaptations to similar environments never take place. One reason this third assumption is made may be because the phenomenon is known by the epithet "parallel evolution." Now if race formation is primarily due to the action of natural selection, neither of these assumptions is necessarily true, and it is more reasonable to assume that: (a) The local environment greatly influences the evolution of the racial type that eventually adapts itself to it. (b) Similar environments will often, or at least occasionally, bring about similar racial adaptations. It is mainly a question of how similar the environments really are and whether a supply of the right mutations is or becomes available.

In the case of the Negritos we can test these opposing points of view by genetical methods. If the Negritos came somehow from Africa,

TABLE 2. BLOOD GROUP GENE FREQUENCIES
IN AFRICAN AND PACIFIC NEGRITOS

Population	Number Tested	Frequency of Gene								
		A	B	O	M	N	R_0	R_1	R_2	r
Pygmies, Belgian Congo	2557	0.198	0.249	0.553	0.468	0.523	0.630	0.074	0.194	0.101
Papuan Pygmies	139	0.075	0.139	0.786	0.102	0.898	0.030	0.850	0.119	0
Andaman Negritos	52	0.54	0.08	0.38	0.61	0.39	0	0.92	0	0

bringing with them their dark skins, frizzy hair, broad noses, and thick lips, they should have brought with them the African blood group frequencies, and no continent has a more characteristic blood group picture than Africa south of the Sahara. Did they?

The blood groups of the Papuan pygmies have been studied by Graydon, Semple, Simmons and Franken (1958:149) and those of the Andaman Negritos by Lehmann (1954). Their results, compared with results found in Africa by Hubinot and Snoek (1949) and Jadin, Julien and Gusinde (see Boyd 1939:113), are shown in Table 2.

The ABO gene frequencies, it is true, do not sharply distinguish the three populations (though the Andaman Negritos certainly seem to have more A than the other two groups); the other two blood group systems show definite and distinctive differences. A characteristic feature of the African pygmies, as of Africans in general, is the high frequency of R_0; this is not reflected in the other two populations shown in Table 2, as it certainly should be if they were of African origin. In fact, the Andaman Negritos seem to have no R_0 at all. The African pygmies, like other Africans, have "normal" frequencies of M and N (M and N about equal); the Papuan pygmies, like other Papuans, have much more N than M. The Rh negative gene (r or cde) is present in the African pygmies (it is almost a monopoly of the Africans and the Europeans), but is absent from the other two populations. Both the Papuan and the Andaman Negritos show the characteristic high Pacific value of R_1, over 10 times as high as that found in the African pygmies.

These numerous and marked genetic differences are not compatible with the supposition that the Pacific pygmies were somehow derived by migration from Africa, or even with the supposition that they are the product of a mixture of some older Pacific stock and some Africans who somehow managed to get to the Pacific and thus contributed their skin, hair, and lips to the hybrids. Any significant African contribution to the ancestry of the Pacific Negritos would have brought also some Rh negative genes, raised the R_0 frequency significantly above zero, lowered the R_1 frequency, and probably raised the M frequency. Graydon *et al.* are being very conservative when they say, "Despite the finding of a high R_0 frequency in Malayan Negritos the present authors believe that the weight of the blood grouping evidence makes it unlikely that the African and Asian pygmies are related. Previously expressed similar views regarding the lack of relationship between the African and Oceanic Negroes" (1958:169).

It would seem that genetical methods, though relatively new, have already made distinctive and worthwhile contributions to physical anthropology.

Summary

Certain recent arguments against the application of genetical methods in physical anthropology are examined, and four examples are offered of contributions made by such methods: (a) confirming the Indian origin of the Gypsies, (b) computing the per cent of White mixture in American Negroes, (c) establishing that the Lapps are a highly distinctive separate race, but European, (d) showing that the ancestors of the Papuan pygmies did not come from Africa.

REFERENCES CITED

ALLISON, A. C., B. BROMAN, A. E. MOURANT and L. RYTTINGER
    1956   The bloodgroups of the Swedish Lapps. Journal of the Royal Anthropological Society 86:87–94.

BERNSTEIN, F.
    1931   Die geographisch Verteilung der Blutgruppen und ihre anthropologische Bedeutung. Comitato Italiano per lo Studio dei Problemi della Populazione. Rome.

BIELICKI, T.
    1962   Some possibilities for estimating inter-population relationships on the basis of continuous traits. Current Anthropology 3:3–8.

BIRDSELL, J. B.
    1951   Some implications of the genetical concept of race in terms of spatial analysis. Cold Spring Harbor Symposia on Quantitative Biology 25:259–314.
    1952   On various levels of objectivity in genetical anthropology. American Journal of Physical Anthropology 10:355–362.

BOYD, W. C.
    1939   Blood groups. Tabulae Biologicae 17:113–240.
    1950   Genetics and the races of man. Boston, Little, Brown.
    1956   Anthropologie und Blutgruppen. Klinische Wochenschrift 34:993–999.

COON, C. S.
    1939   The races of Europe. New York, The Macmillan Co.

DENIKER, J.
    1900   The races of man. New York, Charles Scribner's Sons.

DIXON, R. B.
   1923   The racial history of man. New York, Charles Scribner's Sons.
FLEURE, H. J.
   1945   The distribution of skin color. Geographical Review 35:580–595.
GLASS, B., and C. C. LI
   1953   The dynamics of racial intermixture—an analysis of the American
          Negro. American Journal of Human Genetics 5:1–20.
GRAYDON, J. J., N. M. SEMPLE, R. T. SIMMONS and S. FRANKEN
   1958   Blood groups in pygmies of the Wissellakes in Netherlands New
          Guinea. American Journal of Physical Anthropology 16:149–171.
HADDON, A. C.
   1925   The races of man. New York, The Macmillan Co.
HIERNAUX, J.
   1962   (Discussion of papers by A. Wiercinski and I. Bielicki in Current
          Anthrop.) Current Anthropology 3:29–30.
HIRSZFELD, L. and H. HIRSZFELD
   1919   Serological differences between the blood of different races. The
          Lancet ii, 197:675–679.
HOOTON, E. A.
   1946   Up from the ape. 2nd edition. New York, The Macmillan Co.
HOWELLS, W.
   1959   Mankind in the making. New York, Doubleday and Co.
HUBINOT, P. O. and J. SNOEK
   1949   Reparition des genes Rh (CDE cde) chez les pygmees Batswa des
          Ntomba, Comptes Rendus de la Societé de Biologie, Paris 143:579–
          581.
LAYRISSE, M. and J. WILBERT
   1960   El Antigeno del Sistema Sanguinea Diego. Editorial Sucre, Caracas.
LEHMANN, H. and E. W. IKIN
   1954   Study of Andamanese negritos. Transactions of the Royal Society of
          Tropical Medicine and Hygiene, 48:12–15.
MONTAGU, M. F. A.
   1960   An introduction to physical anthropology. Springfield, C. C Thomas.
MOURANT, A. E.
   1954   The distribution of the human blood groups. Springfield. C. C
          Thomas.
OSCHINSKY, L.
   1959   A reappraisal of recent serological, genetic and morphological re-
          search on the taxonomy of the races of Africa and Asia. Anthro-
          pologica 1:1–25.
STEVENS, W. L.
   1952   Statistical analysis of the A-B-O system in mixed populations. Hu-
          man Biology 24:12–24.
STIRLING, M. W.
   1943   The native peoples of New Guinea. Smithsonian Institution War
          Background Studies No. 9. Washington.

VERZÁR, F. and O. WESZECZKY
   1921   Rassenbiologische Untersuchungen mittels Isohämagglutininnen. Bio-
   chemisches Zeitschrift 126:33–39.
WIERCINSKI, A.
   1962   The racial analysis of human populations in relation to their ethno-
   genesis. Current Anthropology 3:9–10.

CHAPTER 11

The Population
as a Unit of Study

Overview

The Mendelian Population

From the point of view of evolutionary theory, the unit of study is a *population*. In the genetic sense, population—that is, *Mendelian population* —is a technical term, and is defined as a spatial-temporal group of interbreeding individuals who share a common *gene pool*. Although the individual is indispensable to the process of evolution because he is the carrier and transmitter of the genetic material and the source of all mutation, genetic change in a single individual is not looked upon by the population geneticist as constituting "evolution." The forces of evolution are cumulative and operate through time on a gene pool, and only a population can have a gene pool. Thus, populations "evolve," not individuals.

Unfortunately, it is easier to define a Mendelian population theoretically, as the Hardy-Weinberg Law does, than to delimit such a unit accurately from the totality of mankind. Although it is obvious that humans are found clustered into groups which are characterized by the proclivity of their members to mate within the group (*endogamous mating*) rather than outside it (*exogamous mating*), boundaries are not sharply demarcated. As members of the single species *Homo Sapiens*, humans are biologically capable of breeding with one another. This biological fact, when combined with man's tendencies to trade, to migrate, to wage war, and to engage in numerous other activities which bring individuals and populations together, has prevented the complete genetic isolation of any group of hu-

33

mans for an extended period of time. Realistically considered, therefore, human populations are not totally closed systems, because their gene pools are shared to varying degrees by individuals from other populations. Thus, the population geneticist must work with actual population units less pure than those found in theoretical constructions.

Theoretically, the concept of the Mendelian population may be extended to include the entire human species, because all humans share in the common gene pool of the species *Homo sapiens*. In this sense, mankind itself constitutes a "population" and can be considered a proper unit of genetic study, but for one purpose only; namely, understanding the cohesive factors that bind the total human species together. Hence, the term population can refer to all of mankind or to its lesser constituent groups, which are of varying size and distribution. The term *deme* is often used, instead of the term population, with reference to a small endogamous group which is *relatively* self-sufficient and isolated from other such groups. Demes are the smallest basic population units studied by the population geneticist, but they can be organized into complexes of increasing size and inclusiveness—tribes, states, national and supranational entities—until all the demes of mankind are included. The Xavante tribe of Brazil (Salzano, *et al.*, Chapter II), the Tristan da Cunha islanders (Roberts, Chapter V), and the subgroups of the Amish sect of Pennsylvania (McKusick, *et al.*, Chapter V) are examples of demes which are described in this volume.

Factors Maintaining Population Boundaries

Although all human gene pools are open to varying degrees, it is evident that panmixis does not take place within the total species. If it did, there would be, of course, no distinguishable populations. Thus, to the proposition that populations are partially open systems must be added the second proposition that populations are partially closed systems. To the extent that one population's gene pool is closed to genes entering from any other population, the populations are *genetically isolated* from one another. Genetic isolation may be the result of (1) the actual physical separation of populations by spatial distance or geographical barriers such as mountains or rivers, and (2) the social separation of populations which culturally differ in language, religion, dress, and other such characteristics.

In the early history of human populations geographic isolation was more important in keeping populations apart than it is in the modern world. The island population of Tristan da Cunha described by D. F. Roberts (see Chapter V) is an unusual case of geographic isolation in a modern setting. Tristan da Cunha is located in the middle of the Atlantic Ocean and the islanders had little opportunity for social contact and mating with individuals of other populations. Yet even under these almost "ideal" isolated circumstances, the island population was subjected to occasional social and biological intrusion by survivors from shipwrecks or a rare immigrant from the continents. But in 1961 a volcanic eruption abruptly ended 145 years of geographical isolation by forcing the evacuation of the people from the island and their resettlement in England. The eventual breakdown of isolation, such as that experienced by the Tristan da Cunha population, is undoubtedly a common process which has operated throughout the history of the human species.

Today, the isolation of many primitive groups has broken down in the wake of expanding industrialized societies with their technologically advanced transportation and communication. Even the Xavante of remote Central Brazil have been living, more or less permanently since 1951, in proximity to government posts maintained, by the Indian Protective Service of Brazil. The Polar Eskimos, who believed they were the only people in the world at the time they were discovered, and who, in fact, probably had been totally isolated for two hundred years or more, are a rare phenomenon among human populations (Laughlin, 1967). It is highly unlikely that any human population has ever remained in total geographic isolation, and therefore total genetic isolation, for any significant length of time.

The more important mechanisms maintaining genetic isolation of populations today are *cultural* rather than geographical. A large portion of the behavior of any human group is directly or indirectly involved in maintaining its cultural distinction, and consequently its social and genetic isolation, from other groups. The various institutions and customs of a culture therefore operate in such a way as to maximize matings within the group and minimize matings outside the group. The residential segregation of Negroes and whites described by the sociologists R. Farley and K. Taeuber in this chapter operates in exactly this manner. The authors describe the growth and residential isolation of the Negro population in many large central cities of the United States through immigration and high fertility and the accompanying out-migration of the white population from the central cities to the suburbs. Using city-block data Farley and Taeuber have calculated a *dissimilarity index* which "indicates the minimum percentage of Negroes (or of non-Negroes) whose census tract of residence would have to be changed to obtain a really homogeneous distribution of the two groups." Their indices show a pattern of continuing residential segregation in American cities, which in some cities approaches 100 percent.

In another study, Taeuber (1965) examined the factors usually cited as an explanation of residential segregation: choice (Negroes prefer to live with Negroes), poverty (Negroes can't afford to live in other areas of the city), and discrimination. He concluded that only *discrimination* sufficiently explains the residential segregation of American cities. For this reason the various social and legal measures designed to alleviate the problem but which do not deal with patterns of social discrimination have had little impact on residential segregation. Not until these patterns of social discrimination, as exercised particularly through housing discrimination, are altered can any change in residential segregation be expected.

From a population geneticist's point of view, the effect of such residential segregation is to create racial homogeneity in person-to-person contacts on the street, in the stores, schools, and other neighborhood facilities. As a result, social contacts between the two populations are limited, mating across population boundaries is minimized, and mating within each separate population is maximized. Discrimination, as it is exercised by the institutions and individuals of white society, places the Negro and white populations of the United States in varying degrees of social, and consequently genetic, isolation from one another. In a like manner, the cultural and behavioral patterns of the Hopi Indian population and those of the Amish population (Chapter V) operate to enforce these populations' social and genetic isolation in the midst of larger populations of different cultural identity. Although their social isolation is maintained more by choice than that of the American Negroes, the consequences are the same—relatively closed gene pools and genetic interpopulational differences.

Identifying the Breeding Population

In addition to the problem introduced by the biological openness of human-population systems, accurate definition of a human Mendelian population is complicated by the fact that man clusters in social groupings which may or may not serve as biological breeding units. For example, one might mistakenly regard any American city as a Mendelian population, whereas the true structure of the city is that of a political unit occupied by numerous so-called "ethnic" or "racial" groups which are the real biological breeding units—clusters of Japanese, Chinese, Puerto Ricans, Mexican Americans, Negroes, and so on. Only a careful *demographic* investigation of the structure and dynamics of a living group will distinguish between the primarily social or political units and the true biological units. This type of demographic analysis is well-illustrated by the study by Salzano and his colleagues on the Xavante Indian villages of Brazil. The authors describe the Xavante social and political group as the village; yet as they point out, this unit cannot be considered the biological breeding unit. Rather, they say, "the picture which now emerges is of constant, continuing realignment among groups within the population, of such a degree that . . . over a period of several generations there should be so much exchange between 'villages' that the breeding unit approximates the *entire tribe*."

The first problem of the population geneticist, therefore, is to identify and describe, as accurately as possible, the biological population before he can undertake an analysis of the gene pool and the forces acting on it. Because direct analysis of a population's gene pool is impossible, all conclusions regarding its composition are necessarily *inferential*, and must be made on the basis of direct examination of the phenotypes of the reproducing individuals. To infer the composition of a gene pool at a single point in time the population geneticist must first enumerate and describe the *actual progenitors*, that is, the parents in a population. These progenitors constitute the *breeding population* (N), which is always much smaller, usually by about one-third, than the total number of individuals actually living in a group. The breeding number (N) cannot be obtained by census alone, but only by a careful demographic analysis wherein age groups are distinguished, mating behavior is documented, and reproductive histories are elicited. In practice, this type of investigation is very difficult to conduct, especially if the geneticist or demographer must work with primitive populations where no records are kept of such vital events.

On the basis of the demographic analysis of three Xavante villages, Salzano and his co-workers conclude that 44 percent of all individuals enumerated are "progenitors"; this proportion includes all males aged 20 to 45 years and all females aged 15 to 40 years. By applying the figure of 44 percent to the total Xavante population of approximately 1800 persons, the authors calculate that the Xavante breeding population (N) consists of 792 persons. Among the Xavante the age categories can be used to enumerate the breeding population (N), because all women are married after the age of 15 years and sterility is so uncommon that all adult women are considered as contributing to the next generation. Thus, adult age status and parenthood are equated in this population, but in many populations where sterility may be as high as 10 percent among all couples, age and parenthood are totally unrelated in some individuals. In most populations N can be correctly enumerated only by identifying all successfully reproductive adults or "parents," as opposed to identifying all adults of reproductive age.

When identifying the parents or progenitors of a population, the important point is that the next generation's gene pool is determined by the actual reproductive matings that occur in the population, rather than by the social rules governing mating in the population. Although a sharp distinction may be drawn socially between the system of mating approved or legitimatized by a society in the form of "marriage" and that illegitimate mating which takes place outside the socially sanctioned system, the distinction is unimportant genetically. For this reason the matings in a population must be enumerated and analyzed on the basis of actual reproductive performance rather than in terms of the "marriage" system. The need for this approach is documented by Salzano and his colleagues. They report that the Xavante, in comparison to other tribes, "appear to be relatively restrained in extramarital relationships." But even in this "restrained" group, of the 107 children genetically analyzed, the authors report nine apparent "exclusions"; that is, children whose genotypes were not consistent with those expected on the basis of the genotypes of the socially recognized parents. This genetic data, of course, raises questions about the children's biological paternity and the actual matings involved.

A more striking illustration of the genetic consequences of mating outside the socially sanctioned marriage system is provided by the American Negro populations of the United States. The gene pools of these Negro populations contain varying amounts of genes derived from the surrounding white populations (see T. E. Reed and Workman, *et al.*, Chapter VI). Much of the flow of genes from the white into the Negro populations has occurred through socially disapproved interracial matings, primarily of Negro females with white males. Such matings began at some early point in the history of American slavery, probably before 1700, establishing a pattern of interracial mating which continues today. Until 1952, 30 states had statutes prohibiting and punishing interracial marriages (*miscegenation*). Although between 1952 and 1967, 14 states repealed these laws, not until 1967, when the Virginia law against interracial marriage was declared unconstitutional by the Supreme Court of the United States (Loving v. Virginia, 388 U.S. 1, 1967), did the antimiscegenation laws in the other 15 states lose their legal force. Yet in these same states, for example Georgia, the gene pool composition of the resident Negro populations indicates that as much as 12 percent of their component genes are derived from white populations. It is clear, then, that the genetic composition and evolutionary dynamics of the American Negro populations can be understood only within the context of the actual mating behavior practiced by the populations involved, rather than within the context of the legal rules governing mating patterns. In summary, the important fact is that the gene pool of a population is determined by *fertile matings* rather than by fertile marriages alone.

Effective Breeding Size

After demographic analysis of a population has enabled the investigator to enumerate and define the breeding population, he may need to further refine the estimated N. In order to assess the potential effects of the various forces of evolution on the gene pool, especially the force "genetic drift" (see Chapter V), the population geneticist must also determine the *effective breeding size* (N_e) of the population. The effective breeding size (N_e) is the breeding size (N) reduced in such a way that the remaining number is equivalent to the number of individuals in the "ideal" population. The reason that the actual breeding population must be corrected to its effective size is that the population geneticist wishes to assess the impact of the determinate and indeterminate modes of change operating on the gene pool of the population. The assessment is done using mathematical models which predict the direction and magnitude of genetic change expected within the gene pool of the "ideal" population. Therefore, before it is possible to apply formulae to estimate genetic change within a real population, an "ideal" gene pool must be isolated within the real population, to the extent that the investigator is practicably able to do so.

In the "ideal" population's gene pool a random gamete has equal probabilities of having come from any parent, all of whom are unrelated, and of mating with any other gamete. However, every real human breeding population contains numbers of gamete-carrying individuals who are biologically related to one another to some degree. These individuals, to the extent they are related, do not reflect the size, composition, and level of heterozygosity of the "ideal" gene pool, because the actual number and type of unrelated gametes available to the next generation is less than the actual total number of gametes (related and unrelated) present. The chances of a gamete mating with an unrelated gamete are reduced accordingly, and theoretically heterozygosity in the gene pool, with respect to a pair of alleles, will decrease (*decay*) at the rate of $\frac{1}{2}N_e$ per generation. Thus, because of the presence of related individuals, the effective breeding size (N_e) of any human population is generally smaller than its breeding size (N). For example, correcting the Xavante breeding size, $N = 792$ to the "ideal" standard, Salzano and his colleagues arrive at an effective breeding population $N_e = 470$ or 661, depending on which figures are used in the estimate. In other populations, however, particularly in large populations, the difference between N and N_e may be insignificant.

The number of biologically related individuals in a population is increased in the presence of one or more of four situations. Whenever any of the following four situations is encountered the size of the breeding population (N) must be reduced to the size of the effective breeding population (N_e): (1) unequal numbers of the two sexes; (2) cyclic variations in population size; (3) inbreeding; and (4) reproductive inequalities among individuals, that is, differential fertility (Li, 1963). By applying certain formulae which correct the breeding size for these reductive situations, the population geneticist is able to calculate the "ideal" size (N_e) of the population under consideration.[1] For example, in the case of two breeding

---

[1] The four formulas follow.
1. *Unequal numbers of sexes:*

$$N_e = \frac{4 N_f N_m}{N_f + N_m}$$

N_f is the number of breeding females, N_m the number of breeding males, and $N_f + N_m = N$ (the breeding population).

2. *Cyclic variations in population size:*

$$\frac{1}{N_e} = \frac{1}{t} \left[\frac{1}{N_1} + \frac{1}{N_2} \cdots \frac{1}{N_t} \right]$$

N_e is the harmonic mean of the population size of the various generations over the cycle of t generations with breeding population numbers $N_1, N_2, \ldots N_t$.

3. *Inbreeding:*

$$N_e = \frac{N}{1 + F}$$

populations where $N = 80$, and one population has 40 males and 40 females, and the other 10 males and 70 females, the latter population must be corrected for the inequality of sexes. By applying the formula

$$N_e = \frac{4N_f N_m}{N_f + N_m}$$

it can be calculated that the modes of change will operate in this breeding population ($N = 80$) as if there were actually only

$$\frac{4(70)(10)}{70 + 10} = 35$$

breeding individuals in the group, equally divided between the sexes. Clearly, these two populations and their evolutionary dynamics cannot be equated on the basis of their equal breeding size, $N = 80$. In the first population, which contains an equal number of sexes, the breeding population is equal in size to the effective breeding size ($N = N_e = 80$), but in the second population, with its unbalanced sex ratio, the two are unequal ($N = 70$, $N_e = 35$). Not correcting for the unequal proportion of sexes in the second population ignores the genetic consequences of the disproportionate gametic contribution by the 10 males to the next generation's gene pool. Among the Xavante an unequal sex ratio among the progenitors due to the high frequency of polygynous marriages is the most important factor necessitating the reduction of N. For example, the six men with four or five wives produced two or three times more children than other males.

In practice, precise estimation of the effective breeding size is a difficult, if not impossible, task. First, all the situations which reduce the breeding size often operate simultaneously within the population, as they do among the Xavante, making precise mathematical manipulation of the interacting parameters impossible. Second, as Salzano and his co-workers point out, adequate allowance cannot be made in most living populations for other factors which tend to reduce N_e. For example, past fluctuations in population numbers as the result of war, famine, migration, or disease have

---

F is the inbreeding coefficient. In the extreme case where inbreeding is complete, that is, $F = 1$, N_e is one-half the breeding size.

    4. *Reproductive inequalities among individuals:*

$$N_e = \frac{4N - 2}{\sigma_k^2 + 2}$$

Here k is the number of gametes left by a parent and N is the actual number of parents. Variance of k (that is, σ_k^2) measures the distribution of the number of progeny per parent, that is, the reproductive variability in a population. When there is a random sampling of gametes each generation, $N_e = N$. When $\sigma_k^2 > \bar{k}$, as is the case in most human populations, N_e is smaller than N.

(This material has been modified from Li, 1963).

probably had an extremely important impact on the breeding size of populations. Such past events would substantially affect the present size of N_e, but the impact of unrecorded historical events is impossible to assess and include in any estimations of N_e.

The Structure and Dynamics of Populations

The evolutionary forces acting on a gene pool are external to it and derive from the environment within which the gene pool exists. What factors constitute the "environment"? The environment includes not only all natural features, such as temperature, altitude, and like phenomena, but also a population's culture and structure, of which the gene pool is an intrinsic, albeit abstract, part. Thus, in order to understand the dynamics of genetic evolution, one must appreciate the structure and dynamics of a population, which in turn determine the size and composition of the gene pool.

Every population has a unique *structure* which can be partially described in terms of its distribution, size, and age and sex composition. The structure itself is determined by the vital processes of *fertility*, *mortality*, and *migration*. These vital processes, in their differential operation, are responsible for the structural dissimilarities that exist between populations at any given point in time and within a population through time. For example, as Farley and Taeuber point out, the age composition of Negro populations in American cities is more youthful than that of white populations, and therefore more conducive to high rates of natural increase from fertility. The Amish population is also characterized by a high fertility rate and consequent rapid growth, it having multiplied five-fold in the same period of time that the population of the United States merely doubled. The comparative population profiles of the Amish population and rural non-Amish populations show the structural consequences of the higher fertility of the Amish. Here individuals under twenty years of age comprise a significantly greater proportion of the population than in non-Amish groups (see McKusick, *et al.*, Chapter V).

Changes in the vital process through time also account for the changes that have occurred in the structure of a population historically. For example, as a country undergoes increasing industrialization, the distribution of people is radically altered by migration from rural into urban environments. The United States presents a striking case in point, as it has undergone and continues to undergo the demographic changes which accompany industrialization. In 1850, 85 percent of this nation's total population of 31 million people were classified as "rural" (Potter, 1965). By contrast, only 27 percent of over 203 million people were rural in 1970. This altered distribution of people is the result of continuing migration from rural areas into urban and, especially, metropolitan areas. Since 1940, many rural areas have lost between one-fourth and one-half of their population, whereas over 80 percent of the total population growth of the nation was in the metropolitan areas in the ten-year period from 1950 to 1960 (Hunt, 1966). In 1968, 35 percent of the nation's population were residing in the suburbs as compared to the 24 percent who resided there in 1950. In less than two decades the number of people living in the suburbs has increased by 47 million (Wattenberg, 1970)!

This migration has been accompanied by other changes in the vital processes. For example, because urbanization intensifies certain environmental elements that are, or can become, unfavorable to fertility (such as the relative net cost of rearing children in the city), fertility tends to drop in an urbanized society. The demographer Dudley Kirk (see Chapter IV) notes that the reduction in fertility which has occurred in the United States since 1800 has been dramatic in that the average number of children per woman has dropped from seven in that year to three or less today. Such pronounced changes in the vital processes of fertility, mortality, and migration have marked effects not only on the structure of a population, but also on its gene pool, although as Kirk points out, the exact genetic consequences may not be easily detectable or predictable.

It is clear that the evolution of a population can be understood and studied only in terms of that population's structure and dynamics. However, this essential, basic approach to investigating the genetics of human populations was neglected by physical anthropologists and geneticists for many years. Because the earlier investigators were primarily interested in the genetic classification and comparison of populations, a typological approach developed, which focused on the statics rather than the dynamics of populations. In addition, the statistical description and analysis of the structure and dynamics of populations were traditionally the work of demographers who were not concerned with the genetic evolution of populations. For these reasons, the early genetic work done in populations provided little more than gene frequencies and racial taxonomies of human groups throughout the world. The typological approach yielded no information concerning the dynamic forces acting on the gene pool of populations, and consequently contributed little to the understanding of evolution as it operates in human populations. The neglect of demographic data is especially regrettable in view of the fact that the formal mathematical models describing evolution in populations had already been well-elaborated by Fisher, Haldane, and Wright. Fortunately, the gap between human population genetics and demography is now being bridged (see summaries in Bajema, 1967; and Neel, 1968b). Studies by people such as D. F. Roberts on African populations (1956, 1956/1957), J. V. Neel and F. Salzano on South American tribes (Salzano, 1961, 1964; Neel, et al., 1964; Neel, 1968a), B. Bonne on the Samaritans (1963, 1966) and N. Freire-Maia on Brazilian populations (and A. Freire-Maia, 1963; and Krieger, 1963) successfully combine the methods of demography and genetics.

These more recent investigations have confirmed the observable fact that the human species presents an exceedingly complex system of populations. The large, panmictic Hardy-Weinberg population maintained in equilibrium is an ideal that has not been achieved by any human population. Man clusters in groups of fluctuating size and distribution; he reproduces or does not reproduce for reasons that are often strictly nonbiological; and he usually chooses his mate in a very selective (nonrandom) way. One form of nonrandom mate selection is *assortative mating*, which exists when the mates have more attributes in common than would be expected by chance. The most obvious type of assortative mating is by color of the skin, but there are innumerable other more subtle traits by which man chooses his mate. In the paper included here the sociologist Bruce Eckland discusses some of the various psychological, cultural, social, and geographical factors which nullify the randomness of the mating process.

It is obvious that a large amount of assortative mating occurs in all human populations, but the genetic consequences of assortative mating are practically unknown. Assortative mating affects the gene pool of a population only to the extent that the traits used in mate selection have some genetic basis, and there are little data showing this relationship. The physical anthropologist J. Spuhler (1969) has reviewed the results of studies dealing with assortative mating solely as it is related to certain physical characteristics of individuals in human populations. With the exception of age and stature, none of the physical traits investigated were strongly associated with mate selection. In his own studies on the population of Ann Arbor, Michigan, Spuhler (1962) found significant assortative mating for body size and intelligence, as measured by scores on three cognitive tests.

But numerous methodological difficulties are involved in dealing with quantitative physical characters, such as stature, body size, and intelligence, which are often the traits underlying mate selection. First, because these traits are extremely responsive to environmental influences, exact determination of their genetic component is impossible. Second, it is difficult to correlate nonphysical traits, such as social class, that may directly determine mate selection with the more indirectly involved physical traits such as height. In view of these problems, the question regarding the influence assortative mating has had, or will have, on the genetic structure of human populations remains unanswered. Yet, as Eckland speculates, a better understanding of the dynamics of mate selection may become one of the outstanding objectives of future investigators as other factors, such as mortality, become less important in the future evolution of human populations.

In a chapter dealing with population as the basic unit of study, enough emphasis cannot be given to the important role the physical environment plays in shaping the evolutionary path of the population. Yet, although the paramount relationship of human populations and their gene pools with their environments is indisputably recognized, it remains almost totally unanalyzed. Modern human ecology, which deals with the relationship of man to his environment, has developed in relative isolation from the field of population genetics, and a body of theory is therefore lacking which can predict particular biological events in ecological as well as evolutionary time. The mathematical models developed by Wiesenfeld (see Chapter IV) to describe the complex interactions in Africa of population size, density-dependent mortality from malaria, and frequency-dependent genetic fitness as bestowed by the sickle-cell trait represent a unique attempt to examine a single limited ecological-genetic problem. Certainly, one can predict that one of the most promising future developments of human-population biology will be in just such studies, which will bridge the gap between population genetics and population ecology.

Demographic Analysis of a Population

See the following reading: "Demographic Data on Two Xavante Villages: Genetic Structure of the Tribe," by Francisco M. Salzano, James V. Neel, and David Maybury-Lewis, p. 46.

Human Behavior and Population Dynamics

See the following readings: "Population Trends and Residential Segregation since 1960," by Reynolds Farley and Karl E. Taeuber, p. 75; and "Theories of Mate Selection," by Bruce K. Eckland, p. 86.

REFERENCES TO THE LITERATURE CITED IN THE OVERVIEW

1. BAJEMA, C. J., 1967. Human population genetics and demography: A selected bibliography. *Eugen. Quart.*, 14: 205–237.
2. BONNÉ, B., 1963. The Samaritans: A demographic study. *Hum. Biol.*, 35: 61–89.
3. _____, 1966. Genes and phenotypes in the Samaritan isolate. *Amer. J. Phys. Anthropol.*, 24: 1–19.
4. FREIRE-MAIA, N. AND A. FREIRE-MAIA, 1963. Migration and inbreeding in Brazilian populations, p. 97–121. In J. Sutter (ed.), *Les Déplacements Humains; Aspects Methodologiques de leur Mésure.* Paris, Hachette.
5. _____ AND H. KRIEGER, 1963. A Jewish isolate in Southern Brazil. *Ann. Hum. Genet.*, 27: 31–39.
6. HUNT, E. P., 1966. *Recent Demographic Trends and their Effects on Maternal and Child Health Needs and Services.* U. S. Dept. of Health, Educ. and Welfare, Children's Bureau. Washington, D.C., U. S. Government Printing Office.
7. LAUGHLIN, W. S., 1967. Race: a population concept. *Eugen. Quart.*, 13: 326–340.

8. NEEL, J. V., 1968a. The demography of two tribes of primitive relatively unacculturated American Indians. *Proc. Nat. Acad. Sci., 59: 680–689.*

9. _____ (Chairman), 1968b. Symposium on genetic implications on demographic trends. *Proc. Nat. Acad. Sci.,* 59: 649–699.

10. _____, F. M. SALZANO, P. C. JUNQUEIRA, F. KEITER, AND D. MAYBURY-LEWIS, 1964. Studies on the Xavante Indians of the Brazilian Mato Grosso. *Amer. J. Hum. Genet.,* 16: 52–140.

11. POTTER, J., 1965. The growth of population in America, 1700–1860, pp. 631–689. In D. V. Glass and D. E. C. Eversley (eds.), *Population in History.* London, Edward Arnold.

12. ROBERTS, D. F., 1956. A demographic study of a Dinka village. *Hum. Biol.,* 28: 324–349.

13. _____, 1956/1957. Some genetic implications of Nilotic demography. *Acta Genet.,* 6: 446–452.

14. SALZANO, F. M., 1961. Studies on the Caingang Indians. *Hum. Biol.,* 33: 110–130.

15. _____, 1964. Demographic studies on Indians from Santa Catarina, Brazil. *Acta Genet. Med. Gemell.,* 13: 278–294.

16. SPUHLER, J. N., 1962. Empirical studies on quantitative human genetics, p. 241–252. In *United Nations World Health Organization Seminar on the Use of Vital and Health Statistics for Genetic and Radiation Studies,* 1960. Geneva, World Health Organization.

17. _____, 1969. Assortative mating with respect to physical characteristics. *Eugen. Quart.,* 15: 71–84.

18. TAEUBER, K. E., 1965. Residential segregation. *Sci. Amer.,* 213: 12–19.

19. WATTENBERG, B., 1970. The nonsense explosion: Overpopulation as a crisis issue. *New Republic,* April: 18-23.

Demographic Data on Two Xavante Villages: Genetic Structure of the Tribe

FRANCISCO M. SALZANO, JAMES V. NEEL, and
DAVID MAYBURY-LEWIS

The Akwẽ-Xavante are a tribe of Gê-speaking Indians of the Brazilian Mato Grosso, comprising some 1,500–2,000 individuals now divided
into at least seven villages scattered along the Rio das Mortes and its
environs, between latitudes 12° and 16° S. Our contacts with the tribe
date back to 1958, when one of us (D. M.-L.) undertook an anthropological study of the village which is located at Sao Domingos near
the Post Pimental Barbosa of the Indian Protective Service. A preliminary, multidisciplinary study of this same village was carried out in
1962 (Neel *et al.*, 1964). The present series of papers will report
the results of multifaceted studies on two additional villages of this
tribe, the field work having been performed in 1963 and 1964.

The Xavante (we shall omit the prefix) were chosen as subjects for
study because they appeared to meet a rather exacting combination
of prerequisites. All three of the villages studied are at a stage where
their culture, although disturbed, is essentially intact; yet they are sufficiently acculturated as to be amenable to contacts of the type to be
described. However, dietary, disease, and other patterns are being
altered rather rapidly as the Xavante become less nomadic, to the extent that in another ten years we doubt whether these villages would be
suitable for this type of study. We hasten to add a very necessary stric-
ture, that we do not present this group as "untouched," but, rather, as

Reproduced by permission of The University of Chicago Press and Francisco Salzano from
American Journal of Human Genetics, *19:463–489 (1967). Copyright 1967 by Grune & Stratton, Inc. This article originally appeared under the title "I. Demographic Data on Two Additional Villages: Genetic Structure of the Tribe."*

The investigations reported in this and subsequent papers in this series were supported
in part by grants from the Pan American Health Organization, the U. S. Public Health
Service, the U. S. Atomic Energy Commission, and the World Health Organization.

We are grateful to the Serviço de Proteção aos Indios for the necessary authorizations
and facilities. It is a pleasure to acknowledge our debt to Mr. Girley V. Simões for technical
assistance in the field and to Mr. Arthur McIntosh, pilot for Azas de Socorro, without
whose staunch assistance this study might not have succeeded. We further acknowledge the
generous hospitality of Sr. Pedro Vani de Oliveira, agent at the Post Simões Lopes of the
Serviço de Proteção aos Indios, and Padre Mario Panziera, chief of the São Marcos
Salesian Mission. Finally, we are indebted to Drs. C. A. B. Smith, W. J. Schull, F. B. Livingstone, and N. A. Chagnon for their careful reading of various of the manuscripts in this
series.

Francisco M. Salzano, *a population geneticist, was born in Brazil in 1928. He received his B.S. and Sc.Lic. from the Federal University of Rio Grande do Sul in 1950 and 1952 respectively, and his Ph.D. from the University of São Paulo in 1955. He is presently the Director of the Natural Science Institute of the Federal University of Rio Grande do Sul in Brazil. His research in population genetics has focused on the population structure and genetic polymorphisms of primitive and modern populations and the genetic admixture of the main ethnic groups in Brazil. A veteran of numerous field expeditions, he has conducted demographic and genetic investigations among various Indian groups of Brazil, including the Caingang, Cayapo, Yanomama, and, the subject of his article reproduced here, the Xavante of Central Brazil. He is a past president of the Brazilian Society of Genetics, a member of the Special Committee of the International Biological Program which coordinates IBP activities all over the world, and a member of the World Health Organization Expert Advisory Panel on Human Genetics. His books dealing with genetics and the evolution of human populations include* Populações Brasileiras, Aspectos Demográficos, Genéticos e Antropológicos, *with F. M. Freire-Maia and N. Freire-Maia (São Paulo, 1967);* Problems in Human Biology: A Study of Brazilian Populations *(Detroit, 1970); and* The Ongoing Evolution of Latin American Populations, *Ed. (Springfield, Mass., 1970).*

about at that stage of acculturation compatible with achieving our objectives.

This first paper will present data on the pattern of population exchange between Xavante villages and then describe the vital statistics and breeding structure of the two additional villages, in an effort to elucidate the genetic parameters of a tribe enjoying a hunting-gathering-incipient agriculture type of economy. Subsequent papers in the series will describe the physical status of the group, the frequency of occurrence of a variety of genetic traits, and the results of biochemical and serological studies aimed at evaluating the biological pressures on the group. The final paper in the series will attempt a synthesis of the implications of the various papers, in the form of a series of propositions or hypotheses.

As will become apparent, the detailed scrutiny of populations of this type raises a number of far-reaching questions concerning the applicability of various commonly employed genetic statistics and concepts. In this series of papers, we shall be more concerned with defining these questions clearly than in proposing solutions based on the study of a single tribe. We and our collaborators have in progress studies on two other tribes at the cultural level of the Xavantes, namely, the Cayapo of Brazil and the Yanomama (Waica) of Venezuela and Brazil. It is anticipated that when the results of all these studies are combined, not only will we be in a sounder position to develop genetic models of human populations at this level and of their biological pressures, but there will be a corpus of data suitable for (1) defining the heritability in

James V. Neel, *a physician and geneticist, was born in Ohio in 1915. He received his B.A. from the College of Wooster in 1935 and his Ph.D. and M.D. from the University of Rochester in 1939 and 1944 respectively. Presently he is the Lee R. Dice University Professor of Human Genetics, Chairman of the Department of Human Genetics, and Professor of Internal Medicine at the University of Michigan. His major interests are in clinical and population genetics and he has authored or co-authored some two hundred papers on these topics. He has conducted field research among human populations in Central and West Africa, Venezuela, and Brazil. Under the auspices of the Atomic Bomb Casualty Commission, he participated with William J. Schull in the extensive demographic and genetic studies of the Japanese populations of Hiroshima and Nagasaki which were designed to assess the genetic effects of the atomic bombing. Past president of the American Society of Human Genetics, he received that organization's Allen Ward in 1965 for outstanding contributions in the field. He is currently a member of the National Academy of Sciences and an editor for* Human Genetics Abstracts *and* Behavior Genetics. *His published books include* Human Heredity, *with W. J. Schull (Chicago, 1942);* The Effect of Exposure to the Atomic Bombs on Pregnancy Termination in Hiroshima and Nagasaki, *in collaboration with R. C. Anderson, et al. (National Academy of Sciences National Research Council Publication 461, Washington, 1956);* Changing Perspectives on the Genetic Effects of Radiation, *with W. J. Schull in collaboration with Arthur L. Drew, et al. (Springfield, Mass., 1963);* The Effects of Inbreeding on Japanese Children, *(New York, 1965).*

this environment of a range of human characteristics, (2) identifying new genetic traits, and (less surely) (3) contributing, through an analysis of fertility differentials and segregation ratios, to an understanding of the action of selection on specific genetic systems. The apparently greater uniformity of the environment at this cultural level may contribute to the realization of these various objectives, since a significant fraction of the "noise" present in more civilized cultural systems may be absent.

Recent History of the Xavantes

Detailed accounts of the history and culture of the Xavantes, as well as recent changes in their way of life, will be found in Maybury-Lewis (1965a,b, 1967; see also Neel *et al.,* 1964). We will present here only that background material essential to developing the genetic arguments which follow.

It is clear from historical references that at one time the Xavantes were found considerably to the east of their present distribution. It is doubtful that at that time their tribal distribution also included the area they now occupy. It is possible they have retreated into this area no more than a century ago, at that time perhaps displacing other Indian groups westwardly. After an initial period of peaceful contacts in the late eighteenth century, relations with neo-Brazilians deteriorated and were in general hostile during the nineteenth century and the

David Maybury-Lewis, a social anthropologist, was born in Hyderabad, India in 1929. He received a B.A. in 1952 and an M.A. in 1956 from Cambridge University, a M.S. in 1956 from the University of São Paulo, an M.A. in 1956 and a Ph.D. in 1960 from Oxford University. He is currently Professor of Anthropology at Harvard University. His field experience has been among Indian tribes of Brazil. His publications in this area include The Savage and the Innocent (London, 1965); and Akwẽ-Shavante Society (Oxford, 1967).

first 40 years of this century. In 1941, in an effort at pacification, the Indian Protective Service located a Post (now known as Pimental Barbosa) on the Rio das Mortes at a place designated as Sao Domingos (see Fig. 1). Shortly thereafter, the Xavantes killed a party of six from the Post who were attempting to establish peaceful contacts, and it was not until 1946 that amicable contacts were established, although for the next five years these were rare and guarded. Only since 1951 has the Post been in a more-or-less permanent relationship

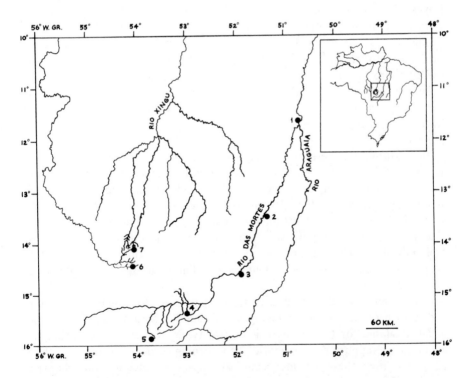

Fig. 1. The present location of the eight Xavante villages. Key: (1) São Felix (one small village near here, one larger some distance off, exact position unknown); (2) São Domingos; (3) Areoes; (4) São Marcos; (5) Sangradouro; (6) Simões Lopes; (7) Batoví. [Redrawn.]

with the Indians. During the last twenty years, peaceful contacts have also been established with some six or seven other Xavante groups; it is difficult to be accurate because of the shifting character of Indian villages. Although there may be groups still uncontacted, this seems doubtful.

One of the two villages studied in 1963–1964 is located near the Simões Lopes Post of the Indian Protective Service (see Fig. 1). This Post originally was established for the Bakairí Indians, who are now well acculturated. The Xavante first made contact here in 1955, but after a brief period of residence at the Post (and uncertain relations with the Bakairí) withdrew and established a village about 12 kilometers off. A few months earlier, another band of Xavantes had made contacts with an Indian Post at Batovı, some 40 kilometers to the north, originally established to attract the Waurá. The group at Simões Lopes was apparently an offshoot of the Batoví group, remaining on such cordial terms with the group that, as we will see, the two villages may be considered one breeding unit. A few of the oldest Xavantes at Simões Lopes claimed kinship with the community at São Domingos. There were approximately 230 persons in the Simões Lopes group in 1964.

The second of the villages studied in 1963–1964 is located in close proximity to the Salesian Mission of São Marcos. The group now settled there made contacts in 1956 with the Salesian Mission at Merure, originally established for the Bororo tribe. The Xavantes and the Bororos did not get along well, so in 1958 the Salesians set up a mission at São Marcos especially for the Xavante. This village now numbers about 400 persons.

In addition to these two villages, there are currently six other villages, as follows: (1) São Domingos, the village studied in 1962, current estimated population, 150; (2) Batoví, already mentioned, current estimated population, 200; (3) Sangradouro, current estimated population, 150; (4) Areoes, current estimated population, 170; (5) two villages near São Felix, one of 50 quite near the settlement and one of 250, the most isolated of all the Xavante villages today, more than 100 kilometers from São Felix, exact position unknown. The location of the villages is shown in Fig. 1.

One of our specific interests in this paper is to document the amount and nature of the genetic exchanges between Xavante Villages. The material available for this purpose is of two types, namely, some rather inadequately supported information concerning events during the first 12 years after these groups began to establish peaceful contacts with neo-Brazilians and some more adequately documented material relating to events since 1958, the year in which one of us (D. M.-L.) first began anthropological studies of the group. In the field work of 1962, we were impressed by the relatively high degree of endogamy and isolation of the village at São Domingos. At the same time, it was recognized that the chief of that village was an unusually dominant figure, who might have discouraged exchange with other villages. In the course of that field work, it was found that the village had undergone a schism since the initial work in 1958, with the formation of an offshoot village (E Tõ) some 15 kilometers up-river from the parent village, and we were able to present some information concerning the biological lines along which such a schism occurs. By good fortune, the present round of field studies has enabled us to document the results of this schism more accurately.

With respect to the inadequately documented events between 1946 and 1958, it should first be noted that Xavante villages are temporary in character and are readily moved as circumstances require. There is reason to believe that most of the villages mentioned above have assumed their present general location only recently. Padre Mario Panziera, head of the Salesian Mission at São Marcos since its founding, states that an aerial reconnaissance of the region north of Merure undertaken by the Salesians in 1946 revealed a very large Xavante village on the shores of a small lake to the north of the Rio das Mortes (the so-called Lagoa Group). The "twin villages" near Batoví and Simões Lopes are thought to be an offshoot of that village; the time of their departure from the mother village is uncertain. An article in the *Bolletino Salesiano* of January 1, 1954, concerns the recent arrival of a group of Xavantes at the town of Xavantina and includes an aerial photograph of the above-mentioned village, allegedly after these departures. Twenty-two houses can be counted, one of which is probably a bachelor's hut. If each house contains on the average 20 persons (a conservative estimate), this implies a population even then of some 400 persons. The band described above at first making contacts in 1956 with the Salesian Mission to the Bororo Indians located at Merure and then settling at São Marcos is thought by the Salesians also to have originated from the Lagoa group, as is a smaller group which made contacts in 1956 with the Salesian Mission at Sangradouro, originally established for the Bororos. In view of the numbers involved, the Lagoa group would have to have been unusually large to give rise to all four of the villages attributed to it, and it seems quite likely that additional groups may be involved in these developments. On the other hand, Sadock de Freitas (1955) states that there were 618 Xavantes at the village near São Domingos when he visited there in 1954, and the possibility of temporary groupings of this size cannot be discounted. This same author reports that while he was in contact with the village near São Domingos, it was visited by a band from the Lagoa group—an independent confirmation of the existence of this group.

Turning now to more adequately documented recent events, we mentioned above the schism involving the village at São Domingos (cf. Neel *et al.*, 1964). It now appears there have been two schisms since 1958, which have reduced the size of the village from approximately 220 persons to approximately 110. The schism reported in our paper of 1964 resulted in the establishment of a new village about 15 kilometers up-river from São Domingos (E Tõ). When we arrived in São Marcos in May, 1964, we determined that São Marcos had recently received an accretion from E Tõ. This latter village had now dissolved, one fragment coming to São Marcos and another joining the Areoes group located near Xavantina. A few individuals may have returned to the parent village at São Domingos. Twenty-six persons from this group were examined at São Marcos; there may have been a few more persons in this group who were not seen.

The São Marcos group also had received another larger addition since its founding, a group of approximately 90 under the leadership of one Sebastião. This group appears to have joined the São Domingos group about 1959, coming from a now dissolved village at Capitariquara, and to have "seceded" (in whole or part) from the São Domingos group in 1960, and therefore results from an earlier schism than that described above. After brief contacts with several other Xavante communities, this group had arrived at São Marcos that same year. Anyone familiar with both the linguistic and cultural problems inherent in obtaining precise histories of population movement under these circumstances will perhaps be sympathetic to our previous failure to recognize two schisms at São Domingos between 1958 and 1962, and to the difficulties in balancing the numbers involved. Furthermore, the one group came and went in the interval between our contacts; the Indians remaining in the village would regard them as "visitors" even if marital partners were exchanged.

The composition of Sebastião's group is presented in detail in Fig. 2 to illustrate the biological interrelationships in such a group, as best they can be reconstructed. Deceased persons are shown only when they establish links of relationship. There may be presumed to be significant biological relationships between the individuals shown in generations 1 and 2 beyond those we were able to elicit. The striking feature of the group is the degree to which one sibship of five (of which one member is still alive) dominates the picture. We had hoped to compute the number of independent genomes present in the group, but the data are inadequate for this purpose.

Much additional data concerning the movement of other Xavante groups not involved in the present round of studies will be found in Maybury-Lewis (1965b, 1967). The picture which now emerges is of constant, continuing realignment among groups within the population, of such a degree that, although at any one moment there are many con-

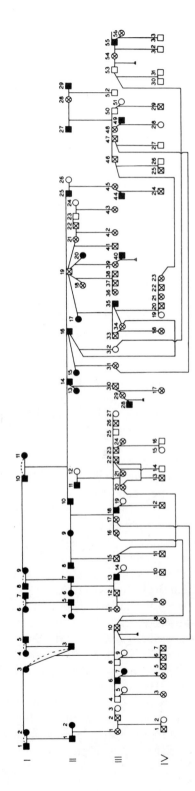

Fig. 2. The pedigree of Sebastião's group, illustrating the biological lines along which village cleavages occur. The dotted lines connecting certain marriages indicate uncertainty as to whether the relationships should be reversed! Individuals indicated by solid symbols are deceased; those indicated by a × were examined.

straints upon the choice of a marriage partner, over a period of several generations there should be so much exchange between "villages" that the breeding unit approximates the entire tribe. *We do not know, to what extent this is a recent phenomenon, in response to the increasing contacts with neo-Brazilians.* Further work on tribes even less disturbed in their social structure is obviously necessary. *In general, we believe that recent events have not created but only increased the internal mobility of the Xavantes.* If this is correct, then it is clear that one may derive a very biased picture of the tribal dynamics of this (and presumably any other) Indian tribe during the course of a brief contact.

Demographic Data

Demographic data on the villages at Simões Lopes and São Marcos were obtained as described by Salzano (1961, 1964) and Neel *et al.* (1964). In São Marcos, genealogies and reproductive histories were obtained independently by F. M. S. and D. M.-L., with a later comparison of the findings and attempt to resolve any discrepancies. Where final differences remained between the two sets of data, the findings of F. M. S. have been utilized, since, aside from the possibility of real discrepancy, the population base is shifting constantly, and the data of F. M. S. correspond to the time of the detailed serological and medical studies. The relatively small discrepancies between the two sets of histories would in no way influence the general conclusions. In Simões Lopes, only F. M. S. collected demographic data. Ages have been estimated by visual inspection and from a knowledge of the age-set system (cf. Maybury-Lewis, 1967). While we believe these data are as accurate as can be obtained under the circumstances, it must be emphasized that the information is undoubtedly less precise, and certainly covers a shorter time space, than would be true in a literate culture. In particular, attention must be directed towards the difficulties which the general cultural milieu of this group, and especially a classificatory kinship system, creates for obtaining precise reproductive data. In a later paper we shall discuss how the extensive blood typings which were carried out enable one to check on the accuracy of the histories obtained.

In the tables to follow, we will routinely compare the findings in the Xavantes with the observations of Salzano (1961, 1964) in six Caingang communities in the States of Rio Grande do Sul and Santa Catarina. The Caingangs are also Gê-speaking "marginal" Indians whose culture formerly had important elements in common with that of the Xavantes (cf. Métraux, 1944). However, although largely intact genetically, they have now completely abandoned a nomadic existence and are settled on reservations where they live primarily by farming. We introduce the comparison in an effort to explore demographic changes in the transition from a hunting-gathering-incipient agriculture economy to a truly agricultural economy.

Isolation

Table 1 summarizes the degree of endogamy for the three villages. Because of the internal mobility of Xavante society described earlier, it is difficult to use the term "endogamy" in a strict sense. As already mentioned, relations between the villages near Batoví and Simões Lopes were close, with the Indians in the latter village appearing to regard themselves as an extension of the former, who for purposes of convenience were living near Simões Lopes. During the period we were there, exchanges between villages were almost daily. Accordingly, marital partners from Batoví were not considered as "outsiders" by the Indians in the Simões Lopes village but treated as from the same group, and we have been forced to do likewise in the tabulation. Where one partner was classified as "from the outside," this was usually a wife acquired peacefully from another Xavante village (other than Batoví). Unfortunately, despite special efforts, it has been impossible to place the origin of these outside spouses with certainty, presumably in large measure because of the young age at betrothal for women and the shifting character of Indian villages. No instance of wife-stealing from other tribes is known to us. The entry "both from outside," applicable only to São Marcos, refers to the marriages within Sebastião's group plus two marriages of persons from São Domingos.

The degree of endogamy of Simões Lopes (-Batoví) appears about equal to that of São Domingos, with São Marcos by contrast quite cosmopolitan. The recent accretions to São Marcos which result in the greater exogamy may reflect the favorable subsistence conditions created by the Mission. In addition to the data summarized in Table 1, eight adults in Simões Lopes reported siblings living in São Marcos (whose presence we did not detect), and in São Marcos, various adults mentioned siblings and other relatives living in São Domingos, Areoes, Sangradouro, and Batoví, for a total of nine persons (two males, seven females).

TABLE 1. NUMBER OF ENDOGAMOUS AND EXOGAMOUS MARRIAGES IN THREE XAVANTE VILLAGES

Only those marriages are listed where one or both marital partners are living.

Tribe and locality		Both partners from same group	One partner from another group	Both from outside	Origin of one partner unknown	Total number of marriages
Simões Lopes	N	76	5	—	—	81
	%	93.8	6.2	—	—	
São Marcos	N	80	11	30	—	121
	%	66.1	9.1	24.8	—	
São Domingos	N	50	4	—	2	56
	%	89.3	7.1	—	3.6	
Total	N	206	20	30	2	258
	%	79.8	7.8	11.6	0.8	

Age and Sex Distribution

As described previously, the estimation of age among the Xavantes presents problems and all ages are approximate. For most of the following comparisons, three age classes are recognized, namely, 0–14 years (the pre-reproductive group), 15–30 years (vigorous young adults), and over 30 years (senior citizens, usually grandparents). Age was estimated only for persons actually interviewed. Persons not seen but reported to be in the village (or temporarily away) were usually siblings (and their children) of informants. These people comprise the "unknown age" column. The information summarized in Table 2 establishes the following points:

1. The base of the age pyramid is broad, suggesting a population replacing itself. This suggestion is borne out by the detailed reproductive data to follow.

2. There appears to be a deficiency of females (or excess of males) in Simões Lopes and São Marcos, involving the younger age groups rather than the older. Although this is statistically significant in only

TABLE 2. AGE AND SEX STRUCTURE OF THE THREE VILLAGES STUDIED, AS ESTABLISHED FROM AN INTERVIEW WITH AS MANY MARRIED COUPLES IN EACH VILLAGE AS POSSIBLE

Figures in parentheses are the per cent of total in each age group.

Tribe and Locality		Age intervals (years)			Unknown age	Total
		0–14	15–30	31–		
Xavantes						
Simões Lopes	♂	60	41	16	7	124
	♀	39	43	15	5	102
	Total	99 (43.8)	84 (37.2)	31 (13.7)	12 (5.3)	226
	Sex ratio	153.8	95.3	106.7	140.0	121.6
São Marcos	♂	83	54	39	42	218
	♀	61	53	42	18	174
	Total	144 (36.7)	107 (27.3)	81 (20.7)	60 (15.3)	392
	Sex ratio	136.1	101.9	92.9	233.3	125.3
São Domingos	♂	29	35	3	17	84
	♀	39	31	8	16	94
	Total	68 (38.2)	66 (37.1)	11 (6.2)	33 (18.5)	178
	Sex ratio	74.4	112.9	37.5	106.3	89.4
Total	♂	172	130	58	66	426
	♀	139	127	65	39	370
	Total	311 (39.1)	257 (32.3)	123 (15.4)	105 (13.2)	796
	Sex ratio	123.7	102.4	89.2	169.2	115.1
*Caingang***						
Total	♂	622	522	418	64	1626
	♀	634	547	367	54	1602
	Total	1256 (38.9)	1069 (33.1)	785 (24.3)	118 (3.7)	3228
	Sex ratio	98.1	95.4	113.9	118.5	101.5

*Data from Salzano (1961, 1964).

one village, we direct attention to it because it may result from phenomena we have not yet identified (rather than chance). The reproductive histories obtained, and on which Table 7 is based, fail to suggest any excess mortality for females prior to age 14. Because of the widespread practice of female infanticide among various American Indians, the possibility must be raised, but no history of such was obtained. In Simões Lopes, where physical examinations were performed on as many persons as possible, it was clear the deficiency involved only the adolescent and preadolescent girls (36 males but only 16 females in the estimated age range 6–15 years). Physical examinations were not routinely carried out at São Marcos and we cannot place the deficiency with equal accuracy. In this latter village, the "young men's house" had recently been located in very close proximity to the Mission and had fallen under the supervision of the Mission. The boys slept there at night and were fed by the Mission. There was thus a much greater degree of "control" over young males than females. It is possible that the latter were more timid and retreated from contacts, but in our opinion the co-operation of the parents and chiefs was such that this is an unlikely explanation. Beyond this, the apparent differences between villages could in part be explained by the movement of betrothed young females between villages, but, if so, this fact did not emerge from the histories taken. Finally, the especially great discrepancy in the numbers of the two sexes in the "age unknown" category may be related to preferential recall of male relatives in a male-dominated society. Unfortunately, we did not detect this inequality in the two sexes in the field, and so failed to pursue certain obvious lines of inquiry.

3. Of the Xavantes whose age is known, 17.8% are 31 years or older, in contrast to 25.2% for the Caingangs. In view of the medical data to be presented later, we view this as reflecting a greater probability of traumatic death in the Xavantes rather than the better health of the Caingangs.

Polygyny

Polygyny is common, as shown in Table 3. There is no apparent difference among villages. Of 184 married men in the three villages studied, 40.2% were polygynous. The most common type of polygynous union involved two wives. The women in these cases were commonly related to each other as sisters, half-sisters, or first cousins. Approximately 70% of the surviving offspring were the issue of polygynous unions, but this figure must be used with caution because it is usually the older men (whose reproductive career is more nearly completed) who have the multiple wives.

In the study of Sao Domingos, "child marriages," involving prepubertal girls, comprised 28.6% of the marriages reported. By contrast, only one such marriage was acknowledged at Simões Lopes, and none at São Marcos. However, the difference in the rate of child marriages in the three

TABLE 3. NUMBER AND RELATIONSHIP OF WIVES REPORTED BY MARRIED XAVANTE MALES

Numbers greater than one are usually accounted for by polygamy, but in some instances successive marriages are involved.

Columns under "Biological relation of wives in polygamous marriages": Sisters · Half sisters · First cousins · 2 half sisters 1 unrelated · 2 sisters 1 unrelated · 2 sisters 1 half sister · 2 sisters 2 half sisters · 2 sisters 2 unrelated · Unrelated · Unknown

Localities	Number of men	Sisters	Half sisters	First cousins	2 half sisters 1 unrelated	2 sisters 1 unrelated	2 sisters 1 half sister	2 sisters 2 half sisters	2 sisters 2 unrelated	Unrelated	Unknown	Number of surviving offspring
Simões Lopes												
Men with one wife	34 (58.6)	—	—	—	—	—	—	—	—	—	—	45 (27.1)
Men with two wives	16 (27.6)	9	1	2	—	—	—	—	—	4	—	61 (36.8)
Men with three wives	7 (12.1)	—	—	—	1	5	1	—	—	—	—	54 (32.5)
Men with four wives	1 (1.7)	—	—	—	—	—	—	—	1	—	—	6 (3.6)
São Marcos												
Men with one wife	55 (61.8)	—	—	—	—	—	—	—	—	—	—	65 (33.3)
Men with two wives	27 (30.3)	6	1	—	—	—	—	—	—	16	4	90 (46.2)
Men with three wives	4 (4.5)	—	—	—	—	3	—	—	—	1	—	18 (9.2)
Men with four wives	2 (2.3)	—	—	—	—	—	—	1	—	—	1	11 (5.7)
Men with five wives	1 (1.1)	—	—	—	—	—	—	—	—	—	1	11 (5.6)
São Domingos												
Men with one wife	21 (56.8)	—	—	—	—	—	—	—	—	—	—	24 (27.0)
Men with two wives	10 (27.0)	6	1	—	—	—	—	—	—	3	—	23 (25.9)
Men with three wives	4 (10.8)	2	—	—	—	—	1	—	—	1	—	13 (14.6)
Men with four wives	1 (2.7)	—	—	—	—	—	—	—	1	—	—	6 (6.7)
Men with five wives	1 (2.7)	1	—	—	—	—	—	—	—	—	—	23 (25.8)
Total												
Men with one wife	110 (59.8)	—	—	—	—	—	—	—	—	—	—	134 (29.8)
Men with two wives	53 (28.8)	21	3	2	—	—	—	—	—	23	4	174 (38.7)
Men with three wives	15 (8.1)	2	—	—	1	8	2	—	—	2	—	85 (18.9)
Men with four wives	4 (2.2)	—	—	—	—	—	—	1	2	—	1	23 (5.1)
Men with five wives	2 (1.1)	1	—	—	—	—	—	—	—	—	1	34 (7.5)

villages is not reflected in the figures on polygyny. We suspect the difference between villages in this respect may be more apparent than real, originating in concealment of such betrothals (or the persons concerned?) in an effort to meet the admonitions of priests and Indian agents, but, if this is so, then it seems unusual that the proportions in the various polygyny classes do not appear to differ in the three villages. On the other hand, if this alleged difference in marital practices is real, it is tempting to relate it to the apparent deficiency of young girls at Simões Lopes and São Marcos, but we have no evidence in favor of this suggestion. Alternatively, of course, we must consider the possibility of a true difference in the polygyny structure of the villages.

In all three villages studied, it was the chief and the heads of clans who enjoyed the highest degrees of polygyny. The six men with four or five wives had 57 surviving offspring, an average of 9.5. Three of these men had probably completed reproduction; three probably had not. As we will see later, the average number of surviving children for Xavante men 40 years of age or over is 3.6. In round terms, then, these highly polygynous individuals produce two to three times as many children as other males. We shall come later to the possible genetic consequences of this mating structure. It may be noted here that polygyny was widespread among Indian tribes, although its quantitative aspects have seldom been documented as here (cf. Steward, 1946–1950).

The betrothal of the women occurs at a very early age (even in the absence of child marriages); the husband is commonly five to ten years the senior of his wife. The data of Table 3 indicate that 184 married men had had a total of 287 wives (some now deceased). Some of these wives are premenarchial, and marriage is a "holding operation" on the part of older males. The young male counterparts of these premenarchial wives are not in fact being deprived of a functional wife.

If an adult male dies leaving wives in the reproductive period—as not infrequently happens—remarriage of these widows occurs almost at once. Some idea of the frequency of death and remarriage can be derived from the fact that among the 175 married women interviewed in the three villages combined, 30 (17.2%) had been married twice and 6 (3.4%) three times. In interpreting this figure, it must be remembered that this, like the figure for polygyny, is a cross section in time—the figure for those who have completed the reproductive period will be higher. Divorce in the usual sense of the word seems uncommon among the Xavantes (cf. Maybury-Lewis, 1967). Thus, among 36 women married more than once, in only two instances was a previous husband alive. We are aware of no true polyandry. No instance was encountered in the three villages of an adult woman who had never been married. Since, as we will see later, sterility is uncommon, one may almost say categorically that all adult women are contributing to the next generation.

Number of Surviving Children Per Marriage

One of the crudest indices to the reproductive structure of a population is the number of surviving children per marriage for all marriages at time of census, but this crude figure, influenced as it is by the age and marriage structure of the population, is often the only one available for comparison. The data on the Xavante and Caingang are presented in Table 4. In the present instance, we observe no difference among the three Xavante villages but a significant difference between Xavantes (1.7 ± 0.1) and Caingangs (2.3 ± 0.1). Only 3.9% of Xavante families contained six or more surviving children, whereas this was the case for 9.9% of Caingang families. Holmberg (1950) reports that in one band of the nomadic Siriono of Eastern Bolivia, the average number of children in 17 "nuclear families" was 2.3, while in a second band the average number in 14 families was 1.6, but adds that "since the latter band had had considerable contact with the whites, a number of their children had been stolen from them." However, seven of these 31 "nuclear families" were polygynous; the data are not given for individual females. The figure for the first band, with intuitive allowance for polygyny, is quite similar to that for the Xavante.

Fertility and Survival

Tables 5, 6, 7, and 8 present data on fertility and survival. In the compilation of these tables, certain difficulties have arisen. For genetic purposes, the most pertinent data concern completed fertility and number of offspring reaching adulthood. The ideal approach, involving a cohort, is not possible under these circumstances. We thus are forced to attempt to approximate completed performance on the basis of cross-sectional data. Each adult was questioned about his or her reproductive performance and that of any deceased spouses. Unfortunately, the culture does not permit the accurate estimation of age at death; nothing more precise than a division into "death before 40" and "death after 39" seems justified, and even in this case we must sometimes be arbitrary.

Table 5 presents data on number of livebirths, subdivided as just described. The data are a mixture of complete and incomplete families. For living women, there is by history about one birth every five years between the ages of 15 and 40 years. We will compare this later with the findings on physical examination. The number of livebirths reported by males 40 years and above and the number reported by spouses for those males estimated to have died after 40 are comparable, lending some credence to the validity of the data for deceased males. On the other hand, no males reported their deceased wives had been 40 or over at the time of death, so that this same comparison is not possible for females. The reason is probably found in the usual age disparity between husband and wife—men whose wives had died at age 40 or greater would themselves often be expected to be 50 years or greater. There are few such males in the village.

TABLE 4. NUMBER OF SURVIVING CHILDREN PER MARRIAGE IN THE THREE VILLAGES, FOR MARRIAGES WHERE AT LEAST ONE MARITAL PARTNER SURVIVES

This figure does not include "child" (prepubertal) marriages.

Tribe and locality		Number of children											Number of families	Average Number of children ($M \pm \sigma_M$)
		0	1	2	3	4	5	6	7	8	10	14		
Xavantes														
Simões Lopes	N	17	30	16	8	2	3	2	2	—	—	—	80	1.7 ± 0.2
	%	21.3	37.5	20.0	10.0	2.5	3.7	2.5	2.5	—	—	—		
São Marcos	N	21	41	25	13	4	4	1	—	—	—	—	109	1.6 ± 0.1
	%	19.3	37.6	22.9	11.9	3.7	3.7	0.9	—	—	—	—		
São Domingos	N	10	12	7	4	2	1	2	1	1	—	—	40	2.0 ± 0.3
	%	25.0	30.0	17.5	10.0	5.0	2.5	5.0	2.5	2.5	—	—		
Total	N	48	83	48	25	8	8	5	3	1	—	—	229	1.7 ± 0.1
	%	21.0	36.2	21.0	10.9	3.5	3.5	2.2	1.3	0.4	—	—		
*Caingang**														
Total	N	148	140	107	101	63	35	31	24	8	1	1	659	2.3 ± 0.1
	%	22.5	21.2	16.2	15.3	9.6	5.3	4.7	3.6	1.2	0.2	0.2		

*Data from Salzano (1961, 1964).

TABLE 5. NUMBER OF LIVEBIRTHS PER MARRIED INDIVIDUAL OVER THE AGE OF 15, BY AGE GROUPS

There were no births to parents less than an estimated age of 15.

Tribe and locality		Ages (years)							Dead	
		15–19	20–29	30–39	40–49	50–59	60–	All ages	Probably < 40	Probably ≥ 40
Xavantes Simões Lopes	Number of females	19	22	12	5	2	—	60	12	—
	Mean number livebirths	0.9	2.9	4.6	5.8	5.0	—	2.9 ± 0.3	3.4	—
	Number of males	3	20	17	4	—	—	44	4	10
	Mean number livebirths	0.3	2.4	4.1	6.5	—	—	3.3 ± 0.4	3.5	5.8
São Marcos	Number of females	10	32	15	15	7	—	79	9	—
	Mean number livebirths	1.1	2.2	3.3	5.2	6.3	—	3.2 ± 0.2	3.6	—
	Number of males	—	27	15	15	—	2	59	11	14
	Mean number livebirths	—	1.7	3.2	5.1	—	12.0	3.3 ± 0.4	1.9	4.9
São Domingos	Number of females	14	9	2	4	1	1	31	4	—
	Mean number livebirths	1.2	2.8	8.5	7.0	8.0	1.0	3.1 ± 0.5	3.2	—
	Number of males	2	13	3	2	—	1	21	7	1
	Mean number livebirths	0.5	2.2	4.0	3.0	—	24.0	3.4 ± 1.1	3.7	11.0
Total	Number of females	43	63	29	24	10	1	170	25	—
	Mean number livebirths	1.1	2.5	4.2	5.6	6.2	1.0	3.1 ± 0.2	3.4	—
	Number of males	5	60	35	21	—	3	124	22	25
	Mean number livebirths	0.4	2.1	3.7	5.1	—	16.0	3.3 ± 0.3	2.8	5.5
Caingang* Total	Number of females	61	189	109	88	40	44	531	25	20
	Mean number livebirths	1.2	3.0	5.4	7.0	6.0	6.1	4.5 ± 0.1	3.6	5.4

*Data from Salzano (1961, 1964).

The three Xavante villages do not differ from one another in the average number of livebirths per married female for all ages, whether for living or deceased persons. The greater number of liveborn offspring reported by males than reported by the spouses themselves of course reflects the polygamy structure of the village. However, it must be remembered that many of these fertilities are incomplete; in a closed system of polygamy where the sexes are equal in number, the mean number of offspring produced by males must equal the mean number produced by females.

In our earlier paper, we commented on the "intermediate" fertilities of the Xavantes. This finding is confirmed by the new statistics. Table 5 reveals that for four of the five age intervals where the numbers are large enough for meaningful comparisons, the number of livebirths was greater in Caingang than in Xavante. Table 6 shows that the average number of livebirths to Xavante women whose reproduction is complete or nearing completion by virtue of an age over 39 or death is (mean and standard deviation) 4.7 ± 2.4; for the Caingang it is 6.1 ± 3.8. This is the mean and standard deviation, as best we can estimate it, for all reproducing women whose reproduction is complete or nearly so. For living women over the age of 39 for the two groups, the figures are 5.7 ± 2.4 and 6.6 ± 3.8. The difference is of course due to the fact that the former figure includes women whose reproductive performance was cut short by death. In future work, we shall attempt to base these figures on women who have completed the menopause, a reproductive milestone more difficult to establish at this cultural level than our own. Also, specific attention will be directed to the possibility of concealed infanticide.

We are aware of only two other sets of data of this latter type for relatively undisturbed primitive groups, those of Grey (1841) on postmenopausal Australian aborigines (mean of 4.6 for 41 women) and of Ranke (1898) for Indian women estimated to be over age 39, in the Upper Xingu region of Brazil (mean of 5.3 for 24 women). Neel (1958) has summarized similar data on completed family size in "simple" agricultural populations in East Pakistan, Ghana, and Liberia; the means and standard deviations are, respectively, 6.5 ± 3.2, 7.0 ± 3.3, and 5.5 ± 3.9. The last figure is depressed by the fact that 11.9% of the women interviewed had been sterile. It seems clear that both the mean and variance of number of children ever born to fertile women who have completed childbearing tends to be smaller in nomadic cultures depending significantly on hunting and gathering than in more settled, agricultural groups.

Turning now to the average number of surviving offspring per woman or man who had at least one liveborn child, again we see no significant difference among the three Xavante villages (Table 7). The average number of surviving children per fertile female is greater for Caingang women than for Xavante women. Since the percentage of children lost when all ages are considered together is very similar for the two groups,

TABLE 6. NUMBER OF LIVEBORN OFFSPRING IN COMPLETED SIBSHIPS

	Number of children																	Number of women	Average number of children
	0	1	2	3	4	5	6	7	8	9	10	11	12	13	14	15	16		
Xavante																			
Alive (≥ 40 years)	—	1	—	6	3	9	7	1	4	1	1	2	—	—	—	—	—	35	5.7 ± 2.4
Dead (est. < 40 years)	—	6	4	3	5	2	2	3	—	—	—	—	—	—	—	—	—	25	3.4 ± 2.1
Total	—	7	4	9	8	11	9	4	4	1	1	2	—	—	—	—	—	60	4.7 ± 2.4
Caingang																			
Alive (≥ 40 years)	10	11	12	7	12	11	25	14	13	12	20	8	7	6	1	2	1	172	6.6 ± 3.8
Dead (est. < 40 years)	1	7	3	2	4	1	1	4	1	1	—	—	—	—	—	—	—	25	3.6 ± 2.7
Dead (≥ 40 years)	2	2	2	2	1	1	1	2	3	1	1	1	—	1	1	—	—	20	5.4 ± 3.9
Total	13	20	17	11	17	13	27	20	17	14	21	9	7	7	1	2	1	217	6.1 ± 3.8

TABLE 7. AVERAGE NUMBER OF SURVIVING OFFSPRING PER FEMALE AND MALE WHO HAD AT LEAST ONE LIVEBORN CHILD, BY AGE GROUPS

Tribe and locality		Alive							Dead	
		15–19	20–29	30–39	40–49	50–59	60–	All ages	Probably < 40	Probably ≥ 40
Xavantes Simões Lopes	Number females	13	21	12	5	2	—	53	12	—
	Average number surviving offspring	0.8	1.6	3.1	3.8	3.5	—	2.0 ± 0.2	2.3	—
	Decrease as % of livebirths	11.1	44.8	32.6	34.5	30.0	—	31.0	32.4	—
	Number males	1	17	16	5	—	—	39	4	10
	Average number surviving offspring	—	1.5	2.5	4.0	—	—	2.2 ± 0.3	2.7	3.8
	Decrease as % of livebirths	100.0	37.5	39.0	38.5	—	—	33.3	22.9	34.5
São Marcos	Number females	10	31	15	15	7	—	78	9	—
	Average number surviving offspring	0.7	1.5	1.9	2.9	4.0	—	2.0 ± 0.2	2.1	—
	Decrease as % of livebirths	36.4	31.8	42.4	44.2	36.5	—	37.5	41.7	—
	Number males	—	27	13	14	—	2	56	11	14
	Average number surviving offspring	—	1.4	1.8	3.1	—	7.0	2.1 ± 0.2	0.8	3.2
	Decrease as % of livebirths	—	17.6	43.8	39.2	—	41.7	39.4	57.9	34.7
São Domingos	Number females	12	9	2	4	1	1	29	4	—
	Average number surviving offspring	0.7	1.8	7.5	4.8	8.0	—	2.3 ± 0.5	3.0	—
	Decrease as % of livebirths	41.7	35.7	11.8	31.4	—	100.0	25.8	6.2	—
	Number males	1	12	3	2	—	1	19	7	1
	Average number surviving offspring	1.0	1.2	3.0	2.5	—	23.0	2.8 ± 1.1	2.6	7.0
	Decrease as % of livebirths	—	45.5	25.0	16.7	—	4.2	17.6	29.7	36.4
Total	Number females	35	61	29	24	10	1	160	25	—
	Average number surviving offspring	0.7	1.6	2.8	3.4	4.3	—	2.0 ± 0.1	2.4	—
	Decrease as % of livebirths	36.4	36.0	33.3	39.3	30.6	100.0	35.5	29.4	—
	Number males	2	56	32	21	—	3	114	22	25
	Average number surviving offspring	0.5	1.4	2.3	3.2	—	12.3	2.2 ± 0.2	1.7	3.6
	Decrease as % of livebirths	20.0	33.3	37.8	37.3	—	23.1	33.3	39.3	34.5
*Caingang** Total	Number females	49	181	106	89	39	39	503	24	19
	Average number surviving offspring	1.1	2.0	3.3	4.1	3.7	3.6	2.8 ± 0.1	2.2	2.8
	Decrease as % of livebirths	8.3	33.3	38.9	41.4	38.3	41.0	37.8	38.9	48.1

*Data from Salzano (1961, 1964).

TABLE 8. SURVIVING OFFSPRING IN COMPLETED FAMILIES

Data based on living individuals age 40 years and older plus recently
deceased individuals for whom information was supplied by surviving spouse.

	Number of children												Number of individuals	Average number of children	Variance (s^2)
	0	1	2	3	4	5	6	7	8	9	11	23			
Xavantes (Total)															
Females	1	7	7	7	7	7	4	3	1	—	—	—	44	3.6	3.9
Males	4	12	14	7	7	6	2	7	—	1	1	1	62	3.6	12.1

it follows that the greater number of living children of the Caingang is due to a higher birth rate rather than better survival of their children. However, Caingang mothers over 29 years of age have lost a higher percentage of children than Xavante mothers, but the reverse is true under this age. Assuming that the recall of the two sets of mothers is the same, this trend, although not statistically significant, suggests that prior to contacts with neo-Brazilians, the Xavante had lower (intermediate) infantile and childhood mortalities than the Caingang, but that the results of these contacts are already reflected in the vital statistics (see section on special mortality statistics).

Of particular genetic interest is the mean and variance in number of surviving offspring of those whose reproductive performance is certainly or probably complete or nearly so. These data are given in Table 8, based on living persons whose age was estimated to be over 39 years plus recently deceased husbands or wives, as the case might be. Although the means are the same for the two sexes—as expected in a polygynous society with equal numbers of the two sexes—the male variance is significantly greater than the female ($F = 3.10$). A single unusual male makes a disproportionate contribution to this variance; with his exclusion, the male variance is 5.9 and the difference between the male and female variances no longer significant ($F = 1.51$). Assuming for the moment that this trend is valid, it follows that the relatively excessive reproductive performance of the few highly polygynous males is offset by the very poor performance of a larger number of males, presumably those who died in early manhood or who had difficulty obtaining wives—although it is rare for a man never to find a wife (Maybury-Lewis, 1965a, 1967).

We will return later to the question of how well this reported structure corresponds to the actual facts as revealed by genetic markers (Shreffler and Steinberg, 1967).

A mean of 3.6 surviving children per completed family implies a rapidly expanding population. This figure is of course an over-estimate of the number reaching sexual maturity, since many of these children are still quite young. On the other hand, it would appear the Xavante are not at this moment a "dying" tribe. Of especial importance in the theory of population genetics is the ratio of variance to mean, here 1.08 for females. If this ratio remains relatively constant as other primitive groups are studied and if the difference between sexes persists, it will seem to constitute a valid parameter throughout temporal fluctuations in the reproductive performance of the tribe as a whole.

The question of stillbirths was not pursued with the same intensity in the early studies of the Caingang that it was in the later studies or for the Xavantes. Thus, although both figures are probably underestimates, the error is greater for the Caingang. Among the Xavantes, eight of 614 births in the three villages reportedly terminated in stillborn infants (1.3 ± 0.5%), whereas among the Caingang, 83 of 2,646 births were stillborn (3.1 ± 0.3%). Accordingly, despite the bias in favor of higher rates in the Xavante, the stillbirth rate is significantly higher in the Caingang ($\chi^2 = 6.9$, df = 1, $0.02 > P > 0.01$). If stillbirths are added to the data on death prior to age of reproduction, the differences between Xavantes and Caingangs in deaths prior to the age of reproduction are accentuated.

Infertility

After the age of 15 years, all Xavante women were married. In the case of multiple (successive) marriages, a woman is scored as fertile if there have been children by any of the marriages. After the second decade, only one of the 195 Xavante women for whom histories are available (0.51 ± 0.51) had not borne a child! The figure for the Caingangs is 23 of 576 women (4.0 ± 0.82). Thus, infertility appears to be less common in the Xavante than in the populations of simple agriculturists for whom we have data, where it averages 3–5% (review in Neel, 1958).

Inbreeding

Every effort was made to establish as complete a pedigree as possible for each marriage, but the combination of a linguistic barrier, short life span, illiteracy, early betrothal with subsequent emphasis on clan affiliation, and a classifactory kinship system render estimations of inbreeding levels unsatisfactory. Thus, for only 18 out of a total of 257 Xavante marriages do we have a satisfactory identification of all four grandparents. Conversely, for 21 marriages we are unsure of the identity of one or more of the parents of the spouses.

Table 9 reveals that, taken at face value, there is more than a ten-fold difference in the frequency of consanguineous marriages in the three villages. However, the figures for São Domingos are based on 39 adult and 16 child marriages, whereas child marriages were reported only once in the other two villages. In São Domingos, consanguinity could be estab-

TABLE 9. FREQUENCY OF CONSANGUINEOUS MARRIAGES IN THE THREE XAVANTE VILLAGES

GPK=all four grandparents known; PK=parents known, incomplete information on grandparents; I=incomplete information about the spouses' parents.

Degree of relationship	Xavante Simões Lopes[1]					Xavante São Marcos					Xavante São Domingos[2]					Xavante Total					Caingang Total[3]	
	GPK	PK	I	No.	%	GPK	PK	I	No.	%	GPK	PK	I	No.	%	GPK	PK	I	No.	%	No.	%
Uncle/niece	—	—	—	—	—	—	—	—	—	—	—	—	—	—	—	—	—	—	—	—	3	0.5
First cousins	—	4	—	4	4.9	1	—	—	1	0.8	—	4	—	4	7.3	1	8	—	9	3.5	26	4.1
First cousins once removed	—	—	—	3	3.7	—	—	—	—	—	—	—	—	—	—	—	—	—	—	—	—	—
Double first cousins	1	2	—	—	—	—	—	—	—	—	2	4	—	6	10.9	3	6	—	9	3.5	13	2.0
Double first cousins once removed	—	—	—	—	—	—	—	—	—	—	1	—	—	1	1.8	1	—	—	1	0.4	—	—
Second cousins	—	—	—	2	2.5	—	—	—	—	—	—	—	—	—	—	1	1	—	2	0.8	42	6.6
Half-double second cousins	1	1	—	—	—	—	—	—	—	—	—	—	—	—	—	—	—	—	—	—	1	0.2
Second cousins once removed	—	—	—	1	1.2	—	—	—	—	—	—	2	—	2	3.6	—	3	—	3	1.1	2	03
Double second cousins	—	1	—	—	—	—	—	—	—	—	—	—	—	—	—	—	—	—	—	—	1	0.2
Third cousins	—	—	—	—	—	—	—	—	—	—	—	—	—	—	—	—	—	—	—	—	1	0.2
Other[4]	—	—	—	—	—	—	—	—	—	—	—	—	—	—	—	—	—	—	—	—	2	0.3
Total	2	8	—	10	12.3	1	—	9	1	0.8	3	10	8	13	23.6	6	18	—	24	9.3	91	14.4
Number marriages	4	73	4	81		9	103	9	121		5	42	8	55		18	218	21	257		630	
Inbreeding coefficient				0.005					0.0005					0.009					0.004		0.005	

[1] Includes one "child" marriage.
[2] Includes sixteen "child" marriages.
[3] Data from Salzano (1961, 1964).
[4] Since the exact degree of relationship is not known in these cases, they were not included in the calculation of the inbreeding coefficient.

lished for five of 39 adult marriages ($F = 0.006$) but for eight of the 16 child marriages ($F = 0.019$). We believe the apparently higher frequency in the latter to be in part misleading, simply because information is usually available for an additional generation in the case of child marriages. On the other hand, the disproportionately large contribution of Chief Apęwę to the young adult generation in São Domingos may have increased the likelihood of consanguinity among the child marriages. We are thus left with the impression that the apparently higher frequency of consanguineous marriages at São Domingos may be in part real (excessive reproduction of Chief Apęwę) and in part spurious or at least of recent origin (absence of reported child marriages in the other two villages).

As noted, there are 18 marriages in the three villages for which all four grandparents can be identified; six of these are consanguineous, the mean coefficient of inbreeding for all 18 being 0.013. Earlier, we suggested that the true coefficient of inbreeding for the São Domingos village might be four or five times the recorded 0.009 (Neel *et. al.*, 1964). This impression was based in part on the apparent high endogamy at that village; on the basis of these more recent data, this estimate would appear high for the entire tribe. We shall for subsequent purposes use a figure of 0.02–0.03. Precise data on other Indian tribes are highly desirable. The coefficient of inbreeding for the three Xavante villages combined (0.004) is essentially the same as that established by Salzano (1961, 1964) for the Caingang.

Mean Coefficient of Relationship

A descriptive parameter which helps characterize a population such as this is the mean coefficient of relationship. Like the coefficient of inbreeding, it suffers in this instance from the obvious defects in the genealogical data, but to a lesser extent, since the proximal generation, predominantly unmarried, enters into this calculation but not into the coefficient of inbreeding.

The mean coefficient of relationship has been calculated in two ways. Firstly, balls representing each person in a village from whom a blood specimen was obtained were placed in a container, and 20 were drawn without replacement. The mean coefficient was calculated on the basis of the 190 possible relationships among these 20 individuals. Secondly, pairs of balls were drawn 190 times, with replacement after each drawing, and the coefficient calculated from these 190 pairs. For São Domingos the resulting value with the first procedure was 0.078, and for the second, 0.082. Since the close relationships are for the most part known, this is much less an underestimate than the coefficient of inbreeding. We conjecture that in view of the inadequate pedigrees the true coefficient of relationship may be as high as 0.125, i.e., a randomly selected pair of individuals would on the average bear a relationship equivalent to first cousins. For Simões Lopes and São Marcos, the observed figures are 0.020 and 0.027, and 0.005 and 0.017, respectively; the true value may be about 0.032 and 0.016.

The differences between villages parallel the differences in coefficients of inbreeding and are in part real and in part spurious. Earlier, we have described the recent decrease in size of São Domingos, as dissident groups slipped away, and the increase in size at São Marcos, as groups joined the village. It seems probable that in São Domingos there has been a tendency for the close clansmen of the chief to remain in the village. On the other hand, on the basis of origin we recognize at least three distinct subpopulations in São Marcos. Within these groups the coefficient of relationship will be much higher than for the village as a whole (see Fig. 1), but across subpopulations, it will usually be quite low. This fact is not reflected in the derivation of a single statistic for the entire village.

Some Special Birth and Mortality Statistics Obtained at Simões Lopes

The agent for the Simões Lopes Post of the Indian Protective Service, Sr. Pedro Vani de Oliveira, has made an unusually conscientious effort to keep a record of births and deaths in the nearby Xavante village since contacts were first made and has also compiled an annual census. He has recorded a total of 90 births, 38 male and 52 female, between 1957 and 1963. Forty-three of these children (47.8%), 16 males and 27 females, have subsequently died, 30 during the first year of life. The observed mortality among these children appears higher than that reported in Tables 7 and 8. Assuming that the probability of ascertaining a live-birth is unrelated to the subsequent death of the child, this permits either of two interpretations, namely, under-reporting on the part of mothers, especially the older ones, or a recent increase in infant and childhood mortality, as already suggested by the data of Table 7.

Both explanations are probably correct. In 1960, the village experienced an epidemic of pertussis, followed by pneumonia, and in 1962, an epidemic of measles. Of 62 deaths recorded by Sr. Pedro between 1958 and 1963 (including these 43), 29 were attributed to measles, 11 to pertussis, seven to pneumonia, seven to diarrhea, two to bronchitis, one to inflammation of the ovary, and five to unknown causes. While it cannot be assumed that none of the children dying of measles and pertussis would have succumbed in the absence of these diseases, it does seem clear that these two diseases of civilization have resulted in an increased mortality in recent years. With respect to the under-reporting of births associated with deaths, this is an ever present possibility, the importance of which is difficult to assess accurately.

Extramarital Conceptions

The sexual practices of Indian tribes vary greatly. From the genetic standpoint, what is important is not the occurrence of extramarital or premarital sexual relationships but the extent to which these result in children. Among Indian tribes, the Xavante would appear to be relatively restrained in extramarital relationships (Maybury-Lewis, 1967). On the other hand, the *wai'a* ceremony, involving the ceremonial rape of selected women by the officiating age-set, is a prominent feature of their culture. We have therefore felt it mandatory in this effort to describe breeding structure in quantitative rather than qualitative terms, to attempt to obtain data on premarital and extramarital conceptions, using the various genetic marker systems. A detailed account of the techniques and findings will be deferred for later papers. Suffice it to say that among 107 children who were studied with respect to phenotype for 12 genetic systems in which variation was present and both of whose parents were similarly studied, there were nine apparent "exclusions," some of whom, however, may be the result of the difficulties in obtaining accurate genealogical data (in the biological sense) in this culture.

Discussion and Summary

We do not present these data as characteristic of the demography and vital statistics of an "untouched" hunting–gathering–incipient agriculture group but only as a reasonable approximation thereto, to be utilized with due caution. On the other hand, we know of no better data from which to begin to derive a picture of the genetic structure of a population at this cultural level. Outstanding among the present findings are (1) the high degree of intratribal mobility, of such a degree that regardless of the situation at any one moment, over a time span of a relatively few generations the entire tribe probably may be regarded as the breeding unit; (2) a high frequency of marriage and a low occurrence of sterility, such that with few exceptions every adult female must be regarded as a member of the reproducing population; (3) a mean and variance for number of livebirths and number of surviving offspring per adult female significantly below those observed at the next cultural levels (simple agriculturalists or pastoralists); (4) a system of polygyny which permits of the disproportionate reproduction of a selected few; and, as a corollary, (5) similar means but a significantly greater variance for number of surviving offspring for males whose reproduction is completed than for similar females.

But, although the main outlines of present Xavante demography appear clear, many details are lacking. For instance, while the structure of the age pyramid and the number of surviving offspring per female whose reproductive performance is complete ($M = 3.6$, $\sigma = 2.0$) suggests an expanding population, some of those surviving children are still very young and will undoubtedly die prior to the age of reproduction. We also have very inadequate data on the death of young males (and females) early in the reproductive period from trauma and war. The present study provides the basis for a cohort-type approach to a variety of demographic questions—but, unfortunately, it is clear that any cohort we might define at present will be subjected to so many new influences that we can never hope to extrapolate with any precision to the past.

One ultimate objective of these studies is to develop realistic genetic models for populations at this level, both as a basis for insights into population structure during the course of human evolution and as a basis for understanding the changes introduced by recent cultural developments. Such models are prerequisite to attempts to define the relative roles of genetic drift and selection in human evolution. A basic parameter in many of the formulations is effective population size (N_e) in the sense of Wright (1931). This number, so useful in the abstract, is extremely difficult to derive in actual life, both because of gaps in our knowledge of populations and mathematical difficulties in combining the effects of all the various factors which reduce the estimate of N_e below that derived from a single count of the reproducing individuals at any time point. A critical question concerns the limits to place on the distribution of the breeding unit. In the case of the Xavantes, it appears that on any kind of time span the unit must be considered to be the tribe as a whole. Interchange with adjacent tribes, for which we have no evidence, would of course extend the boundaries of the unit. In the three villages studied in detail so far, of 691 persons for whom age is known, the breeding fraction may be equated to the 139 males age 20–45 years and the 167 females age 15–40 years. Thus, 44% of the head count is reproductive. A first approximation to N_e results from multiplying this per cent by the estimated tribal size, say 1,800 persons. The resulting figure is 792.

Kimura and Crow (1963; formulae [18] and [20]) have derived formulae for correcting this crude figure for the effects of inbreeding, unequal numbers of the two sexes in the breeding population, and different variances of number of offspring surviving to reproduction for males and females. In order to obtain a general idea of how much the demographic data on the Xavante will reduce N_e below the figure obtained by a simple count, we make the following approximations:

1. α, the measure of departure from Hardy-Weinberg proportions, is here equated to the estimated F value for the population of 0.03. We entertain a number of objections to this approximation, but regard it as sufficiently accurate for present purposes.

2. The numbers of the sexes in the breeding unit are as given above.

3. Mean surviving offspring = 3.6 for males and females (Table 8).

This is too high as, probably, are the variances given below because some of the children on whom this figure is based will die prior to reproduction. Further, if number of sexes in the breeding pool is unequal, the mean number of offspring cannot be the same for both, but we shall ignore this in the approximation.

4. Variance of offspring surviving to reproduction is 3.9 for females but either 5.9 or 12.1 for males, the lower figure resulting from the exclusion from the calculations of the chief with 23 offspring described earlier. This variance has been equated to the S_k^2 of Kimura and Crow (1963).

With these assumptions and approximations, the figure of 792 reduces to either 470 or 661, depending on which variance is employed for the males. Inbreeding contributes relatively little to this reduction; the important factor is polygyny. While at the tribal size of the Xavante, this reduction does not greatly alter the role one assigns to deterministic factors in gene frequencies, at the smaller size which must obtain in the early period of the evolution of a tribe, this much reduction would lower N_e to a point where nondeterminstic factors are of greater relative importance.

There are four additional factors which would tend to reduce N_e for which not even approximate allowance can be made. The first is the restriction on mate selection within the population. Although over a time span of, say, ten generations, the internal mobility of the tribe will permit a high degree of exchange between villages, at any one instant in time an inhabitant of village 1 has very limited access to an inhabitant of village 2 (cf. Fig. 1). Furthermore, any one of these "tribal pseudopods" may be snipped off by war or other calamities. The effect of this is surely to reduce N_e, probably considerably, but neither the necessary data nor the precise formulation is available. Secondly, fluctuations in population numbers, as a result of war, famine, or disease, probably overshadow in their impact on N_e those factors we can now measure. Thirdly, the probable concentrations of relatives in the founding group will also reduce N_e. Finally, as Nei and Murata (1966) have recently demonstrated, if fertility per se is inherited, this also reduces effective population size. Little is to be accomplished in efforts to refine estimates of N_e until more is known both about these four factors and the final ratio of variance in surviving (reproducing) offspring to mean surviving offspring for males and females separately.

REFERENCES

GREY, G. 1841. *Journals of Two Expeditions of Discovery in North-West and Western Australia, 1837–1839,* Volumes I and II. London: T. and W. Boone.

HOLMBERG, A. R. 1950. *Nomads of the Long Bow: The Siriono of Eastern Bolivia.* Smithsonian Institution, Institute of Social Anthropology Publication No. 10. Washington: U. S. Government Printing Office, 104 pp.

KIMURA, M., and CROW, J. F. 1963. The measurement of effective population number. *Evolution* 17: 279–288.

MAYBURY-LEWIS, D. 1965a. *The Savage and the Innocent.* London: Evans Bros., Ltd., 270 pp.

MAYBURY-LEWIS, D. 1965b. Some crucial distinctions in Central Brazilian ethnology. *Anthropos* 60: 340–358.

MAYBURY-LEWIS, D. 1967. *Akwẽ-Shavante Society.* Oxford: Clarendon Press. 356 pp.

MÉTRAUX, A. 1944. The Caingang. In *Handbook of South American Indians,* J. H. Steward (ed.), Volume I. Washington: Smithsonian Institution Bureau of American Ethnology Bulletin 143, pp. 445–475.

NEEL, J. V. 1958. The study of natural selection in primitive and civilized human populations. *Human Bio.* 30: 43–72.

NEEL, J. V., SALZANO, F. M., JUNQUEIRA, P. C., KEITER, F., AND MAYBURY-LEWIS, D. L. 1964. Studies on the Xavante Indians of the Brazilian Mato Grosso. *Amer. J. Hum. Genet.* 16: 52–140.

NEI, M., AND MURATA, M. 1966. Effective population size when fertility is inherited. *Genet. Res.* (Camb.) 8: 257–260.

RANKE, K. E. 1898. Beobachtungen über Bevölkerungszustand und Bevölkerungsbewegung bei Indianern Centralbrasiliens. *Deutsch. Ges. f. Anthr.* 29: 123–134.

SADOCK DE FREITAS, A. 1955. Inquerito médico-sanitário entre os Indios Xavantes. In *Report of Indian Protective Service for 1954,* H. T. Simões (ed.). Rio de Janeiro: Ministerio do Agriculture.

SALZANO, F. M. 1961. Studies on the Caingang Indians. I. Demography. *Human Biol.* 33: 110–130.

SALZANO, F. M. 1964. Demographic studies on Indians from Santa Catarina, Brazil. *Acta Genet. Med. Gemel.* 13: 278–294.

SHREFFLER, D. C., AND STEINBERG, A. G. 1967. Further studies on the Xavante Indians. IV. Serum protein groups and the SC_1 trait of saliva in the Simões Lopes and São Marcos Xavantes. *Amer. J. Hum. Genet.* 19: 514–523.

STEWARD, J. H. (ed.). 1946–1950. *Handbook of South American Indians,* Volumes 1–7. Washington: U. S. Government Printing Office.

WRIGHT, S. 1931. Evolution in Mendelian populations. *Genetics* 16: 97–159.

Population Trends and Residential
Segregation since 1960

REYNOLDS FARLEY and KARL E. TAEUBER

"A great tide of migration is segregating American life, as most of us live it, faster than all of our laws can desegregate it" (*1*). A national concern with civil rights developed in the late 1950's in part as a response to the problems engendered by momentous demographic change, but the change itself was largely unrecognized. The 1960 census eventually produced evidence of the absolute loss of white population and gain of Negro population in many large central cities (*2*). In many other cities, there was net out-migration of whites, particularly in the young adult ages, but the natural increase prevented decline in total numbers and masked the magnitude of change. Census results also documented the wider spread of Negro urbanization. As news stories were subsequently to reveal, Negro population was increasing rapidly, not only in New York and Chicago, but in Los Angeles, Syracuse, Boston, Milwaukee, and most other large cities.

The 1960 decennial census provided the most recent reliable basis for detailed assessment of population trends. No comprehensive data for localities are available for any subsequent date, and results of the 1970 census are several years in the future. Fortunately, the Bureau of the Census from time to time conducts special censuses in various cities. Some are taken at the request and expense of local areas which need current data; some are conducted to pretest census methodologies; and some are conducted under congressional mandate (for example, the Voting Rights Act of 1965). These special censuses provide the best available information about population change, migration patterns, and trends in residential segregation since 1960.

We have assembled data for all 13 cities in which a special enumeration conducted after 1960 reported a total population of at least 100,000 and a Negro population of at least 9000, and for which the 1960 and later census tract grids are reasonably comparable (*3*). These cities, their populations, and their growth rates are shown in Table 1.

75

Karl E. Taeuber, a demographer, was born in 1936 in Washington, D.C. He received his B.A. from Yale University in 1955, and his M.A. and Ph.D. from Harvard University in 1957 and 1960 respectively. His interests in population dynamics focus primarily on migration and urbanization patterns, especially those of the American Negro. His publications dealing with this topic include "The Changing Character of Negro Migration," with Alma F. Taeuber, American Journal of Sociology, 70:429–441 (1965); "The Negro Population of the United States in 1966," in John P. Davies, Ed., The American Negro Reference Book (Englewood Cliffs, N.J., 1966); "Recent Immigration and Studies of Ethnic Assimilation," Demography, 4:798–808 (1967); and Migration in the United States: An Analysis of Residence Histories, Public Health Monograph No. 77 (Washington, D.C., 1968).

TABLE 1. POPULATION CHANGE AND RACIAL COMPOSITION, 1960 TO MID-DECADE. THE DATA ARE TAKEN FROM REFERENCE 3.

City	Date of special census	Total population (thousands)		Change 1960 to later (%)		Negroes (%)	
		1960	Later	White*	Negro	1960	Later
Buffalo	4-18-66	535	481	− 13.5	15.7	13.2	17.0
Providence	10- 1-65	208	187	− 11.9	24.5	5.4	7.4
Rochester	10- 1-64	319	306	− 7.1	34.6	7.4	10.4
Cleveland	4- 1-65	876	811	− 14.7	10.2	28.6	34.1
Des Moines†	4-28-66	209	206	− 1.5	6.3	4.9	5.3
Evansville†	10-20-66	142	143	0.4	6.2	6.6	6.9
Fort Wayne	1-24-67	155	160	0.2	39.8	7.5	10.2
Greensboro	1-25-66	120	132	8.8	13.9	25.8	26.7
Louisville†	5-14-64	391	387	− 3.0	11.7	17.9	20.2
Memphis†	3-27-67	491	497	− 6.8	14.8	37.6	42.6
Raleigh	1-25-66	94	105	12.2	12.7	23.4	23.4
Shreveport†	6-15-66	158	147	− 9.1	− 2.2	33.1	34.7
Sacramento†	10- 9-64	189	192	− 0.8	29.7	6.5	8.3

*Includes "other races." †Areas annexed after 1960 are excluded.

Population Change

Seven of the 13 cities experienced a decline in total population, as much as 10 percent in Providence and Buffalo. In each city the Negro population grew more rapidly or—in the case of Shreveport—decreased less rapidly than the white population. As a consequence the percentage of Negroes rose after 1960. This occurred in the southern cities and in Sacramento as well as in the northern cities.

National sample surveys conducted by the Bureau of the Census document on an aggregate basis the prevalence of the demographic change observed in the 13 cities (4):

In the first six years of the 1960's, the Negro population in large cities increased by more than 2 million while the white population in the same areas decreased by 1 million. The survey of March, 1966, confirms that, to an in-

Reynolds Farley, *a demographer, was born in 1938 in Harrisburg, Pennsylvania. He received his B.A. from the University of Notre Dame in 1960 and his M.A. and Ph.D. from the University of Chicago in 1963 and 1964 respectively. At present he teaches in the Department of Sociology, University of Michigan. He is a member of the Population Research Advisory Committee, Center for Population Research, National Institute of Health, and a member of the Advisory Committee on Problems of Census Enumeration, National Academy of Sciences. His major interest in demographic research is the population dynamics of American Negro populations. His publications in this area include "The Demographic Rates and Social Institutions of the Nineteenth Century Negro Population: A Stable Population Analysis," Demography, 2:386–398 (1965); "Recent Changes in Negro Fertility," Demography, 3:188–203 (1966); The Growth of the Black Population: A Study of Demographic Trends (forthcoming); "The Changing Distribution of Negroes within Metropolitan Areas: The Emergence of Black Suburbs," American Journal of Sociology, 75:512–529 (1970); and "Fertility Among Urban Blacks," Milbank Memorial Fund Quarterly, XLVIII, Part 2:183–233 (1970).*

creasing extent, Negroes are living in metropolitan areas, and, within these areas, in the central cities. Between 1960 and 1966, the Negro population living in metropolitan areas increased by 21 per cent, from 12,198,000 to 14,790,000, and almost all of this increase occurred within central cities. The white population living in metropolitan areas increased by 9 per cent, from 99,688,000 to 108,983,000, and all of this metropolitan increase occurred outside central cities.

Migration Patterns

Special census tabulations, like those from the decennial census, show the population by age, sex, and color. From these data estimates of net migration were calculated. As a first step, survival ratios from a national life table for 1962 were applied to the 1960 population of each city (specifically for age, sex, and color) to estimate its population at the special census date (5). This estimated population was then compared to the population enumerated by the special census and the difference represented net migration. Table 2 presents the estimated net migration, by color, and the net migration per 100 original population by age and color. Except for Sacramento, at least 94 percent of the nonwhites in each city are Negroes.

From 11 of the 13 cities there was a substantial net out-migration of whites in the post-1960 period. Cleveland had the highest rate of migration loss, 110,000 people or 18 percent of the white population, in a 5-year span. Buffalo's migration loss was 15 percent during a 6-year period, and Shreveport, Memphis, and Providence also had net out-migration of more than 10 percent in the 5- to 7-year period.

Migration losses were proportionately greatest among whites aged

TABLE 2. ESTIMATED NET MIGRATION, BY COLOR AND AGE, 1960 TO MID-DECADE. THE DATA ARE TAKEN FROM TABLE 1 AND REFERENCE 5.

Color	Net migra-tion	Net migration per 100 persons in age group in 1960							
		Total	0–9	10–19	20–29	30–39	40–49	50–64	65+
				Buffalo					
White	− 68,565	− 15	− 19	− 9	− 26	− 16	− 11	− 12	− 13
Nonwhite	+ 574	+ 1	+ 3	0	+ 6	+ 1	− 6	− 2	− 7
				Providence					
White	− 24,292	− 12	− 19	+ 3	− 31	− 17	− 11	− 8	− 8
Nonwhite	+ 1,125	+ 9	+ 13	+ 23	+ 13	+ 2	− 1	+ 5	*
				Rochester					
White	− 22,477	− 8	− 12	+ 8	− 19	− 14	− 9	− 8	− 16
Nonwhite	+ 4,210	+ 17	+ 12	+ 29	+ 23	+ 10	+ 3	+ 3	*
				Cleveland					
White	− 110,893	− 18	− 33	− 5	− 38	− 32	− 19	− 18	− 20
Nonwhite	+ 1,878	+ 1	+ 3	+ 3	+ 4	− 5	− 2	+ 2	− 3
				Des Moines					
White	− 12,973	− 7	− 12	+ 6	− 17	− 11	− 5	− 4	− 2
Nonwhite	− 306	− 3	+ 1	− 9	− 3	− 1	− 1	− 2	*
				Evansville					
White	− 6,824	− 5	− 7	− 7	− 4	− 5	− 5	− 2	− 5
Nonwhite	− 363	− 4	+ 1	− 24	*	+ 1	− 2	+ 26	*
				Fort Wayne					
White	− 9,311	− 6	− 11	+ 8	− 15	− 10	− 7	− 5	− 6
Nonwhite	+ 1,806	+ 15	+ 16	+ 29	+ 19	+ 7	+ 3	*	*
				Greensboro					
White	+ 941	+ 1	− 1	+ 15	0	− 2	− 5	− 6	− 4
Nonwhite	+ 954	+ 3	+ 6	+ 26	− 24	+ 1	− 14	− 1	*
				Louisville					
White	− 23,030	− 7	− 14	+ 3	− 20	− 11	− 5	− 4	− 2
Nonwhite	+ 2,483	+ 4	+ 4	+ 9	+ 5	+ 3	+ 5	+ 6	− 2
				Memphis					
White	− 35,991	− 12	− 20	+ 4	− 22	− 16	− 11	− 7	− 5
Nonwhite	+ 2,266	+ 1	+ 4	− 5	− 10	0	− 3	+ 7	− 7
				Raleigh					
White	+ 3,875	+ 5	0	+ 39	− 27	− 2	− 4	− 6	− 5
Nonwhite	+ 1,333	+ 6	+ 5	+ 23	− 6	+ 3	+ 1	+ 4	*
				Shreveport					
White	− 13,351	− 13	− 21	− 6	− 23	− 16	− 11	− 4	− 1
Nonwhite	− 9,420	− 17	− 18	− 26	− 24	− 16	− 14	+ 1	− 11
				Sacramento†					
White	− 5,264	− 3	− 4	+ 1	− 11	+ 2	− 1	− 4	− 6

*Rate not calculated because denominator is less than 1000. †No migration rates for non-whites were calculated for Sacramento since 40 percent of the nonwhites in this city were Orientals.

20 to 29 at the start of the period. The 30- to 39-year-olds also had high migration losses, and the 0- to 9-year-olds migrated along with their parents. For eight cities there was a net migration balance into the city among whites aged 10 to 19. These results are consistent with a variety of other migration data indicating a continued attractiveness of central cities to young adults, but a marked out-movement during the family-expansion stage of the life cycle (6). In all cities there was a

net out-migration of older white population, giving no evidence of a "back-to-the-city" movement among those whose children are grown.

Migration patterns for nonwhites are diverse. Among those aged 20 to 29 in 1960, there tends to appear the pattern long thought to be typical: net out-migration from southern cities and net in-migration to northern cities. But there is no simple way to summarize the patterns among other age groups. In some northern cities with large Negro populations (Buffalo and Cleveland), net migration during the early 1960's was slight. In northern cities with smaller Negro populations, net migration was sometimes large (Rochester and Providence) and sometimes small or negative (Des Moines and Evansville). Some southern cities had a net gain of Negro population through migration (Greensboro, Memphis, Louisville, and Raleigh), but some lost (Shreveport). The early 1960's may represent a transitional period in Negro migration. As Negro migrants seek out a variety of urban destinations, the earlier pattern of movement from southern cities to a few large northern cities may no longer be a dominant feature (7).

Growth of Negro population and increases in Negro percentages are not dependent on continued in-migration of Negroes. Negro populations in most cities are youthful, with many women in the childbearing ages and many more about to enter those ages. White populations not only have a significant out-migration, but their more elderly age structures are less conducive to high rates of natural increase. For instance, in Providence in 1965 the median age of whites was 35 years, of Negroes 19 years. In Buffalo in 1966 the median age of whites was 35 years, of Negroes 21 years. In Rochester in 1965, 17 percent of the whites, but 44 percent of the Negroes, were under age 15. Differential natural increase and white out-migration from cities are sufficient for continued increases in Negro percentages regardless of the pace of Negro migration to cities.

Trends in Residential Segregation

The growing Negro populations in many cities have expanded into housing outside the previously established Negro residential areas. Inspection of census tract data reveals this type of change. Census tracts are small areas, containing on the average about 4000 persons, for which basic census data are tabulated. In Buffalo, for example, in 1960 most Negroes lived in a belt of tracts extending south and west of downtown. By 1966 this belt had grown to include several more tracts. In Cleveland tracts were added to the principally Negro areas on the east side. Almost all of Cleveland's Negroes, in both 1960 and 1965, lived east of the Cuyahoga River in a broad belt stretching from downtown to the city limits. Local estimates indicate the development of several predominantly Negro residential areas in the eastern suburbs

(8). Few Negroes lived on the other side of downtown: in 1965 the special census counted 300,000 Clevelanders west of the Cuyahoga, of whom more than 99 percent were white.

The other cities were lacking in such extensive established Negro areas in 1960, but solidly Negro residential areas have developed. In Rochester, areas southwest and immediately north of the central business district became increasingly Negro. In Providence, Negroes replaced whites in tracts in the Federal Hill area and south of downtown along the Providence River. In each of the 13 cities the development and spread of predominantly Negro residential areas can be traced.

It is also possible to use these data to calculate summary indices of the degree of residential segregation. In contrast to the detailed descriptions of Negro residential patterns obtained from maps, such indices facilitate comparisons among cities and through time.

Using city block data for a large number of U.S. cities, the Taeubers assessed trends in residential segregation from 1940–1960 (9). In cities of all sizes and in every part of the country, Negroes and whites were found to be residentially segregated. From 1940 to 1950, the housing market was very tight. Existing segregation patterns were maintained and additional white and Negro population was housed in a highly segregated pattern. Residential segregation generally increased. During the 1950's there was an increased availability of housing. A multiple regression analysis for 69 cities relating changes in segregation to changes in other characteristics suggested that in many northern cities "the growing Negro populations, together with the demand for improved housing created by the improving economic status of Negroes, were able to counteract and in many cases to overcome the historical trend toward increasing residential segregation. In southern cities Negro population growth was slower and economic gains were less. The long-term trend toward increasing segregation slowed but was not reversed" (10).

Dissimilarity Index

The Taeubers' measure of segregation was the dissimilarity index, calculated from city block data on the number of housing units occupied by whites and by nonwhites. For assessment of post-1960 trends, we shall use the dissimilarity index calculated from census tract data on the number of Negroes and non-Negroes. The magnitude of the index depends on the areal units from which it is calculated, and the indices shown here are not directly comparable with the Taeubers' (9). Calculation of the index requires a percentage distribution of Negroes across all the census tracts of a city, and a similar percentage distribution of non-Negroes. The index is one-half the sum of absolute differences between the two percentage distributions. The numerical value of the index indicates the minimum percentage of Negroes (or of non-Negroes) whose census tract of residence would have to be changed to obtain an areally homogeneous distribution of the two groups. A value of 100 indicates complete segregation; of zero, no segregation.

Dissimilarity indices for the 13 cities for 1960 and the special census dates are shown in the first two columns of Table 3. The differences indicate a pattern of increasing residential segregation. Only in Fort Wayne and Sacramento did Negroes and non-Negroes become less segregated from one another during the early 1960's. There is no evidence in these data of an acceleration or even continuation of the trend toward decreasing segregation observed for northern cities from 1950–60.

The 13 cities are not a random sample, and we cannot claim to show that residential segregation in American cities is generally increasing. Putting these results together with those for 1940–60, there is strong evidence that the pervasive pattern of residential segregation has not been significantly breached. Whether the temporal trend for a particular city has been up, down, or fluctuating, the magnitude of the change has usually been small. Stability in segregation patterns has been maintained despite massive demographic transformation, marked advances in Negro economic welfare, urban renewal and other clearance and resettlement programs, considerable undoubling of living quarters and diminished room-crowding, high vacancy rates in many of the worst slums, and an array of federal, state, and local anti-discrimination laws and regulations.

TABLE 3. INDICES OF RESIDENTIAL SEGREGATION, 1970 AND MID-DECADE. THE DATA ARE TAKEN FROM REFERENCE 3.

| City | Dissimilarity index | | Replacement index | | Homogeneity index | | | |
| | | | | | Negro | | White* | |
	1960	Later	1960	Later	1960	Later	1960	Later
Buffalo	84.5	85.1	19.4	24.0	65	74	95	95
Providence	64.2	70.3	6.6	9.6	23	30	96	94
Rochester	76.7	79.3	10.5	14.8	44	53	96	95
Cleveland	85.2	87.2	34.8	39.2	81	86	92	92
Des Moines	76.7	77.3	7.1	7.8	35	40	97	97
Evansville	76.9	80.5	9.5	10.3	54	61	97	97
Fort Wayne	79.8	79.2	11.1	14.5	38	52	95	95
Greensboro	83.8	89.1	32.1	34.9	83	88	94	96
Louisville	78.6	81.2	23.1	26.2	68	73	93	93
Memphis	79.3	83.7	37.2	40.2	79	86	88	89
Raleight†	75.0	78.0	26.9	28.0	72	74	92	93
Shreveport	82.5	85.1	36.5	38.6	81	85	90	92
Sacramento	58.2	57.2	7.1	8.7	24	29	95	94

*Includes "other races." †Indices were calculated from data for 19 tracts lying entirely within the city and 11 tracts lying across the city boundary.

The analysis of census data points to stability in segregation patterns, with some preponderance recently of small increases in a segregation index. How may these results be reconciled with the rapidly increasing segregation perceived by most civil rights groups and many other observers? Such observers are likely to be looking at something more than simply the patterns of housing segregation.

For example, consider the problems of *de facto* educational segregation faced by a city with a small but completely segregated Negro population. Negroes live in only a few areas of the city, but there are not enough Negroes to make an extensive "ghetto." One or two elementary school districts may be solidly Negro, but no high school district is solidly Negro. If the Negro population increases, and continues to be housed in a segregated manner, additional elementary school districts will become all Negro, and one or more high school districts may become all Negro. If the white population is declining (the total city population is constant or declining), the Negro percentage in the city will be increasing rapidly. Hence the magnitude of the desegregation task will increase. Yet the basic segregated residential pattern is merely persisting, not worsening. If this example is modified slightly so that the initial segregation pattern is one of great rather than complete racial segregation, we have an approximation to the actual situation in many U.S. cities. It is the increasing number and proportion of Negroes in most central cities that account for the increasing visibility of segregation-induced problems, not any change in the residential pattern.

Composite indices may be formulated which combine measures of the proportion Negro with measures of residential segregation. Because the two components are not highly correlated, the Taeubers argued against use of a composite index for comparisons between cities (9, p. 195). Nevertheless, we believe there may be heuristic value to the calculation of selected composite indices. From among the many that have been proposed, two seem particularly well formulated to represent, respectively, the magnitude of the desegregation problem and magnitude of the segregation problem.

Desegregation Problem

By the desegregation problem, we refer to the proportion of the population that would have to be moved to effect complete residential desegregation. The index of dissimilarity gives a superficial answer to this problem. It specifies the desegregation problem on the assumption that persons of only one race are to be moved, from areas in which they are overrepresented to areas of underrepresentation. Moving persons of only one race is unrealistic in the sense that it would depopulate many areas and require substantial additional housing in others. More realistic is a series of exchanges of white and Negro households, accomplishing desegregation while maintaining existing housing stock. The minimum percentage of the total population that would be moved by such a precedure is given by

$$2q(1 - q)D$$

where q is the proportion Negro in the total population, and D is the index of dissimilarity. This measure has been called the replacement index (*11*).

By the segregation problem, we refer to the tendency of residential segregation to create racial homogeneity among neighborhood contacts (on the street and in stores, schools, and other neighborhood facilities). For an objective, census-based measure of this type, it is necessary to assume that contacts within an area (census tract, city block, school district) are made at random from among the resident population. For a Negro chosen at random from the city's population, the probability of residing in tract i is n_i/N, where n_i is the number of Negroes in tract i and N is the total number of Negroes in the city. The probability that another individual randomly chosen from tract i is also a Negro is $(n_i - 1)/(t_i - 1)$ where t_i is the total population in tract i. For convenience, this term may be approximated by n_i/t_i. If we take the joint probability of the two events, sum over tracts, and express the result in percentage scale, we have (*12*)

$$(100/N) \sum n_i^2/t_i.$$

This index may also be interpreted as the average percentage Negro in census tracts, weighted by the number of Negroes in the tract. From the Coleman report, some evidence may be adduced for the proposition that the educational achievement of Negro pupils is less the higher the percentage of Negroes in their schools (*13*). Calculating the index for schools would provide a measure of the average Negro percentage faced by Negro school children. More generally, the social-psychological consequences of residential segregation might be hypothesized to be some function of the average Negro percentage encountered by Negroes in their neighborhoods. It is in this sense that the index may be regarded as measuring the segregation problem. We designate it the Negro homogeneity index.

The dissimilarity and replacement indices are racially symmetrical. Negroes and non-Negroes are equally segregated from each other. The Negro homogeneity index is racially specific. The average Negro percentage encountered by Negroes may differ from the average white percentage (non-Negro) encountered by whites (the white homogeneity index). The complements of these measures are also of interest: the weighted average white percentage encountered by Negroes in tracts and the weighted average Negro percentage encountered by whites in tracts.

Values of the replacement and homogeneity indices are shown in Table 3. In contrast to the dissimilarity index, there is a wide range in magnitude of these indices. This reflects the wide range in values of q (the proportion Negro) and the additional variance introduced by the squared terms appearing in each composite index. The replacement and Negro homogeneity indices are highly correlated. Both indices increased for each of the 13 cities between 1960 and the later date. For most cities, both segregation (D) and proportion Negro (q) increased, but the relative increase was small in the former compared to the latter. Trends in the composite indices are largely determined by trends in the Negro proportion.

We examined special census data for 13 cities to assess trends in population, migration, and residential segregation from 1960 to mid-decade. In these cities, the demographic trends of the 1950's are continuing. There is a net out-migration of white population, and in several cities a decline in total population. Negro population is growing rapidly, but natural increase rather than net in-migration increasingly is the principal source. The concentration of whites in the suburbs and Negroes in the central cities is continuing. Within the cities, indices of racial residential segregation generally increased. The combination of small increases in residential segregation and large increases in the Negro percentage has greatly intensified the magnitude of the problems of segregation and desegregation of neighborhoods, local institutions, and schools.

REFERENCES AND NOTES

1. EDITORIAL, Washington Post, 28 Dec. 1966.
2. L. F. SCHNORE, The Urban Scene (Free Press, New York, 1965), p. 255.
3. The 13 cities are listed in Table 1. Data for the later date were published as follows: U.S. Bureau of the Census, Current Population Reports, Ser. P-28, Nos. 1376, 1377, 1386, 1390, 1393, 1411, 1413, 1430, 1431, 1435, 1441, 1446, and 1453 (1964–7); data for 1960: U.S. Bureau of the Census, Censuses of Population and Housing: 1960 PHC(1) (1961), Parts 21, 28, 39, 45, 49, 57, 83, 89, 122, 124, 127, 129, and 143.
4. U.S. BUREAU OF THE CENSUS, Current Population Reports, Ser. P-20, No. 157 (1966), p. 1.
5. U.S. NATIONAL CENTER FOR HEALTH STATISTICS, Vital Statistics of the United States: 1962, II, Section 5 (1964).
6. H. S. SHRYOCK, Population Mobility within the United States (Community and Family Study Center, Chicago, 1964), p. 424.
7. K. E. TAEUBER AND A. F. TAEUBER, Amer. J. Sociol. 70, 429 (1965).

8. REGIONAL CHURCH PLANNING OFFICE (Cleveland, Ohio), *Newsletter*, "Changes in the Non-white Population," No. 21 (1965).
9. K. E. TAEUBER AND A. F. TAEUBER, *Negroes in Cities* (Aldine, Chicago, 1965).
10. K. E. TAEUBER, *Sci. Amer.* **213**, 17 (1965).
11. D. WALKER, A. L. STINCHCOMBE, M. J. McDILL, *School Desegregation in Baltimore* (Johns Hopkins Center for the Study of Social Organization of Schools, Baltimore, 1967), p. 5.
12. W. BELL, *Social Forces* **32**, 357 (1954).
13. J. S. COLEMAN, E. Q. CAMPBELL, A. M. MOOD, *Equality of Educational Opportunity* (Government Printing Office, Washington, D.C., 1966), p. 21.
14. Support for this study was provided by a faculty research grant from Duke University to R.F. and by the Computation Laboratories of Duke University and the University of Michigan. Taeuber's participation was facilitated by appointment as expert, U.S. Commission on Civil Rights Race and Education Project, with support from funds granted to the Institute for Research on Poverty at the University of Wisconsin by the Office of Economic Opportunity. The conclusions are the sole responsibility of the authors.

Theories of Mate Selection

BRUCE K. ECKLAND

This paper is devoted to a review and clarification of questions which both social and biological scientists might regard as crucial to an understanding of nonrandom mate selection. Owing to the numerous facets of the topic, the diverse nature of the criteria by which selection occurs, and the sharp differences in the scientific orientations of students who have directed their attention to the problem, it does not seem possible at this time to shape the apparent chaos into perfect, or even near-perfect, order and, out of this, develop a generalized theory of mate selection. Nevertheless, it is one of our objectives to systematize some of our thinking on the topic and consider certain gaps and weaknesses in our present theories and research.

Before embarking on this task, it would be proper to ask why the problem is worth investigating, a question which other speakers no doubt also will raise during the course of this conference. If the social and biological scientists had a better understanding of mate selection, what would happen to other parts of our knowledge or practice as a result? Despite the fact that our questions arise from quite different perspectives, there is at least one obvious point at which they cut across the various fields. This point is our common interest in the evolution of human societies, and assortative mating in this context is one of the important links between the physical and cultural components of man's evolution.

Looking first from the geneticists' side, at the core of the problem lies the whole issue of natural selection. Any divergence from perfect panmixia, i.e., random mating, splits the genetic composition of the human population into complex systems of subordinate populations. These may range from geographically isolated "races" to socially isolated caste, ethnic, or economic groups. Regardless of the nature of the boundaries, each group is viewed as a biological entity, differing statistically from other groups with respect to certain genes. To the extent that different mating groups produce more or fewer children, "natural" selection takes place.

In the absence of differential fertility, assortative mating alone does not alter the gene frequencies of the total population. Nevertheless, it

Bruce K. Eckland, *a sociologist, was born in Chicago in 1932. He received his B.S. in 1957, his M.A. in 1960, and his Ph.D. in 1964 from the University of Illinois. His major research interests are social biology, social stratification, and the sociology of education. In 1968 he served as co-chairman of the Task Force on Psycho-Social Deprivation for the National Institutes of Health. He is on the editorial board of Behavior Genetics and on the Board of Directors of the American Eugenics Society. His publications focusing on problems of human-population genetics include* "Genetics and Sociology: A Reconsideration," *American Sociological Review, 32:173–194 (1967);* "Evolutionary Consequences of Assortative Mating and Differential Fertility in Man," *in Th. Dobzhansky, Ed., Evolutionary Biology (forthcoming); and* "Educational Selection and Population Genetics," *in H. V. Muhsan, Ed., Education and Demography (forthcoming).*

does change the distribution and population variance of genes (Stern, 1960) and this, itself, is of considerable importance. Hirsch (1967), for example, has stated:

As the social, ethnic, and economic barriers to education are removed throughout the world, and as the quality of education approaches a more uniformly high level of effectiveness, heredity may be expected to make an ever larger contribution to individual differences in intellectual functioning and consequently to success in our increasingly complex civilization. Universally compulsory education, improved methods of ability assessment and career counseling, and prolongation of the years of schooling further into the reproductive period of life can only increase the degree of positive assortative mating in our population. From a geneticist's point of view our attempt to create the great society might prove to be the greatest selective breeding experiment ever undertaken. (p. 128)

Long-term mate selection for educability or intelligence increases the proportion of relevant homozygous genotypes which over successive generations *tends* to produce a biotic model of class structure in which a child's educability and, therefore, future social status are genetically determined. Since these propositions hold whether or not everyone has the same number of children with exact replacement, assortative mating would seem to have consequences just as relevant as any other mechanisms involving the genetic character of human societies.[1]

Also from the biological point of view, it is probable that assortative mating is becoming an increasingly important factor relative to others affecting the character of the gene pool. Infant mortality, for instance,

---

[1] I have attempted in the early part of this paper to place mate selection in an evolutionary perspective. The discussion later will focus on explanatory theories, treating assortative mating as the dependent variable. In another paper, I shall discuss in much greater depth than outlined here the social-evolutionary consequences of mate selection. See Bruce K. Eckland, "Evolutionary Consequences of Assortative Mating and Differential Fertility in Man," in Theodosius Dobzhansky (ed.), *Evolutionary Biology*, Vol. IV, Appleton-Century-Crofts, in press.

does not appear to exert the same kind of selection pressure on the populations of Western societies today as it did a hundred, or even fifty, years ago. Likewise, accompanying the rise of mass education and spread of birth control information, fertility differentials appear to have narrowed markedly, especially in this country (Kirk, 1966). For example, the spread is not nearly as great as it once was between the number of children in lower and upper socio-economic families. It is not altogether clear, of course, just how the relaxation of selection pressures of this kind would, in the long run, affect future generations. Yet, assuming, as some have suggested, that these trends will continue, then a broader understanding of the nature and causes of mate selection may eventually become one of the outstanding objectives of population geneticists. One reason is that the more the assortative mating, the greater the rate of genetic selection. If nearly all members of a society reproduce and most reproduce about the same number of children, and these in turn live to reproduce, it might then be just as important to know who mates with whom as to know who reproduces and how much.

The interest of social scientists in mate selection has been more uneven and much more diffuse. Some anthropologists undoubtedly come closest to sharing the evolutionary perspective of geneticists, as indicated by their work in a variety of overlapping areas which deal in one way or another with mating, e.g., genetic drift, hybridization, and kinship systems. In contrast, sociologists have been less sensitive to genetic theories. We share with others an evolutionary approach, but one that rests almost wholly on social and cultural rather than physical processes. Nonetheless, mate selection lies at the core of a number of sociological problems. These range, for example, from studies of the manner in which class endogamy is perpetuated from one generation to the next to studies in which endogamy is conceived as a function of marital stability. While sociologists have helped to ascertain many facts as well as having developed a few quasi-theories about assortative mating, it is rather difficult when reviewing our literature on the subject to distinguish between that which is scientifically consequential and that which is scientifically trivial. The general orientation of social scientists, in any case, is far from trivial and can be used instructively in the region of mate selection and in ways heretofore neglected by population geneticists. Some of their "theories" will be reviewed later in this paper.

Evolution in Parallel and Interaction

Differences in the basic theoretical orientations of the social and biological sciences with respect to human evolution and assortative mating perhaps can best be understood in terms of the set of diagrams

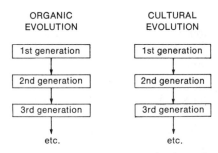

Fig. 1.—Evolution in parallel. [Redrawn.]

that follow. Fig. 1 illustrates the usual manner in which investigators in either field approach their subject matter. The course of human development is traced on separate but parallel tracks. Some textbooks and elementary courses in sociology begin with a brief treatment of genetics, but it is soon forgotten. In a like manner, students in a course in genetics are told that the expression of the genetic character of an individual depends largely on environmental influences, after which no further reference to environment seems to be necessary (Caspari, 1967).

Evolution viewed in parallel has allowed each field to articulate its own theories and perspectives. Mate selection is only one case in point, but a good one. The anthropologist or sociologist typically begins with some universal statement to the effect that in no society is mate selection unregulated and then he may proceed to analyze the cultural controls that regulate the selection process. As he has defined his problem, there perhaps has been no need to consider physiological processes. The geneticist, on the other hand, typically introduces the topic with some statement about how mate selection alters the proportion of heterozygotes in the population (as we have done) and then proceeds to a discussion of allele frequencies or consanguinity. Because he is concerned almost exclusively with the nature of the genetic material, he does not care, for example, why tall people seem to prefer to marry tall people. I doubt that sociologists especially care either. There are, however, traits far more relevant than these, like education, which serve as a basis of assortative mating and to which sociologists have given considerable attention and the geneticists relatively little.

The gap about which I am speaking also can be illustrated by the manner in which some geneticists define assortative mating. To repeat a definition which appeared recently in the *Eugenics Quarterly,* assortative mating is "the tendency of marriage partners to resemble one another as a result of preference or choice" (Post, 1965). The reference to individual "preference or choice" illustrates one of the major weaknesses in the geneticist's understanding of the nature of culture and society.

(It is not just this particular statement that is troublesome, but many others like it throughout the literature.)

Mate selection is *not* simply a matter of preference or choice. Despite the increased freedom and opportunities that young people have to select what they believe is the "ideal" mate, there are a host of factors, many *well* beyond the control of the individual, which severely limit the number of eligible persons from which to choose. As unpalatable as this proposition may be, it rests on a rather large volume of data which suggests that the regulatory systems of society enforce in predictable ways a variety of norms and sometimes specific rules about who may marry whom. Perhaps the most important point I will have to make in this paper is that geneticists must begin to recast their assumptions about the nature of culture and society, just as sociologists must recast their thinking about genetics (Eckland, 1967).

Assuming that both geneticists and sociologists do reconsider their positions and assuming, too, that each discipline has a hold of some part of the truth, there still remains the unfilled gap in the kinds of knowledge needed to develop a set of interlocking theories between the social and biological sciences with regard to mate selection. I do not question that organic and cultural evolution can and, in many ways, must be studied as separate phenomena. The point is, however, that they do interact and this, too, should be studied; and to do so will require a much broader historical perspective than most geneticists and social scientists have exhibited up to now.

An interaction model of organic and cultural evolution must specify the precise nature of the relationships between the hereditary factors and environmental influences. Although certainly a very old idea, the notion of *interaction* has laid relatively dormant until recent years, probably largely due to the nature-nurture controversy and the racist arguments that covered most of the first half of the twentieth century. The expanded model in Fig. 2 suggests a more elaborate system of causal paths along which there is continuous feedback between the genetic and cultural tracks from one generation to the next. As before, we are dealing with the pro-

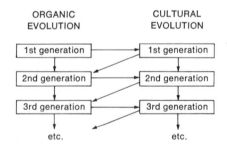

Fig. 2.—Evolution in interaction. [Redrawn.]

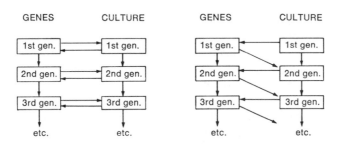

Fig. 3.—False models. [Redrawn.]

cesses by which generational replacement and change occur. However, in addition to the duplication of most genes and most cultural traits in each succeeding generation, new patterns invariably emerge through the interaction of heredity and environment. Briefly, and with no intent on my part to intimate either purpose or consciousness, (*a*) genes restrict the possible range of man's development and (*b*) within these limits he alters his environment or cultural arrangements in such ways as to change the frequencies or distribution of genes in the next generation which (*c*) enables him to carry out further changes.

It is important to note here that the interaction of heredity and environment does not occur within the duration of a single generation, a point that social scientists, in particular, need to recognize. Holding for inspection a very short segment of the life span of a single cohort, as so often we do, it is not possible to observe, even to logically think about, heredity and environment in interaction. Within the span of one generation, the relationship appears only as a one-way process, with the genetic make-up of individuals determining the norms of reaction to the environment. The path from environment *back* to genetics which actually allows us to speak in terms of *inter*action appears only *between* generations, as in the above model. In other words, models of the sort abbreviated in Fig. 3 do not fit reality. The cultural environment, of course, may have an immediate and direct effect upon an individual's endocrine system, as well as other physiological and morphological structures, but it cannot, as far as we know, alter his genes. Environment can only alter their phenotypic expression and, owing to selective mating, the genes of one's progeny in the next generation.

We have now moved into a position whereby we might raise two rather crucial questions regarding the search for significant variables in mate selection, that is, significant in the context of an interaction model. The first is: What genotypes have social definitions attached to their behavioral manifestations or, conversely, what physical, personality, and social traits depend on our genes? The answer requires determining how much, if any, of the variance of a particular trait is due to heredity (and how much to environment). For example, taking the operational definition of intelligence we now employ, if none of the variance can be attributed to genetic sources, then no matter how intense assortative mating is for intelligence, we most certainly would exclude it from any further consideration in our model. Objections sometimes have been raised against partitioning the variance on the grounds that there is a strong interaction component in the development of most traits. It will be recalled, however, that our general model permits no interaction of this form between heredity and environment in the development of the intelligence or any other phenotype of an *individual*. Every character is determined during the lifetime of that individual, with genotypes determining part of the course of development and not the other way around. There are other problems to be encountered in any analysis of variance which attempts to sort out the hereditary component, but this is not one of them.

The second question is: What criteria for mate selection are *functionally* relevant within a particular population at a particular time? This question, of course, raises some long-standing issues in genetics regarding the "adaptive" quality of characteristics which are genetically variable. It appears, for example, that some traits like the O, A, B, and AB blood types for the most part are adaptively neutral or, at least, it is not known how they affect the biological or social fitness of their possessors in any significant way. Likewise, there are traits like eye color which apparently have no clear functional value and yet seem to be involved in the sorting which unites one mate with another. By this, I do not mean that the search for socially relevant traits in mate selection should be directed toward putting the science of genetics to the service of human welfare. Rather, it is my belief that the discovery of socially relevant biological dimensions of human variation is likely to be of the sort, such as intelligence, which may be treated simultaneously as Mendelian mechanisms in the reproductive process and as sorting and selecting mechanisms in the allocation of social status and in the maintenance of boundaries between social groups, the discovery of which may serve to further our general understanding of human evolution. Any delimiting, therefore, of the class of mate selection variables we eventually must take into account should deal, on the one hand, with traits which are understood in terms of genetic processes and partly in terms of social and other environmental processes and, on the other hand, with traits whose survival or social value is at least partly understood.

Notes on Terminology

Two basic forms of nonrandom mate selection are *assortative* mating and *inbreeding*. Assortative mating usually encompasses all character-specific mate selection which would not be expected to occur by chance. Inbreeding, on the other hand, encompasses all mating where departures from perfect panmixia involve the relatedness or ancestry of individuals. While some authors have used the terms in essentially this manner (e.g., Spuhler, 1962; Post, 1965), others have not (e.g., Allen, 1965, Warren, 1966). The latter have not restricted assortative mating to refer only to character-specific situations but have included inbreeding as one of its forms. Another variation is that some authors have used the labels *genotypic* assortative mating to refer to inbreeding and *phenotypic* assortative mating to refer to the nonrandom, character-specific form (e.g., Fuller and Thompson, 1960). Also, the terms *consanguine* and *conjugal* sometimes are used to make the same distinction.

Attention to the rules governing the selection of a spouse has led to another set of terms: the first, representing conformity to the norms, called *agathogamy*; the second, involving prohibited deviations from the norms, called *cacogamy* (Merton, 1964). *Incest*, a special case of inbreeding, involves prohibited deviations from the rules controlling matings between closely related persons and is also a special case of cacogamy since the latter includes other forms of socially disapproved matings as well, such as *mesalliance*, a marriage with one of an inferior position. Special cases of mesalliance are *hypergamy* to denote the pattern wherein the female marries upward into a higher social stratum (the male marries the one in the inferior position) and *hypogamy* wherein the female marries downward into a lower social stratum.

In common use are the more general terms *endogamy* and *exogamy* which refer to in-group marriages of almost any kind. Inbreeding is a special case of endogamy; *hybridization* and *admixture* are special cases of exogamy in which "racial" features are the implied criteria. *Interbreeding* and *intermarriage* also have about the same meaning as above, except the latter term is more frequently used in reference to traits dealing with categories other than race, such as *interfaith* marriages. Miscegenation, another form of exogamy, is the term usually applied to interbreeding between white and Negro or other intergroup matings (legitimate and illegitimate) wherein the contractants have violated cultural proscriptions; and, in this respect, miscegenation is also a form of cacogamy, as well as a form of mesalliance.

Still another term commonly employed to describe assortative mating is *homogamy* which denotes something about the likeness or similarity of the married couples, with or without specific reference to any particular set of characteristics. Thus, one may speak in terms of racial homogamy or social homogamy, or simply, homogamous marriages. The antonym, *heterogamy*, is not widely used but could logically refer to mixed matings, the tendency toward random mating, or selection for "dissimilar" traits. The latter, however, is more often called *negative* assortative mating; all other forms are called *positive* assortative mating.

The above discussion probably comes close to exhausting the arsenal of terms we employ. However, with few exceptions, the concepts which arise from their meaning do not appear to be especially useful for classifying mating patterns in such a manner as to provide a sound basis for bridging the gap between the organic and social models presented earlier. It is quite probable that not only do we need more knowledge of assortative mating upon which to base more generalized theories, but we very well might find it necessary either to develop a new set of concepts (and terms) or to undertake a major revision of those now used. At present, they are confusing and often redundant, many do not appear particularly relevant to our problem, and few perhaps mean the same thing to both the geneticist and social scientist.

In the remainder of this paper, I shall review briefly some of the current theories of mate selection. By no means a complete review, I have neglected, for example, the very large body of work of anthropologists and population geneticists dealing with inbreeding. Studies of consanguineous marriages provide important information about genetic processes, such as the mutation load which is especially sensitive to inbreeding. Also reported in this literature, but not here, are a number of theories that attempt to explain the cultural development of kinship systems in which inbreeding is permitted or prescribed. However, most, although not all, of this work tends to deal with small populations which have been isolated for many generations. It is not convenient for explaining assortative mating in large, relatively open, and highly mobile cultures. The following discussion, therefore, involves a search for those psychological and structural features which best show how assortative mating operates in contemporary societies.

Individualistic Theories

The disappearance of unilineal kinship systems in Western societies has led to a decline of kinship control over mate selection. The resulting freedom which young people now enjoy has brought about an enormously complex system. No doubt, the selection process actually begins long before the adolescent's first "date." Moreover, under conditions of serial monogamy where it is possible to have many wives but only one at a time, the process for some probably never ends. Determining the "choice" are a myriad of emotional experiences and it is these experiences, along with a variety of subconscious drives and needs, upon which most psychological and other "individualistic" theories are based.

The Unconscious Archetype

Some of the earliest and perhaps most radical theories of mate selection suggested that what guides a man to choose a woman (it was seldom thought to be the other way around) is instinct. Scholars believed that there must be for each particular man a particular woman who, for reasons involving the survival of the species, corresponded most perfectly with him. A modern rendition of the same idea is Carl Jung's belief that falling in love is being caught by one's "anima." That is, every man inherits an anima which is an "archetypal form" expressing a particular female image he carries within his genes. When the right woman comes along, the one who corresponds to the archetype, he instantly is "seized" (Evans, 1964). However, no one, as far as we know, has actually discovered any pure biologically determined tendencies to assortative mating.

The Parent Image

A psychoanalytic view, based on the Oedipus configuration, has been that in terms of temperament and physical appearance one's ideal mate is a parent substitute. The boy, thus, seeks someone like his mother and the girl seeks someone like her father. While it admittedly would seem reasonable to expect parent images to either encourage or discourage a person marrying someone like his parent, no clear evidence has been produced to support the hypothesis. Sometimes striking resemblances between a man's wife and his mother, or a woman's husband and her father, have been noted. Apparently, however, these are only "accidents," occurring hardly more frequently than expected by chance.

Another generally unproven assumption, at least with respect to any well-known personality traits, involves the notion that "likes attract." Cattell and Nesselroade (1967) recently found significant correlations between husband and wife on a number of personality traits among both stably and unstably married couples. The correlations, moreover, were substantially higher (and more often in the predicted direction) among the "normal" than among the unstably married couples. As the authors admit, however, it was not possible to determine whether the tendency of these couples to resemble each other was the basis for their initial attraction ("birds of a feather flock together") or whether the correlations were simply an outgrowth of the marital experience. Although the ordering of the variables is not clear, the evidence does tend to suggest that the stability of marriage and, thus the number of progeny of any particular set of parents, may depend to some extent on degrees of likeness.

The Principle of Complementary Needs

Probably as old as any other is the notion that "opposites attract"; for example, little men love big women, or a masochistic male desiring punishment seeks out a sadistic female who hungers to give it. Only in the past twenty years has a definitive theory along these lines been formulated and put to empirical test. This is Winch's theory of complementary needs which hypothesizes that each individual seeks that person who will provide him with maximum need gratification. The specific need pattern and personality of each partner will be "complementary" (Winch, 1958). Accordingly, dominant women, for example, would tend to choose submissive men as mates rather than similarly dominant or aggressive ones. The results of a dozen or so investigations, however, are inconclusive, at best. More often than not, researchers have been unable to find a pattern of complementary differences. No less significant than other difficulties inherent in the problem is the discouraging fact that the correlation between what an individual thinks is the personality of his mate and the actual personality of his mate is quite small (Udry, 1966). Nevertheless, the theory that either mate selection or marital stability involves an exchange of interdependent behaviors resulting from complementary rather than similar needs and personalities is a compelling idea and perhaps deserves more attention.

No firm conclusions can yet be reached about the reasons for similarity (or complementariness) or personality and physical traits in assortative mating. (Even the degree of association or disassociation on most personality characteristics is largely unknown.) To state that "like attracts like" or "opposites attract," we know are oversimplifications. Moreover, few attempts to provide the kinds of explanations we seek have thus far stood up to empirical tests.

Sociocultural Theories

In a very general way, social homogamy is a critical point in the integration or continuity of the family and other social institutions. It is a mechanism which serves to maintain the status quo and conserve traditional values and beliefs. And, because marriage itself is such a vital institution, it is not too difficult to understand why so many of the social characteristics which are important variables generally in society, such as race, religion, or class, are also the important variables in mate selection. Thus, most studies in the United States report a very high rate, over 99%, for racial endogamy, an overall rate perhaps as high as 90% for religious homogamy, and moderately high rates, 50% to 80% for class homogamy, the exact figures depending on the nature of the index used and the methods employed to calculate the rate.

One possible way of illustrating the conserving or maintenance function of social homogamy in mate selection is to try to visualize momentarily how a contemporary society would operate under conditions of *random* mating. Considering their proportions in the population, Negroes actually would be more likely to marry whites than other Negroes, Catholics more often than not would marry Protestants, and a college graduate would be more apt to marry a high school dropout than to marry another college graduate. In a like manner, about as often as not, dull would marry bright, old would marry young, Democrats would marry Republicans, and teetotalers would marry drinkers. What would be the end result of this kind of social heterogamy? A new melting pot, or chaos?

It seems that, in the absence of "arranged marriages," a variety of controls govern mate selection and, in the process, substantially reduce the availability of certain individuals as potential mates. Many structures in society undoubtedly carry out these functions, sometimes in quite indirect ways, such as, the subtle manner which the promotion of an "organization man" may be based, in part, on how well his mate's characteristics meet the qualifications of a "company wife." Thus, despite the "liberation" of mate selection and the romantic ideals of lovers who are convinced that social differences must not be allowed to stand in their way, probably one of the most important functions of both the elaborate "rating and dating" complex and the ceremonial "engagement" is to allow a society to make apparent who may "marry upward" and under what conditions exogamy is permitted. We are referring here, then, not merely to a society's control over the orderly replacement of personnel, but to its integration and the transmission of culture as well.

Rather than reviewing any very well-formulated theories (since there may be none) in the remaining discussion, I have attempted to touch upon a fairly broad range of conditions under which homogamy, as a social fact, relates to other aspects of contemporary societies.

Propinquity and Interaction

Whether we are speaking about place of residence, school, work, or such abstruse features of human ecology as the bus or streetcar routes along which people travel, propinquity obviously plays a major part in mate selection since, in nearly all cases, it is a precondition for engaging in interaction. (The mail-order bride, for instance, is one of several exceptions.) A person usually "selects" a mate from the group of people he knows. Findings which illustrate the function of distance have been duplicated in dozens of studies. In Columbus, Ohio, it was once found that more than half of the adults who had been married in that city had actually lived within sixteen blocks of one another at the time of their first date (Clarke, 1952). Cherished notions about romantic love notwithstanding, the chances are about 50–50 that the "one and only" lives within walking distance (Kephart, 1961).

As many authors have pointed out, people are not distributed through space in a random fashion. In fact, where people live, or work and play, corresponds so closely with one's social class (and race) that it is not quite clear whether propinquity, as a factor in mate selection, is simply a function of class endogamy or, the other way around, class endogamy is a function of propinquity. Ramsøy's (1966) recent attempt to resolve this issue, I want to note, misses the mark almost completely. Investigating over 5,000 couples living in Oslo, Norway, she concludes that propinquity and social homogamy are "totally independent of one another" and, therefore, rejects the long-standing argument that "residential segregation of socioeconomic and cultural groups in cities represents a kind of structural underpinning both to propinquity in mate selection and to homogamy." More specifically, the author shows that "couples who lived very near one another before marriage were no more likely to be of the same occupational status than couples who lived at opposite sides of the city." This is astonishing, but misleading. The author equated the social status of the bride and, implicitly, her social class origin with *her* occupation at the time of marriage. No socioeconomic index other than the bride's occupation unfortunately was known to the investigator and, thus, it was a convenient although poorly considered jump to make. To most sociologists, it should be a great surprise to find in any Western society, including Norway, that the occupations young women hold before marriage give a very clear indication of their social status, relative either to the occupational status of men they marry or to their own places of residence.

Exchange Theory

An explanation often cited in the literature on mate selection, as well as in that on the more general topic of interpersonal attraction, deals in one form or another with the principle of exchange. A Marxian view, marriage is an exchange involving both the assets and liabilities which each partner brings to the relationship. Thus, a college-educated woman seldom brings any special earning power to the marriage, but rather she typically enters into contract with a male college graduate for whom her diploma is a social asset which may benefit his own career and possibly those of his children. In exchange, he offers her, with a fair degree of confidence, middle-class respectability. Norms of reciprocity might also help to explain the finding that most borderline mentally retarded women successfully marry and even, in some cases, marry upward, if they are physically attractive. This particular theory, however, has not been well-developed in regard to mate selection, despite its repeated usage. Also, it may be a more appropriate explanation of deviations from assortative mating or instances of negative mate selection than of positive selection.

Values and Belief Patterns

In contrast to the inconclusive evidence regarding assortative mating in terms of personality characteristics, numerous studies do indicate that married couples (and engaged couples) show far more consensus on various matters than do randomly matched couples. Even on some rather generalized values, as in the area of aesthetics or economics, social homogamy occurs. Apparently, our perception that other persons share with us the same or similar value orientations and beliefs facilitates considerably our attraction to them (Burgess and Wallin, 1943).

The importance of norms and values in mate selection, part of the social

fabric of every society, also can be illustrated in a more direct way by looking at some of the specific sanctions that we pass along from generation to
generation. Without really asking why, children quite routinely are brought
up to believe that gentlemen prefer blondes (which may be only a myth
perpetuated by the cosmetic industry), that girls should marry someone
older rather than younger than themselves (which leaves most of them
widows later on), and that a man should be at least a little taller than the
woman whom he marries (which places the conspicuously tall girl at an
enormous disadvantage). Simple folkways as such beliefs presently are, they
nevertheless influence in predictable ways the "choice" of many individuals.

Social Stratification and Class Endogamy

We have already noted that the field of eligible mates is largely confined
to the same social stratum to which an individual's family of orientation
belongs. Social-class endogamy not only plays a significant part in the
process of mate selection, it may also help to explain other forms of assortative mating. For example, part of the reason why marriage partners or
engaged couples share many of the same values and beliefs no doubt is because they come from the same social backgrounds.

There are at least five explanations which can be offered for the persistence of class endogamy, each of which sounds reasonable enough and probably has a hold on some part of the truth.

First, simply to turn the next to last statement around, persons from
the same class tend to marry *because* they share the same values (which
reflect class differences) and not because they are otherwise aware or especially concerned about each other's background.

Second, during the period of dating and courtship most young people
reside at the home of their parents. (Excluded here, of course, are the large
minority in residential colleges and those who have left both school and
home to take an apartment near their place of work.) The location of parental homes reflects the socioeconomic status of the family and is the general basis for residential segregation. With respect to both within and between communities, the pattern of segregation places potential mates with
different backgrounds at greater distances than those with similar backgrounds. Thus, to the extent that the function of distance (or propinquity)
limits the field of eligibles, it also encourages class endogamy by restricting
class exogamy.

Third, class endogamy in some cases is simply a function of the interlocking nature of class and ethnicity. A middle-class Negro, for example,
probably is prevented from an exogamous marriage with a member of the
upper-class not so much because class barriers block it but because he (or
she) is Negro. The majority of the eligible mates in the class above are
whites and, in this instance, what appears to be class endogamy is really
racial endogamy.

Fourth, ascriptive norms of the family exert a great deal of pressure on persons, especially in the higher strata, to marry someone of their "own kind," meaning the same social level. The pressures that parents exert in this regard sometimes are thought to have more than anything else to do with the process and certainly are visible at nearly every point at which young people come into meaningful contact with one another. Norms of kinship regarding the future status of a child may be involved, for example, in the parent's move to the right community, sending a child to a prep school, or seeing that he gets into the proper college.

Fifth, and an increasingly convincing argument, even as the structure of opportunities for social mobility open through direct competition within the educational system, class endogamy persists owing to the educational advantages (or disadvantages) accrued from one's family of orientation. Most colleges, whether commuter or residential, are matrimonial agencies. As suggested earlier, despite whatever else a woman may gain from her (or, more often, her parents') investment in higher education, the most important thing she can get out of college is the proper husband or at least the credentials that would increase her bargaining power in an exchange later on. Given the fact that men generally confer their status (whether achieved or ascribed) upon women and not the other way around (female proclamations to the contrary notwithstanding), marriage as a product of higher education has far more functional value for women than vocational or other more intrinsic rewards.

To carry this argument a bit further, access to college depends in large measure on the academic aptitude (or intelligence) of the applicants. Moreover, the hierarchical ordering of colleges which is based on this selectivity has led to a system of higher education which, in many ways, replicates the essential elements of the class structure. Differentiating those who go to college from those who do not, as well as where one goes to college, are *both* aptitude and social class. These two variables correspond so closely that despite the most stringent policies at some universities where academic aptitude and performance are the central criteria for admissions and where economic aid is no longer a major factor, students still come predominately from the higher socioeconomic classes. For whatever the reason, genetic and environmental, this correspondence facilitates the intermarriage of individuals with similar social backgrounds, especially on American campuses where the sex ratio has been declining. It is interesting to note in this context that Warren's recent study of a representative sample of adults showed that roughly half of the similarity in class backgrounds of mates was due to assortative mating by education (Warren, 1966).

Ethnic Solidarities

While intermarriage is both a cause and consequence in the assimilation of the descendants of different ethnic origin, various writers claim that the American "melting pot" has failed to materialize. Religious and racial lines, in particular, are far from being obliterated. In fact, the very low frequency of exogamous marriages across these lines itself underscores the strength of the cleavages. Most authors also agree that nationality is not as binding as either race or religion as a factor in mate selection. Nation-type solidarities are still found among some urban groups (Italian and Poles) and rural groups (Swedes and Finns), but our public school system and open class structure have softened considerably what were once rather rigid boundaries. There is some evidence, too, that religious cleavages have been softening somewhat, and perhaps are continuing to soften as the functions of this institution become increasingly secular and social-problem oriented. On the other hand, racial boundaries, from the view of mate selection, appear to be as binding today as at any previous point in history; at least I have found no evidence to the contrary. The gains that Negroes have made in the schools and at the polls during the past ten years apparently have not softened the color line with respect to intermarriage.

Explanations of racial endogamy in America, some of which would take us back several centuries in time, are too varied to discuss here. It might be well to point out, however, that cultural and even legal prohibitions, probably have relatively little to do with the present low rate of interracial marriage. As one author has stated, "the whole structure of social relationships between whites and Negroes in the United States has been organized in such a way as to prevent whites and Negroes from meeting, especially under circumstances which would lead to identifying each other as eligible partners. . . . Under these circumstances, the few interracial marriages which do occur are the ones which need explaining" (Udry, 1966).

For the population geneticist, too, it would seem that the deviant cases are the ones which require attention. Elsewhere I have suggested, for example, that genes associated with intelligence may simply drift across the white and Negro populations since it appears that only certain morphological features, like skin color, actually operate to maintain the color line (Eckland, 1967). In other words, if the skin of an individual with Negro ancestry is sufficiently light, he may "pass" (with no strings attached) into the white population. Even just a lighter-than-average complexion "for a Negro" probably enhances his changes of consummating what we socially define as an "interracial" marriage. In neither the first or second case, however, is intelligence necessarily involved.

If intelligence *were* associated in any predictable way with racial exogamy, the drift would not be random and we would then have a number of interesting questions to raise. For instance, do only the lighter *and* brighter pass, and, if so, what effect, if any, would this be likely to have on the character of the Negro gene pool? What, too, is the character of the inflow of genes from the white population? We do know that the great majority of legally consummated interracial marriages involve Negro men and white women. Does this information provide any clues? And, what about the illegitimate progeny of white males and Negro prostitutes? How often are they placed for adoption in white households and with what consequences? Before taking any of these questions too seriously, we would want to have many more facts. For obvious reasons, our knowledge is extremely meager.

Precautionary Notes

In conclusion, five brief comments may be made upon the present state of research and theories of mate selection as revealed in the foregoing discussion.

First, there is a great deal of evidence of homogamous or assortative mating but relatively few theories to explain it and no satisfactory way of classifying its many forms.

Second, nearly all facts and theories regarding mate selection deal with engaged or married couples and hardly any attention has been given to illegitimacy (including adultery) and its relationship to assortative mating. It may be, such as in the case of miscegenation, that some of the most important aspects of mate selection occur outside the bonds of matrimony.

Third, our heavy emphasis upon courtship and marriage has obscured the fact that people often separate, divorce, and remarry. Mate selection may be a more or less continuous process for some individuals, affecting the character of the progeny of each new set of partners.

Fourth, the relationships between fertility and assortative mating still must be specified. Are there, for example, any patterns of assortative mating on certain traits, like education, which affect the number of children a couple will have?

Fifth, most of the factors in mate selection appear to covary. We discussed some of the more obvious problems in this regard, such as the relationship between residential segregation (propinquity) and class endogamy. It would appear that much more work of this sort will need to be done.

In regard to the last point, it would also appear that it is precisely here that social scientists, and sociologists in particular, may best serve the needs of population geneticists. Through the application of causal (chain) models and multivariate techniques, it may eventually be possible to sort out the relevant from the irrelevant and to specify in fairly precise terms not only the distribution of assortative mating in the social structure with regard to any particular trait, but also the ordering of variables and processes which restrict the field of eligibles.

REFERENCES

ALLEN, GORDON. Random and nonrandom inbreeding. Eugen. Quart. **12**:181–198.
BURGESS, ERNEST W., and PAUL WALLIN. 1943. Homogamy in social characteristics. Amer. J. Sociol. **49**:109–124.

CASPARI, ERNST. 1967. Genetic endowment and environment in the determination of human behavior: Biological viewpoint. Paper read at the annual meeting of the American Educational Research Association, February 17, 1967.

CATTELL, RAYMOND B., and JOHN R. NESSELROADE. 1967. "Likeness" and "completeness" theories examined by 16 personality factor measures on stably and unstably married couples. (Advanced Publication No. 7.) The Laboratory of Personality and Group Analysis, University of Illinois.

CLARKE, ALFRED C. 1952. An examination of the operation of residential propinquity as a factor in mate selection. Amer. Sociol. Rev. **17**:17–22.

ECKLAND, BRUCE K. 1967. Genetics and sociology: A reconsideration. Amer. Sociol. Rev. **32**:173–194.

EVANS, RICHARD I. 1964. Conversations with Carl Jung. Van Nostrand, Princeton.

FULLER, J., and W. THOMPSON. 1960. Behavior genetics. Wiley, New York.

HIRSCH, JERRY. 1967. Behavior-genetic, or "experimental," analysis: The challenge of science versus the lure of technology. Amer. Psychol. **22**:118–130.

KEPHART, WILLIAM M. 1961. The family, society and the individual. Houghton Mifflin, Boston.

KIRK, DUDLEY. 1966. Demographic factors affecting the opportunity for natural selection in the United States. Eugen. Quart. **13**:270–273.

MERTON, ROBERT. 1964. Intermarriage and the social structure: Fact and theory, p. 128–152. *In* Rose L. Coser (ed.), The family: Its structure and functions. St. Martin's, New York.

POST, R. H. (ed.). 1965. Genetics and demography. Eugen. Quart. **12**:41–71.

RAMSØY, NATALIE ROGOFF. 1966. Assortative mating and the structure of cities. Amer. Sociol. Rev. **51**:773–786.

SPUHLER, J. N. 1962. Empirical studies on quantitative human genetics, p. 241–252. *In* The use of vital and health statistics for genetics and radiation studies. United Nations and World Health Organization, New York.

STERN, CURT. 1960. Principles of human genetics. W. H. Freeman, San Francisco.

UDRY, J. RICHARD. 1966. The social context of marriage. J. B. Lippincott, Philadelphia and New York.

WARREN, BRUCE L. 1966. A multiple variable approach to the assortative mating phenomenon. Eugen. Quart. **13**:285–290.

WINCH, ROBERT. 1958. Mate selection. Harper and Row, New York.

Mutation

Overview

Mutation occurs when there is a change in the genetic material (DNA) of the cell, which constitutes the chromosomes of the nucleus. Mutation, therefore, necessarily involves chromosomes, either whole or in part, and accordingly it can affect groups of genes or a single gene. Mutations may occur in the somatic or the germinal cells of any organism, but *germinal-cell mutations* alone account for human evolution. Only they can be transmitted by sexual reproduction to succeeding generations, thereby providing a population with the capacity for and perpetuation of genetic change. Without mutation there would be no change of genetic material in individuals, and there would therefore be no evolution of human populations.

Although we now know that mutation provides the basic building blocks of evolution, Charles Darwin, writing in 1859, was ignorant of the process of mutation; consequently, his theory of natural selection was fundamentally weak. Because of his ignorance of mutation, the variation within a species, which was an essential element in his theory, remained unexplained with regard to both its origin and the means by which it was maintained (see Chapter IV). Not until fifty years later, in 1909 to 1910, did knowledge about mutation become available. At that time, the Dutch botanist Hugo de Vries described in *The Mutation Theory* the numerous plant mutants which had appeared during his extensive breeding experiments with the evening primrose. De Vries concluded that mutation causes drastic, nongradual changes in an organism, thereby abruptly forming a new species without intermediate gradations. His view of mutation and speciation was therefore contrary to Darwin's explanation of the evolution of species; de Vries believed that evolutionary change and speciation was discrete and radical, whereas Darwin held it to be continuous and gradual. On the other hand, the two men were in close agreement on the random nature of individual changes or variations. Although ignorant of their source, Darwin had postulated all kinds of individual changes—some advantageous, some detrimental, and others of no significance to a species. De Vries also saw no direction or pattern inherent in the range of mutations which had appeared in plants, concluding that "nature . . . furnishes every possibility, so to speak, and leaves it to the environment to choose what suits it." On this issue de Vries and Darwin stood as one, and today the *randomness* of the mutation process is generally accepted.[1]

The basic difference between de Vries's mutation theory of abrupt speciation and Darwin's natural-selection theory of gradual speciation was resolved by the work of T. H. Morgan and his associates with the fruit fly *Drosophila*. They reported the appearance of numerous mutations, some effecting drastic changes, others resulting in barely detectable changes, but none resulting in a new species of fly (for example Morgan, 1910). Their research showed that mutations do, in fact, possess the characteristics requisite for the hereditary variation basic to Darwin's theory of gradual evolution of the species. And although it is possible that the extreme variations necessary for the instantaneous formation of a new species may arise suddenly through mutations having drastic phenotypic effects, modern geneticists agree that such so-called "macromutations" have not been significantly involved in evolutionary change and speciation.

The investigation of the dynamics of naturally occurring (*spontaneous*) mutation is difficult, and presently little is known about the causes of spontaneous mutations in natural populations. They have primarily been inferred from laboratory experiments which show that certain agents may artifically induce the mutations of individual genes or chromosomes in various organisms. These mutation-inducing agents (*mutagens*) include high and low temperatures, ionizing radiations, and certain natural and manufactured chemical substances, including mustard gas, formaldehyde, caffeine, and many others. But it has been impossible to prove that any of these agents, as they occur in their natural states in the environment, account for the rates of spontaneous mutation estimated for various plant and animal populations. In addition, studies on humans have shown statistical associations of mutations with other variables such as viral infections and increased parental age (see the following articles by Vogel, and Court-

---

[1] "Randomness" refers here only to the random nature of where and when a mutation will occur, and the lack of correspondence between the phenotypic effect and the mutation-inducing stimulus. In other respects mutations are nonrandom; for example, mutations in a gene are probably channelized along certain lines predetermined by the existing gene structure (Grant, 1963).

Brown and Smith). However, the biological mechanisms which underlie the associations have not been clearly demonstrated. In general, the causes of spontaneous mutations remain unaccounted for.

Ionizing Radiation as a Mutagen

Ionizing radiation, because of its availability, its ease of application in the laboratory, its consistently high mutagenic effectiveness, and its increasing use by man for war and peace, is the most extensively studied mutagenic agent. The famous geneticist H. J. Muller was the first to demonstrate in *Drosophila* that mutations could be induced by irradiation (Muller, 1927). For this work and his subsequent extensive research on mutations, which opened a new epoch in mutation studies, he received the Nobel Prize in 1946. One result of the extensive radiation research is that the process by which ionizing radiation causes mutations is now known. The process is one in which ionizing radiation strips electrons from (ionizes) the atoms of a molecule, thereby breaking its chemical bonds. The DNA of a chromosome is a molecule, and the irradiation and breakage of the chemical bonds of the DNA, therefore, breaks the chromosomes. Numerous experiments have demonstrated that the chromosomes, in fact, are the cell sites most sensitive to radiation damage, and the broken chromosomes of irradiated cells are clearly visible microscopically.

Radiation damage to the genetic material results in mutation, and/or destruction of a cell's reproductive capacity. Radiation-induced mutations may take one of several different forms: First, the broken chromosomes are themselves mutations in that the original genetic material has been changed by its separation. Second, although chromosomes possess the power to self-repair by rejoining broken ends, if the broken segments of different chromosomes join, a new abnormal combination occurs, that is, a mutation. Finally, even when segments of the same chromosome rejoin correctly, there may have been chemical changes in or a loss of the gene or genes located at the site of the break. Such chromosomal breakage or abnormal junction may also destroy the cell's capacity to reproduce (Puck, 1960).

Importantly, laboratory experiments suggest that the doses of radiation sufficient to cause chromosomal damage may be very low. *In vitro* irradiation of human cell cultures shows that the average dose needed to cause a single chromosomal break per cell is 20 to 25 rads, and the average lethal dose for cellular reproduction is in the range of 50 rads. However, doses of 20 to 50 rads are not enough to destroy the metabolic functions of a cell; doses 10 to 100 times those capable of inhibiting reproduction are required to destroy cellular metabolism, and a cell may be reproductively dead, but metabolically alive. For this reason a human victim of a radiation accident who has been exposed to the lethal total body dose of 400 to 500 rads suffers a prolonged period prior to death. Doomed because 50 percent of its cells have lost their reproductive capacity but not their metabolic function, the body gradually ceases to function as first one tissue and then another fails from lack of cellular multiplication. The slowly manifested cellular reproductive failure explains the typical lag between a man's initial exposure to a lethal dose of radiation, the development of pathological symptoms, and his ultimate death (Puck, 1960).

Extensive laboratory work has conclusively demonstrated that artificially ionizing radiation may destroy the capacity to reproduce in some cells and induce mutations in others. However, the important fact for human populations is that ionizing radiation with identical destructive capacities exists outside the laboratory. It occurs naturally in the form of cosmic rays and radioactive elements in the earth's crust, both of which are incorporated by plants and eventually ingested by man in his food. Naturally occurring ionizing radiations are estimated to contribute an average dose of 3 rads per generation averaged over a whole population (Newcombe, below). However, it is not with natural ionizing radiation that man is, or should be, most concerned, but rather with the increasing amounts of ionizing radiation that are being produced and released into the atmosphere artificially.

The potential dangers of irradiation to human germinal cells was recognized as early as 1927 by H. J. Muller, who began at that time his lifelong crusade to reduce and control the exposure of human populations to irradiation. Initially he emphasized the hazards of medical and dental X-rays and their role in increasing man's load of lethal and detrimental genes, and called upon physicians to use all possible measures to keep radiation exposure of their patients at an absolute minimum. With the advent of the atomic bomb, Muller broadened his concerns to include the dangers of irradiation from atomic-explosion radioactive fall-out. Until his death in 1967 he relentlessly led the fight for full public disclosure by the federal government of all information concerning the possible damaging effects of radiation. Muller was unquestionably the one man most responsible for public awareness of the health hazards of radiation and the present governmental policy controls over permissible levels of radiation.

Human Populations and Radiation Research

The decisive factor involved in determining radiation controls is the ratio of the estimated benefits to be derived from radiation to the expected risks. On the basis of the "ratio factor" the present standard for maximum permissible dose levels for humans has been set at 10 rads of man-made radiation per individual per generation. This dose is that which is expected to increase the natural spontaneous mutation rate, but which "necessarily involves a compromise between deleterious effects and social benefits" (Report UNSCEAR, 1958). The precisely quantified "hazard" of radiation to human populations is an elusive factor to arrive at under any circumstances; the lack of data on it makes the establishment and exercise of controls more susceptible to political policy at times than to biological considerations. (Consider, for example, the political policies and arguments underlying the Amchitka bomb tests.) Numerous research efforts have been directed toward quantitatively measuring the hazardous effects of radiation, especially its mutagenic effects, on humans. But this type of research is fraught with methodological problems, for it is practically impossible to measure the relationship between the dosage and duration of irradiation received by human populations and the rates of mutation induced.

The problem of measurement stems from the kind of data that the geneticist must rely upon to demonstrate the relation between radiation and mutation. Laboratory experiments must be done either on tissue-cell cultures or on nonhuman organisms, and extrapolation from the laboratory results to humans must always be done cautiously. It is highly questionable whether the responses of human germinal or somatic cells irradiated *in vivo* are identical to the laboratory responses of either human germinal cells irradiated *in vitro* or the germinal cells of nonhuman organisms irradiated *in vivo*. For example, the radiation dosage required to destroy the cellular reproductive capacity of the common bacterium *E. coli* is in the range of 5,000 to 15,000 roentgens, a dosage which contrasts strikingly with the dosage of 50 roentgens which is lethal to human tissue cells (Puck, 1960). The best data would, of course, come from controlled laboratory radiation experiments on actual humans, but as this course of action is morally and ethically unthinkable, the next best data come from studies of humans who have been exposed to high doses of radiation under a variety of conditions. They include people who work constantly with radiation, such as radiologists, research workers, certain factory workers; patients exposed to large doses of medical X-rays; and victims of radiation accidents or atomic bombings.

Of special interest from a genetic perspective are the studies dealing with the reproductive performance of women who have been irradiated prior to or during reproductive age. Supposedly their offspring should manifest the mutagenic effects of the radiation on the maternal germinal cells. The various phenomena which may be due to irradiation, as recorded in such studies, are changes in the sex ratio of the offspring and increased rates of abortion, fetal death, and prematurity (for example, Kaplan, 1957). One group of investigators studied the reproductive performance of 1500 women whose own germinal cells had been irradiated *in utero* prior to their birth. The mothers who themselves had been exposed *in utero* before they attained 30 weeks of gestation had significantly increased numbers of male babies, longer gestation periods, lower placental weights, and more precipitate labors (Meyer, *et al.*, 1969). Although such studies appear to confirm radiation damage to human somatic and germinal cells *in vivo*, they provide no means whereby the observed damage may be translated into the result of germinal mutation per se. Is an abortion, for example, to be attributed to a single mutation, multiple mutations, or to the poor uterine environment of a woman already being therapeutically irradiated for medical reasons?

It was hoped that the extensive study of the survivors of the 1945 Hiroshima and Nagasaki atomic bombings would overcome some of the methodological defects of other human-radiation studies. Begun in 1958 by the Atomic Bomb Casualty Commission, the study was designed to specifically assess, among other things, the short-term and long-term genetic effects of measurable amounts of radiation on a large human population. However, the evaluation of 70,000 children who were conceived after the bomb explosion has failed to reveal to date unequivocal genetic effects of radiation. In a publication reviewing the published evidence relating to delayed radiation effects in the bomb survivors, the author concludes:

No influence of radiation was demonstrable in this study, which, statistical tests have shown, was likely to detect a 2-fold increase in rates of malformation or a 1.8-fold increase in rates of stillbirths and deaths of newborns. Moreover, there was no effect on birth weight or on anthropometric values at 8 to 10 months attributable to radiation exposure. The sex ratio (the proportion of males to females) for children conceived after exposure of one parent to radiation will, in theory, be diminished if the mother was irradiated, and increased if the father was. In a study of about 120,000 births, such shifts in the sex ratios were found to occur in the first 10 years following detonation of the atomic bombs but not thereafter. No effect has been found on the mortality of children conceived after exposure of their parents to the bombs in Hiroshima or Nagasaki. Thus, though laboratory experimentation leaves no doubt that irradiation is mutagenic, the effect could not be demonstrated in the F_1 generation studied by ABCC (Miller, 1969).

A significant increase in chromosomal abnormalities was observed among persons directly exposed to the bomb and those *in utero* at the time of the bombing whose mothers received a dose of at least 100 rads. However, no chromosomal effects were observed in children conceived after parental radiation exposure, so it would appear either that the germinal cells had not undergone mutation, or that any mutations present had not been transmitted to the offspring.

With reference to the Japanese bomb study, Muller (1955) has argued that the available data cannot be used to support the idea that the atomic radiation has had no mutagenic effects. He estimated on the basis of laboratory experiments that the dose of 200 rads received by many Hiroshima survivors would have caused their offspring to inherit on the average at least one new mutation. But such radiation-induced mutations would be spread thinly over a large number of generations, and their over-all impact would be too insidious to affect the population noticeably as a whole. The statistical methods, as used in the Japanese study, are too insensitive to detect the occurrence of such radiation-induced mutations. In addition, the ascertainment of any increase in human-mutation rates due to increased radiation is stymied by the lack of information about the natural spontaneous-mutation rate. As Vogel points out in the included article, the basic problem of measuring spontaneous-mutation rates in human populations remains unsolved. Hence, in the absence of a standard mutation rate for comparison, any fluctuations or deviations from the natural rates which might be due to increased levels of radiation are not detectable.

Aside from the provisional conclusions reached by radiation studies, two generally accepted premises do emerge: (1) There is no threshold below which mutations fail to be induced by radiation, and (2) there is a simple linear relationship between the frequency of mutation and the radiation dosage. The two premises lead to the inevitable prediction that any rise in the levels of radiation to which a population is exposed will increase the frequency of mutation. There is no doubt that modern human populations are being exposed to increasing amounts of radiation from the fall-out of atomic- and hydrogen-bomb testing, the peaceful uses of atomic energy, and medical and dental X-rays. The genetic consequences of this increased exposure must be carefully considered.

In his article included here, Howard Newcombe deals with some of the methodological and moral problems involved in determining the risk of genetic damage to populations exposed to increased radiation. He cites studies showing that the human suffering and the costs to society already produced by the present rate of mutation are significantly burdensome: "In Northern Ireland, for example, more than one-quarter (26.5%) of the hospital beds and institutional places for the handicapped are occupied by persons with hereditary or partially hereditary conditions; and about 2 per 1000 of the population, exclusive of relatives, are employed full-time looking after them." He further estimates that under present maximum permissible doses, 800 additional individuals affected with a serious hereditary condition will be born in Canada each year. Newcombe then concludes, "Such an effect, while not to be regarded as castastrophic, is nevertheless substantial enough to merit serious consideration."

Scientists are also becoming increasingly concerned that man may be increasing the mutation rate with his extensive use of various common nontoxic chemicals such as drugs, pesticides, and the like. Clearly, under present conditions, the rate of human mutation is of immediate concern not only to the population geneticist, but to all mankind.

Determining the Human Mutation Rate

Determination of the spontaneous human mutation rate is, as Vogel discusses in his article, fraught with difficulties. Putting aside for the moment the interpolations made to human populations from the study of laboratory populations, the population geneticist must rely on a single method for the estimation of mutation rate in human populations. This so-called *direct method* consists of counting all new cases of a dominant genetic trait appearing from the mating of two normal parents, and comparing this number with the whole population number to obtain a *mutation rate* (μ) expressed in terms of the number of gametes involved.[2] Vogel gives tables of selected mutation rates as determined by the direct method. The estimates range from 1×10^{-4} (neurofibromatosis, that is, development of a large number of neurofibromas in the body) to $6.3\text{-}9.1 \times 10^{-6}$ (diaphyseal aclasis, that is, imperfect formation of bone in the diaphyseal cartilage). These figures, however, should not be taken to mean that all mutations in man are of this order of magnitude. Rather, the figures are representative only of mutations which lead to clearly discernible or disadvantageous phenotypical conditions, which are, therefore, easily detected and counted by the population geneticist.

Interestingly, the estimates of the average mutation rates for genes producing lethal or semilethal effects on several species of *Drosophila* are also of the order 10^{-5} per generation. However, given the differences in generational time between man (years) and *Drosphila* (days), these data indicate that possibly human genes are much more stable than those of *Drosphila* (Dobzhansky and Spassky, 1954).

Reliance on the type of data traditionally used in mutation studies has led to several conclusions regarding the rate of mutation and the nature of mutations which, in the light of recent research, must now be regarded as provisional. For example, the conclusion that most mutations are lethal or obviously injurious in some way to the organism is primarily based on the results of studies done on obvious dominant or recessive disorders, the vast majority of which are undeniably inadaptive. Such data are then supplemented by the unproved conclusion that any mutation tending to improve adaptation would have already been incorporated into the gene pool of a species. However, questions regarding this conclusion are raised, for example, by experiments on *Drosophila* where of all the mutations induced by radiation only about 20 percent exerted drastic fatal or deleterious effects on the flies, whereas 45 percent lowered viability only slightly from 100 percent to 95 percent, and 10 percent actually raised viability from 100 to 105 percent (Grant, 1963). On the basis of similar experimental evidence Muller (1950) suggested that "small" mutations[3] are about five times more frequent in organisms than those producing drastic phenotypic effects.

[2]In Denmark in 1941 there were 79 chondrodystrophic dwarfs who had normal parents. Because this trait is assumed to be due to the presence of a dominant gene, the 79 are considered to be primary mutants. In a total Danish population of approximately 3,800,000, the mutation had occurred among 79 of the 7,600,000 gametes which had been inherited, putting the mutation rate (μ) at about 1:100,000 (1×10^{-5}) gametes per generation (Slatis, 1955).

[3]There is no qualitative distinction to be made between "types" of mutation, that is, "macromutation," "major," "small," and the like. The magnitude of the phenotypic effect differs, but the underlying genes are not different kinds of genes.

The proposition that "small" mutations are more frequent in man is, of course, difficult to investigate directly, for if the effects of numerous mutations are so small as to be undetectable, then the mutations escape enumeration and their rates remain undetermined. However, there is other evidence which indirectly bears on the question concerning the frequency of "small" mutations. Viewed retrospectively, all the genetic variability that is present in any human gene pool as alleles originated in past germinal-cell mutations which have been occurring throughout the evolutionary history of *Homo sapiens*. The introduction and refinement of various new biochemical techniques are now enabling geneticists to identify a vast array of alleles in human populations which do not exert drastic effects on the individual phenotype and have therefore gone undetected until recently. The relatively new data showing the ubiquity and multiplicity of such alleles in human populations raise important questions concerning some of the traditional views about (1) the rate of mutation in man, (2) the deleterious effects of the majority of mutations, and (3) the factors which have determined the allele frequencies observed.

In his article reproduced in Chapter IV, the English geneticist Harry Harris discusses some of the questions raised by the ubiquity and frequency of the many human protein alleles revealed by the biochemical technique of electrophoresis. Biochemical analysis has shown that the great majority of these variants differ from the "normal" gene by only a single amino acid substitution in one or another of the constituent polypeptide chain(s) of the protein in question. The assumption is that the amino-acid substitution is the result of mutational events which have caused a single base change in the corresponding gene of the DNA helix. This type of mutation is referred to as a *point mutation*. Theoretically, point mutations are responsible not only for the rare genetic disease entities which are used in the direct estimates of mutation rates, but also for the common alleles such as those of the blood groups and various enzymes or proteins which the population geneticist uses in his calculation of gene frequencies. Electrophoretic enzyme surveys of various populations are revealing that in any given population there must be a significant fraction of loci at which two or more relatively common alleles and a considerable number of rare alleles occur. "In fact," as Harris states, "it is beginning to appear that, if virtually any enzyme or protein is examined by sufficiently sensitive methods in a large enough number of individuals, one or more rare variants are likely to be detected."

On the basis of such data, Harris suggests with respect to the mutation rate that, on an average, every newborn infant may be "expected, as a result of a new mutation in either of its parents, to synthesize at least one structurally variant enzyme or protein." He concludes:

In general . . . it would appear likely that the mutation rates often cited for rare abnormalities in man are probably higher than the rate at which a single base at a particular site in a gene is altered by mutation. On the other hand, they probably under-estimate the total mutation rate per locus per generation, since this would include base changes at all possible sites within the gene.

With respect to the assumed deleterious nature of the majority of mutations, the properties of the rare biochemical variants in healthy heterozygotes indicate that the proportion which are markedly deleterious in the homozygous state may, in fact, be quite small. Finally, the biochemical data are beginning to bring into question the traditional assumption that the incidence of rare variants in a gene pool is primarily determined by a balance between recurrent mutations and their elimination by natural selection (see discussion in the introductory essay to Chapter IV). Clearly, research of this nature on biochemical genetic variation has opened new avenues of theory and research with respect to mutation.

Human Cytogenetics

The mutations discussed thus far are point mutations, and although point mutations are most frequently used by the population geneticist, mutations involving greater segments of the chromosomes or whole chromosomes also occur frequently in all human populations. However, human *cytogenetics*—that is, the study of chromosomes—is a relatively young field, having been made possible by technical advances only within the last twelve years. The correct diploid number of 46 chromosomes in human tissue cells was not determined until 1956 (Tjio and Levan, 1956), and the association between chromosomal abnormalities and genetic syndromes, such as Down's syndrome (Mongolism), was not proven until 1959 (Lejeune, Turpin, and Gautier, 1959). Their work was quickly followed by the discovery of many other autosomal and sex-chromosomal abnormalities, and human cytogenetics is now a well-defined discipline. However, because cytogenetics is so young and some of its methods are so different from classical genetics, it has not yet been closely integrated with established quantitative genetic methodology.

Much of the work in human cytogenetics to date has been devoted to identifying and examining persons with obviously abnormal *karyotypes* or screening human abortion material for the presence of abnormalities. Knowledge about the incidence and prevalence of chromosomal abnormalities in human populations is almost totally lacking. The paper included here by William Court-Brown and P. G. Smith reviews the work done on the frequency of chromosomal abnormalities, and examines some of the problems involved in doing human population cytogenetics. The authors point out that the lack of information about chromosomal abnormalities in normal unselected populations is unfortunate for several reasons. If total reliance were to be placed in the types of studies done to date, such as those which show that nearly a third of early spontaneous human abortions involve chromosomal abnormalities of various types (see, for example,

Polani, 1966), or that an estimated one out of every one hundred children born alive suffer from conditions determined by chromosomal aberrations (Report, UNSCEAR, 1962), the obvious conclusion would be that chromosomal abnormalities are either incompatible with life or significantly reduce the viability and reproductive capacity of the individual affected. But in the absence of complete population cytogenetic surveys the conclusion is untenable. In addition, it is impossible for the medical geneticist to offer counseling to the parents of chromosomally affected offspring, because there are no data concerning the developmental history of chromosomal abnormalities in large numbers of *unselected* babies. Possibly the prognosis for some of the children might be encouraging; a few chromosomally abnormal individuals who are phenotypically healthy and normal have been reported. Finally, the lack of accurate figures on the incidence of certain chromosomal abnormalities, such as XYY, in the general population bears importantly on certain sociolegal problems.

It has been suggested, for example, that the presence of an extra Y chromosome may produce criminal propensities in a male. The suggestion stems from studies which show a higher incidence of XYY individuals than would be expected in certain British maximum-security hospitals (cf. Court-Brown and Smith). These statistics have been used to raise an important question about the legal status of an XYY person: May a chromosomal abnormality be a valid defense for a crime in that the person may not be responsible for his criminal acts? In two cases of accused murder, one in France and one in Australia, testimony of the defendants' 44 + XYY chromosomal constitution was admitted at the trial. In the Australian case the defendant was acquitted on the grounds of insanity; in the French case the genetic testimony may have been a strong factor in the reduced sentence given (Housely, 1969). Yet it is obvious that unless an indisputable cause-and-effect relationship can be established between an individual's chromosomal complement and his social behavior, the association is a purely statistical one, derived, as Court-Brown points out, from figures on "populations" which have not been rigorously defined. Although the present material seems to suggest that an XYY male, in comparison with a normal XY male, may incur some increased risk of developing a behavioral disturbance, it does not prove that such a person is genetically destined to do so. A recent estimate suggests that the XYY chromosome abnormality may be more common than is believed, and that in any population, a number of XYY individuals may be alive and pursuing normal lives, not genetically destined for an institution. For example, in a Canadian hospital one of every 250 newborn males was 44 + XYY, a figure which would put the number of XYY's in the population far beyond the total population of all penal institutions (Court-Brown, 1968). Only complete cytogenetic population surveys will provide the information needed to clarify and resolve some of the questions raised by the hospital and institutional data;

meantime, some mutants will continue to present a dilemma to the courts of law.

One important evolutionary consequence of all types of mutations is that they provide the vast variety of genetic material which segregates and recombines during meiosis and sexual reproduction, creating an almost infinite number of different genotypes in any population. The resulting uniqueness which sets each individual apart from any other (with the exception of monozygotic twins) is obvious in both the external physical traits and those revealed only by laboratory analysis of blood and other body tissues. Harris calculates that, with respect to only ten enzyme loci, the probability that any two randomly selected people in an English population will have exactly the same enzyme phenotype is about 1 in 200. If probability estimates are extended to 17 blood genetic loci commonly found in Western Europeans, the chances of two randomly selected people having the same phenotype are less than 1 in 350,000 (Giblett, 1969). In fact, in any population the potential number of different genetic combinations far exceeds the number of individuals who are born in any one generation, or all generations combined.[4] "One may plausibly conclude that, in the last analysis, every individual will be found to have a unique enzyme constitution" (Harris, below).

The Genetic Load

It has been suggested that the genetic variability present in any gene pool in the form of recessive mutations imposes a "hereditary burden" on a population in the form of mortality and morbidity suffered by the homozygotes. The hereditary burden is referred to as the *genetic load*, a term which is defined in a number of ways. Muller (1950) originally used the term to express the proportion of a population undergoing genetic elimination or the amount of disability suffered by the average individual. Crow (1960) defined genetic load as the proportion by which average fitness in the population is decreased in relation to what it could be if the factor under consideration (mutation) were absent. Wallace and Dobzhansky (1959) have maintained that all deleterious genes in a gene pool are the genetic load.

Granting that any population contains in its gene pool numbers of potentially deleterious or lethal mutations, the question arises, How can we estimate the number of such genes carried by individuals and thereby calculate the so-called genetic load of the population? One approach currently used to

---

[4]The number of genotypes (g) that can be assembled by any number of alleles (n) at a given locus in the gene pool of a population is given by the formula

$$g = \frac{n(n + 1)}{2}.$$

The total number of possible genotypic combinations is then

$$g_1 \times g_2 \times g_3 \times \cdots g_i.$$

provide load estimates uses the data provided by *consanguineous matings* (see Chapter V). The assumption underlying the use of this approach is that the deleterious mutations are supposedly recessive genes and therefore their presence in the gene pool of a population should be revealed in the reproductive performance of consanguineous matings. Because closely related individuals have a greater chance of carrying identical alleles than unrelated or distantly related individuals, the children of consanguineous matings have a greater chance of being homozygous for rare recessive alleles than do children from nonconsanguineous matings. Thus, some population geneticists have argued that the genetic load of a population is susceptible to analysis through studies of the reproductive performance of consanguineous matings, and that such analysis may provide insight into the means by which genetic variability is maintained in human populations.

The genetic impairment or "load" revealed by such studies may be of several types, including (1) the *mutational load*, due to the presence of disadvantageous mutants at the so-called "classical loci"; (2) the *incompatibility load*, due to maternal-fetal interactions; (3) the *segregational load*, due to stable polymorphisms maintained by heterozygote advantage or varying selective coefficients at the so-called "balanced loci" (see Chapter IV), and (4) the *disequilibrium load*, due to departure of gene frequencies in a given generation from expected equilibrium values. These categories are not mutually exclusive in any population, and although the geneticist would like to isolate each component to evaluate its total contribution, this type of analysis is not possible. Presently much of the argument between the geneticists working on the problem of the genetic load centers on two issues: (1) whether or not genetic variability is maintained by the "classical" or "balanced" loci (that is, mutational vs. segregational load), and (2) which statistical methods should be applied to which demographic data to obtain estimates of the size of the load carried by a population.

The majority of the work on genetic load has been done using the so-called semidirect statistical methods involving consanguinity-data analysis. Presently there are six semidirect methods available; two are applicable to morbidity and four are applicable to mortality data. In general, the latter four methods are based on the theory that a linear relationship, expressed by the B/A ratio, exists between mortality and inbreeding (Freire-Maia, 1964). However, the population data and statistical methods by which the B/A ratio is calculated vary according to the choice of the investigator, and the results obtained may be significantly different, as illustrated by the two articles included here (Schull and Neel; Freire-Maia and Azevedo). Schull and Neel compare the mortality of the offspring of consanguineous matings in Japan with a control group of children born of unrelated parents. Using the method of Morton, Crow, and Muller, Schull and Neel express the B/A ratio as the ratio of the regression coefficient (B) to the intercept (A). Their method assumes that the natural logarithm of the frequency of survivors (S_i) is linearly correlated with the coefficient of inbreeding (F_i). The weighted regression coefficient of this correlation (B) represents an estimate of the average number of lethal equivalents per gamete. The assumption is that the ratio B/A will be large if the genetic load is a "mutational load," whereas the ratio will be small if the load is a "segregational" one. Applying this method to their inbred- and control-group data, Schull and Neel obtain low B/A ratios and no statistically significant effect of inbreeding. They conclude that "the data lead to no firm conclusions," but that "there is no convincing evidence that the effects of inbreeding on mortality differ in the major racial groups nor that these effects are large."

The Brazilian population geneticists N. Freire-Maia and B. C. Azevedo dispute the findings of Schull and Neel in their article included here. They are of the opinion that because the ratio B/A includes a certain fraction of deaths attributable to nongenetic factors, the calculations should be done, if possible, on populations where high standards of living may insure that prenatal and infant-juvenile deaths are largely due to genetic factors. Therefore, from Schull and Neel's data they take only the control population ($F = 0$) and the inbred population, which is socioeconomically the most similar to the control and which coincidently shows the lowest inbreeding coefficients. Applying a slightly different formula[5] to these subsamples, Freire-Maia and Azevedo obtain significantly higher B/A ratios for the Japanese populations, suggesting an inbreeding effect on mortality. On the basis of their results they conclude that more precise estimates of lethal equivalents may be obtained if population data with the lowest inbreeding levels ($F \leq 1/64$) are used, and that many previously reported interpopulational differences in magnitudes of the inbred load may be spurious, because of poor selection of data.

It is obvious that the theoretical and methodological problems involved in genetic-load studies are enormous. In fact, the general validity of the concept of genetic load and what the B/A ratios actually reveal has been severely questioned (cf. Sanghvi, 1962). Yet it can be generally stated that numerous studies show increased mortality in children of consanguineous

[5]The formula used where a control sample ($F = 0$) is available is:

$$B = \frac{\log (S_i/S_c)}{2n \log [1 - (F_i/2n)]},$$

where F_i is the mean coefficient of inbreeding of the whole group of inbred subsamples, $S_i = (1 - M_i)$, $S_c = (1 - M_c)$ where M_i is the mortality rate in the inbred group and Mc, the rate in the control group, and n is the average number of common ancestors per consanguineous couple. In practice the equivalent simplified equation

$$B = \frac{\log (S_i/S_c)}{-0.4343 F_i}$$

may be used (Freire-Maia, 1964).

marriages, and that some of this may be due to homozygosity from the meeting of two recessive mutations. But no species, including man, will ever be without a certain number of deleterious mutations; they are the small price a population pays for the genetic variability so essential to long-term survival. All one can hope is that the mutation rate of lethal or deleterious genes will be kept as close as possible to the natural rate, and not be increased by the irrationality of mankind.

Sources and Rates of Mutation in Populations

See the following readings: "Problems in the Assessment of Genetic Damage from Exposure of Individuals and Populations to Radiation," by Howard B. Newcombe, p. 121; and "Mutations in Man," by Friedrich Vogel, p. 131.

Products of Mutation

See the following readings: "Human Population Cytogenetics," by W. M. Court-Brown and P. G. Smith, p. 153; "Some Observations on the Effect of Inbreeding on Mortality in Kure, Japan," by William J. Schull and James V. Neel, p. 171; and "Extraneous Variation In Inbreeding Studies," by Newton Freire-Maia and J. B. C. Azevedo, p. 182.

REFERENCES TO THE LITERATURE CITED IN THE OVERVIEW

1. COURT-BROWN, W. M., 1968. Males with an XYY sex chromosome complement. *J. Med. Genet.*, 5: 341–359.
2. CROW, J. F., 1960. Mutation and selective balance as factors influencing population fitness. In L. I. Gardner (ed.), *Molecular Genetics and Human Disease.* Springfield, Ill., Charles C. Thomas.
3. DOBZHANSKY, T. AND B. SPASSKY, 1954. Rates of spontaneous mutation in the second chromosomes of the sibling species, *Drosophila pseudoobscura* and *Drosophila persimilis. Genetics*, 39:899–907.
4. FREIRE-MAIA, N., 1964. On the methods available for estimating the load of mutations disclosed by inbreeding. *Cold Spr. Har. Sympos. on Quant. Biol.,* 29:31–40.
5. GIBLETT, E. R., 1969. *Genetic Markers in Human Blood.* Philadelphia, F. A. Davis.
6. GRANT, V., 1963, *The Origin of Adaptations.* New York, Columbia University Press.
7. HOUSLEY, R., 1969. The XYY chromosome complement and criminal conduct. *Okla. Law Rev.*, 22: 287–301.
8. KAPLAN, I. I., 1957. Treatment of female sterility with X-rays to ovaries and pituitary. *Canad. Med. Assoc. J.*, 76: 43–46.

9. LEJEUNE, M. J., R. TURPIN, AND M. GAUTIER, 1959. Chromosome diagnosis of mongolism. *Arch. Franc. Pediat.*, 16: 962–963.

10. MEYER, M. B., T. MERZ, AND E. L. DIAMOND, 1969. Investigation of the effects of prenatal X-ray exposure of human oogonia and oocytes as measured by later reproductive performance. *Amer. J. Epidemiol.*, 89: 619–635.

11. MILLER, R. W., 1969. Delayed radiation effects in atomic-bomb survivors. *Science*, 166: 569–574.

12. MORGAN, T. H., 1910. Sex limited inheritance in *Drosophila. Science*, 10: 120–122.

13. MULLER, H. J., 1927. Artificial transmutation of the gene. *Science*, 66: 84–87.

14. ———, 1950. Our load of mutations. *Am. J. Hum. Genet.*, 2: 111–176.

15. ———, 1955. The genetic damage produced by radiation. *Bull. Atomic Sci.*, 193: 58–68.

16. POLANI, P. E., 1966. Chromosome anomalies and abortions. *Develop. Med. and Child Neurol.*, 8: 67–70.

17. PUCK, T. T., 1960. Radiation and the human cell. *Sci. Amer.*, 202: 142–153.

18. *Report of the United Nations Scientific Committee on the Effects of Atomic Radiation*, 1958. General Assembly Official Records; Thirteenth Session, Suppl. 17, A/3838.

19. *Report of the United Nations Scientific Committee on the Effects of Atomic Radiation*, 1962. General Assembly Official Records; Seventeenth Session, Suppl. 16, A/5216.

20. SANGHVI, L. D., 1963. The concept of genetic load: a critique. *Am J. Hum. Genet.*, 15: 298–309.

21. SCHULL, W. J. AND J. V. NEEL, 1965. *The Effects of Inbreeding on Japanese Children*. New York, Harper and Row.

22. SLATIS, H. M., 1955. Comments on the rate of mutation to chondrodystrophy in man. *Am. J. Hum. Genet.*, 7: 76–79.

23. TJIO, J. H. AND A. LEVAN, 1956. The chromosome number of man. *Hereditas*, 42: 1–6.

24. WALLACE, B. AND T. DOBZHANSKY, 1959. *Radiation, Genes and Man.* New York, Henry Holt.

Problems in the Assessment of Genetic Damage from Exposure of Individuals and Populations to Radiation

HOWARD B. NEWCOMBE

For considerably more than a decade special interest has been shown in the possible effects of exposure to levels of x-rays and other ionizing radiations that would at an earlier period have been considered quite harmless. Two kinds of risk are involved, that to the exposed individual himself and that to his descendants or, for short, the *somatic* and the *genetic* risk. Interest in both sorts of effect stems from the fact that exposure in greater or lesser degree to some form of man-made radiation is now widespread, and the risks, even if small to any one person, may sometimes be regarded as substantial when summated over large numbers of persons or whole populations. Thus, even such a small personal risk of death or disability as, for example, one in ten thousand, would nevertheless seriously affect some 300,000 persons if all the world's population of approximately three billion people were exposed to it. Furthermore, in the case of the genetic effects, an essentially new element has been injected into our thinking, namely a concern which is primarily for the well-being of individuals who are as yet unborn and who may not be conceived for some time to come.

Although it is now quite generally accepted that even very low doses involve some increase in risk, and especially some increase in genetic risk, attempts to estimate the magnitudes of the possible effects are still far from satisfactory and the best quantitative assessments currently available are subject to uncertainties of perhaps an order of magnitude in either direction. Thus, if we were to estimate that some 10,000 individuals from the Canadian population would be seriously affected over a 30-year period, the true figure might well be anywhere in the range from 1000 to 100,000. In spite of this degree of uncertainty, it is necessary that such estimates serve as guides in making administrative decisions which will limit the exposures arising out of industrial practices and, to a lesser extent, from medical practices as well.

From the Biology Branch, Chalk River Nuclear Laboratories, Chalk River, Ont.

Reproduced by permission of the publisher and Howard B. Newcombe from The Canadian Medical Association Journal, *92:171–175 (1965). Originally published under the title "Radiation Protection in Canada. Part VI. Problems in the Assessment of Genetic Damage from Exposure of Individuals and Populations to Radiation."*

Howard B. Newcombe, a geneticist, was born in Nova Scotia in 1914. He received his B.S. from Acadia University in 1935 and his Ph.D. from McGill University in 1939. At the present time he is head of the Biology Branch of the Atomic Energy Control Commision of Canada, Limited. His major field of interest is human-population genetics, with special emphasis on the genetic effects of radiation on populations. He is a member of numerous scientific committees, including the World Health Organization Advisory Panel on Human Genetics and the International Commission on Radiological Protection. From 1956 to 1966 he served as a member of the Canadian delegation to the United Nations Scientific Committee on the Effects of Atomic Radiation. He is past president of the American Society of Human Genetics and the Genetics Society of Canada. Some of his more recent publications include "Environmental versus genetic interpretations of birth order effects," Eugen. Quart., 12:90–101 (1965); "Present state and long term objectives of the British Columbian population study," in Proceedings of the 3rd International Congress of Human Genetics (Baltimore, 1967); "Multigeneration pedigrees from linked records," in E. Livingstone and S. Livingstone, Record Linkage in Medicine (Edinburgh, 1968); and "Pooled records from multiple sources for monitoring congenital anomalies," Brit. J. Prev. and Soc. Med., 23:226–232 (1969).

The present account will deal with some of the problems of arriving at predictions of the probable genetic harm done by exposing large numbers of people to relatively low levels of radiation.

Uses of Risk Estimates

Concern on the part of various national and international bodies with the likely effects of low levels of radiation has been stimulated in part by public reaction over radioactive fallout from nuclear weapons testing. More important, however, has been a requirement to set sensible limits for the occupational exposures of radiological technicians and employees of atomic energy establishments. This has led inevitably to consideration being given, in the planning of future nuclear power developments, to the levels of radiation to which it is justifiable to expose local populations in the vicinity of nuclear reactors, and to the maximum which might be set for exposure averaged over the populations of whole countries.

In this latter connection, an upper limit of 5 rads per generation from the peaceful uses of atomic energy, averaged over a whole population, has been suggested on the basis of genetic considerations by the International Commission on Radiological Protection, for purposes of planning future developments. No such limit has been set for medical exposures, since these must be dictated by the needs of patients, but, for the present at least, the gonadal doses from this source may be taken as being in the vicinity of 3 rads per generation if fluoroscopies are included. (The rad is a unit of absorbed radiation approximately equivalent to the roentgen; the lethal dose for whole-body exposures delivered in a short period of time is in the vicinity of 500 to 600 rads.) Nature contributes another 3 rads per generation. In the future, a further substantial contribution to the overall total may result from exposures of air crew and passengers of high-flying aircraft, operating at 70,000 feet and above, to galactic cosmic radiation and to the sometimes intense particulate radiation from solar prominences. (Cosmic rays from outside the solar system include rapidly moving particles ranging in size from protons to the atomic nuclei of iron; cosmic rays from flares on the sun consist almost wholly of protons. In both cases, however, the effects are essentially similar to those of x-rays.)

Limits set for the exposures of populations from future peaceful uses of atomic energy will presumably influence the design of reactors and the methods for management of radioactive wastes. Thus, the considerations entering into the choice of an acceptable maximum represent a balancing, albeit a subjective balancing, of the needs for safety against the costs of achieving a given level of safety. As in many other human activities, absolute safety in the purist's sense is economically unattainable, but an exceedingly high level of safety, as judged by present standards, would be a reasonable aim in any such new and major undertaking.

The reasoning has its analogy in clinical practice, in that attempts to reduce the gonadal exposures from diagnostic procedures raise questions of how great are the risks one seeks to avoid. In both situations it is reasonable to ask, is the risk such as to warrant the added trouble and efforts of reducing the gonadal exposure?

Special Reasons for Inferring that Low Doses are not Wholly Safe

Concern over a possible increase in the numbers of harmful hereditary traits in individuals, due to the inheritance of radiation-induced defects in the germ plasm, has arisen because of the apparently simple linear relationship between the radiation dose administered and the yield of induced mutations. This relationship has been demonstrated for a wide range of laboratory organisms, including microbes, fruitflies, plants and mice. There is thus no reason to suppose that the yield of mutations in man varies in other than a linear manner with the magnitude of the radiation exposure. It has also been shown that the mutagenic effects of repeated exposures to the same individual organisms are cumulative.

This means that there is no strictly safe dose; smaller doses produce smaller numbers of heritable changes, but no dose is so low that it can be said to carry no risk at all of producing a mutation. It follows, also, that one rad administered to 100 individuals would involve the same risk of genetic harm to future generations as 100 rads given to one person, and so on, the damage being proportional to the total number of man-rads no matter how low the exposure to any one individual. To illustrate this, suppose that 1,000,000 man-rads were required to cause the transmission of 1000 newly defective genes into the next generation; the harm to future generations would be the same whether 10,000 people received 100 rads each, 20 million people 1/20th of a rad each, or 3 billion people 1/3000th of a rad each.

This implication of the linear dose-mutation relationship has, for some reason, given rise to an immense amount of popular confusion. And yet, the principle of linearity has not been seriously challenged on scientific grounds, nor has the above inference been questioned on logical grounds. The reasoning has, in fact, been employed by all of the major national and international committees which have published detailed technical statements on the subject.

The evidence for this "linearity principle" is based on approximately 40 years of painstaking research, and the principle itself has not been seriously challenged. Recent studies of radiation induction of mutations in immature germ cells of mice, i.e. spermatogonia and dictyate oocytes, have indicated a dose-rate dependence in the form of an enhanced production per unit dose when an exposure is given at high intensities. However, for spermatogonia, one-third of the mutations induced at high intensities can be shown to be intensity-independent and, for these, the linearity principle would be expected to hold down to zero dose.

Even the claim that it is theoretically impossible to prove linearity of a response curve down to zero dose has been disproved by experiments with micro-organisms in which the dose-effect curve has been plotted down to doses so low that less than one ion cluster or "hit", is sustained per nucleus ("hit" being the name applied to the chemically active ion pairs, or clusters, that result from absorption of x- or gamma-ray photons). It follows rigorously from such a finding that the curve must be linear to zero dose, for this particular organism and for the particular changes studied.

Similar elegance of experimental design is not possible, or else is too expensive, in studies with higher organisms. However, one very laborious investigation using the fruitfly, Drosophila, has come close to proving the point; and the exceedingly expensive and painstaking studies with mice have done nothing to cast doubt on the linearity principle.

Radiation-induced gene mutations have now been studied in such a wide range of plants and animals, from the microbial forms to the highest groups, and with such consistent results, that there is no reason for supposing that man may be exceptional and perhaps immune to this sort of radiation effect. A related kind of change that is likewise induced in genetic materials by irradiation, namely chromosome breakage and gross chromosomal rearrangement, has moreover been observed in human cells following exposure in artificial cell culture, and also following accidental exposure of marrow cells *in situ*.

Neither is there reason to believe that the bulk of the induced mutations are other than harmful in their effects. There is an abundance of evidence on this from studies of living materials ranging from fruitflies to cereal plants, but the most strictly relevant information comes from work with laboratory mammals. Of the radiation-induced mutations studied in mice, which were chosen for this purpose solely because of their readily detectable but seemingly innocent effects on coat colour: (a) many mutations have been found to be harmful even when inherited from only one parent; (b) the majority are lethal whenever an individual has the misfortune to inherit the same gene from both his parents; and (c) many of these so-called recessive lethals kill at stages of development that would cause them to be regarded as tragedies in human experience (e.g. one group of them causes infant death at about the time of weaning). Studies of a different sort, carried out at Chalk River, Ontario, on groups of rats descended from many generations of irradiated ancestors, have shown a resulting increase in the rate of infant death and in the occurrence of dwarf or "runt" animals, and a substantial reduction in learning ability as measured in tests using one of the standard designs of maze.

While it is recognized that mutations may occasionally be beneficial, and that evolution must have occurred in the past through a process of selection for such beneficial effects by means of differences in fertility and mortality, such selection must be regarded as costly in terms of human suffering. It should not be thought that we might speed up the evolutionary process simply by increasing the mutation rate, because, if we were to attempt to do this, a corresponding increase would be required also in the rate at which the resulting misfits are weeded out. This, of course, would be contrary to our traditions and instincts.

For physicians who, as a group, are necessarily preoccupied with treating the ills of people who are now with us, the unborn descendants of these people, and especially those descendants removed by a number of generations, may seem remote indeed. Nevertheless, in assessing the genetic harm from radiation exposures there is no logical course but to attach the same weight to a given degree of personal injury of a human individual regardless of the lapse of time between the causal event and the eventual expression of the harm done. It should not lessen the importance of the injury that the defective individual may occur in some other generation, or that the causal exposure and the resulting defect will never be identified with one another.

Thus, in assessing the magnitude of the hereditary effects of a given exposure it is necessary to think in terms of the numbers of affected individuals, regardless of the time over which they may be distributed. There are a number of ways of going about this task of assessment.

Approaches to the Assessment of Genetic Damage

One possible line of reasoning starts with the known amount of hereditary disease already present in the population, and attempts to distinguish that fraction of it that remains prevalent because of mutations of natural origin. It is this fraction which will be expected to increase in direct proportion to any artificially induced elevation of the mutation rate, such as occurs when a population is exposed to ionizing radiation.

An alternative approach involves a simple extrapolation to man from the numbers of recessive lethal changes known to be induced in experimental organisms such as Drosophila or the mouse. The simplicity of this procedure stems from the fact that many recessive lethals are known to have harmful effects even when inherited in single dose *via* just one parent. The severity of the effect in any one individual may be ignored because, where the effect is not immediately lethal, a succession of carrier individuals may be expected to perpetuate the gene until it is eliminated through death or failure to reproduce. The less severe the effect, the longer the line of descent and the greater the number of carrier individuals who will be slightly affected before the gene eventually causes the elimination of a line of descent. Even where a recessive lethal mutation is neutral in its effect when inherited in single dose, the elimination, although delayed will eventually occur when the same gene is inherited *via* both parents, in double dose. In other words, unless a population is to acquire an ever-increasing "load" of genetic damage, the influx of harmful mutation must be balanced by a corresponding number of deaths or failures to reproduce.

From this reasoning has come the dictum that, on the average, each harmful mutation leads eventually to a "genetic death", i.e. the elimination of a line of descent, regardless of the severity of its usual effect. Not all such eliminations are to be regarded as painful because some will take place through early fetal losses. A substantial proportion, however, must also certainly be associated with deaths and handicaps of liveborn individuals, as indicated by studies of mutations in mice.

Both of the above approaches have serious limitations so that any estimate of the numbers of affected individuals must be regarded as perhaps very considerably greater, or very considerably less, than the true value.

We will deal with these two methods in turn.

The Amount of Hereditary Disease Already Present in the Population

Interest in the genetic effects of radiation has done much to foster recognition of the contribution from hereditary factors to the ills and handicaps that are already present in the population.

By far the best documented estimate of the total amount of genetic and partially genetic disease has been made for the population of Northern Ireland. The most recent revision of this is given in the 1962 Report of the United Nations Committee on the Effects of Atomic Radiation (General Assembly Official Records: Seventh Session. Supplement No. 16, A/5216). From this survey it appears that at least six out of every 100 children born alive will be more or less severely affected at some time in their lives by conditions that are largely or in part hereditary in origin. A number of broad categories of defect contribute to the 6%.

Conditions that are inherited in a simple fashion and are determined by single gene differences are known to affect seriously about 1% of liveborn persons. Examples include phenylketonuria, albinism, Huntington's chorea, multiple neurofibromatosis, cystic fibrosis. the muscular dystrophies, and a number of sorts of blindness and deaf mutism. Many of these conditions are individually rare, but something like 400 known heritable traits of man are determined by single genes.

A further 1% of liveborn infants suffer from conditions that are likewise inherited in simple fashion but which are determined by the presence or absence of whole chromosomes or parts of chromosomes. Mongolism (Down's syndrome) is the most common of these, but the group includes Klinefelter's and Turner's syndromes, and others which have been named to indicate the particular chromosome involved.

Among the conditions that are not inherited according to any simple pattern, one broad category consists of the congenital malformations (other than those known to be caused by single gene or chromosome differences). For the majority of these, the effect is probably determined by an unfortunate combination of an adverse fetal environment with an inherited susceptibility. There is considerable evidence of this as relating to a number of such conditions and, although the relative importance of the environmental and hereditary components may vary from one kind of malformation to another, it is unlikely that any substantial proportion of these anomalies is caused wholly by environment, or by heredity, acting alone. Examples include clubfoot, congenital dislocated hip, pyloric stenosis, spina bifida and hydrocephalus. Such malformations are recognizable in about 1% of liveborn individuals at birth, and in a further 1.5% by 5 years of age.

Another broad category, consisting of the so-called "constitutional" disorders of adult life, like-wise exhibits irregular inheritance. The U.N. Committee has included in this group of conditions: epilepsy, diabetes, pernicious anemia, exophthalmic goitre, schizophrenia and manic-depressive reaction. For all of these there is strong evidence both of environmental and of genetic factors in the causation. Collectively, the above six conditions affect about 1% of liveborn individuals at some time in their lives.

The overall total of 6% of liveborn persons seriously affected by conditions from one or other of the above four broad categories somewhat underestimates the genetic component in human morbidity and mortality. Omitted from the list are some quite common and important diseases, such as gastric and duodenal ulceration, gout, essential hypertension, and coronary artery disease, that show family groupings indicative of a genetic component in the causation. Also excluded is the apparently substantial genetic contributions to human sterility and pregnancy wastage.

Other measures of the social burden of hereditary and partially hereditary diseases are possible. The percentage of physicians' consulting time spent on cases of these kinds is fairly modest, being, for Northern Ireland at least, in the vicinity of 6 to 8%. However, since many of the genetic conditions tend to be chronic, the demands which they create for hospital space and special facilities are large. In Northern Ireland, for example, more than one-quarter (26.5%) of the hospital beds and institutional places for the handicapped are occupied by persons with hereditary or partially hereditary conditions; and about 2 per 1000 of the population, exclusive of relatives, are employed full-time looking after them. No similar estimates have been made for Canada.

Effect of an Increase in Mutation Rate

There is no certainty that all of these conditions would increase in frequency with an artificially induced rise in the mutation rate.

The harmful single-gene traits, however, are believed to be maintained in the population by repeated mutations of natural origin which balance the losses of the causal genes that occur through deaths and failures to reproduce. These conditions would therefore be expected to increase in frequency in direct proportion to any artificially induced rise in mutation rate such as that caused by exposure to radiation. The same must also be true for syndromes that are determined by gross chromosomal differences, since the affected individuals are usually much less fertile than the rest of the population and the incidence rates would inevitably depend on the *de novo* production of the underlying genetic changes.

Less is known about the probable importance of mutations as affecting the frequencies of the less regularly inherited genetic conditions. Thus, for only about one-third of the harmful expressions, i.e. a third of the total 6%, is there much assurance that an upward shift in incidence would, in fact, occur.

A doubling of the overall mutation rate would therefore be expected, from these considerations, to increase the number of affected persons from six per 100 liveborn to eight per hundred, and lesser changes in the mutation rate would have proportionate effects.

Prediction of the amount of damage from a given radiation exposure depends, however, upon knowledge of the dose-mutation relationships. Study of these is virtually impossible in man and is exceedingly laborious and expensive even in the most convenient of laboratory mammals. Nevertheless, extensive mutation studies with mice have been carried out, and they indicate that for chronically exposed males about 100 rads per generation are required to produce additional mutations equalling in number those that occur naturally. For females this so-called "doubling dose" for low-intensity exposures is considerably higher, but since only a small fraction of the mutations occurring normally in each generation arise in the female germ cells we will not be far wrong if we take 100 rads as the doubling dose for the total mutation rate.

There is no similar figure from mice for the dose required to double the frequency of changes leading to the loss or gain of whole chromosomes and parts of chromosomes. However, for the fruitfly, Drosophila, the doubling dose for this kind of change, although somewhat larger, is of the same order of magnitude. We will therefore not be wrong by more than twofold if 100 rads is taken as a representative doubling dose for gene and chromosomal mutations combined.

A simple calculation is sufficient to indicate the magnitude of the effect that would be expected of this line of reasoning from a given exposure. If, for example, whole populations were in the future to receive 5 rads per generation from peaceful uses of atomic energy (the maximum suggested by the International Commission on Radiological Protection) plus another 3 rads from medical radiology, i.e. a total of 8 rads per generation from man-made sources, this might be expected to raise the mutation rate by 8%, and to increase the present six individuals with serious hereditary conditions per 100 liveborn by about one-third of this percentage, i.e. by an additional 0.16 per 100 liveborn. For a country such as Canada in which there are 0.5 million births per year this would amount to an increment of 800 affected individuals annually.

Such an effect, while not to be regarded as catastrophic, is nevertheless substantial enough to merit serious consideration. It should be emphasized again that there are uncertainties in extrapolating from mice to men, and a present lack of adequate information on the importance of natural mutations in maintaining the prevalence of the hereditary ills which we already have. The true increase might therefore be perhaps as much as an order of magnitude smaller, or larger, than estimated.

Figures arrived at in the alternative fashion fall well within this broad range. Recessive lethal changes are known to be induced in male mice at a rate of about 20 per 100,000 gametes per rad of acute exposure and, presumably, at one-third this rate for chronic exposures; the rate in females being substantially lower. This 8 rads per generation to both sexes would result in approximately 50 new recessive lethals being transmitted to each 100,000 liveborn individuals, which would be eliminated from the population only through deaths and failures to reproduce. For a country with 0.5 million births per year, 250 eliminations per year would be required to balance the increased mutation rate.

To some extent these two methods measure different things, and it is perhaps surprising that the independently derived estimates of genetic harm are as similar as they are.

Conclusions

While little quantitative reliance can be placed upon current numerical estimates of risks to future generations from radiation-induced hereditary changes, there is every reason to believe that such changes are real and that they are predominantly harmful. Furthermore, we do not know whether the best available estimates are too high, or too low.

Nevertheless, attempts at estimating genetic risks from ionizing radiations do serve a practical purpose. They indicate that the hereditary consequences of current medical exposures involving the gonads, and of the suggested upper limits for exposures of whole populations from the future peaceful uses of atomic energy, are sufficient to justify continued scrutiny. The genetic effects of the radiation from these two sources, when stated in terms of the numbers of seriously affected people per generation, or per year, might perhaps approach within an order of magnitude of the present highway fatality rate for this country. Thus, by two essentially independent estimates, in the vicinity of 250 or 800 individuals might be seriously affected each year in a population of 20 million people, as compared with a present rate of fatalities from motor vehicle accidents of approximately 4000 per year.

More precise estimates of the amount of genetic harm done by a given increase in the average level of exposure to a whole population will depend much upon a better understanding of the role played by mutations of natural origin, as compared with other factors, in maintaining the prevalence of the hereditary and partially hereditary diseases that are now with us.

Knowledge of the role of mutations of natural origin is required, in any case, because of its bearing on the larger problem of genetic disease in general. In this connection, it is perhaps not generally realized that in the one country for which the amount of such disease is best known, about a quarter of all hospital beds and of the institutional places for the handicapped are occupied by persons with hereditary or partially hereditary conditions.

Mutations in Man

FRIEDRICH VOGEL

Research on human mutations in general has slowed down somewhat during the last years. One of the reasons might be that the activity of the human geneticists interested in the mutation problem has shifted towards visible chromosomal changes. The traditional methods, on the other side, seem to be more or less exhausted, while new methods, which were welcomed with great hopes, have not turned out to be thoroughly successful to date.

1. Mutation Rates for Single Rare Mutations

In spite of all difficulties and sources of error in practical use which have been repeatedly reviewed [13, 43, 44, 45, 46, 50, 77, 85] the most reliable data available are mutation rate estimates for a few autosomal dominant or sex-linked recessive hereditary diseases. Since Haldane gave a review at the International Congress of Genetics 1948,[24] the number of more or less reliable estimates has increased somewhat, and in general estimates for the same traits have decreased.

Before we turn to the available estimates, two principal methods of estimation have to be reviewed.

The first one is the so-called direct method which simply consists in counting all sporadic carriers of a dominant disease in a population (patients with two healthy parents). The sporadic carriers are looked upon as dominant mutations and their number is compared with the whole population number. The direct method was used first by Penrose (1935)[21] in tuberous sclerosis.

The second method is the so-called indirect one, which was introduced by Haldane (1935)[22] in a mutation rate estimate for haemophilia. This method is based on the assumption that for rare deleterious mutations, there is an equilibrium between selection and mutation, which leads to selective elimination of exactly as many alleles as are produced by mutation from the wild allele. In mutations with a strong selective disadvantage, Haldane has demonstrated that a stable equilibrium is reached within a few generations, provided that no disturbing influences are at work.

This leads us to the difficulties and sources of error in mutation rate estimates. One of the greatest difficulties in practical organization is the neces-

Friedrich Vogel, a human geneticist, was born in Berlin in 1925. He received his M.D. degree in 1952 from the University of Berlin. Presently he is Professor in and Director of the Institute for Anthropology and Human Genetics at the University of Heidelberg. His major research interests are problems of mutation and natural selection in man and the genetics of the electroencephalogram (EEG). He has done population-genetics research among populations in Germany, India, and Thailand. He is coeditor of the journal German Humangenetik and author of Über die erblichkeit des normalen Elektro-enzephalogramms (Stuttgart, 1958); Lehrbuch der allgemeinen Humangenetik, with W. Fuhrmann (Berlin, 1961); Genetisch Familienberatung, with G. Röhrborn, et al. (Heidelberg, 1968); and Strahlengenetik des Säugers (Stuttgart, 1969).

sity to ascertain the trait concerned completely in a population of some millions. Many mutation rate estimates are unreliable simply because their basis of actual data is insufficient. One certain research project which could be carried out theoretically has not been put into practice to date only because the problems of organization are prohibitively difficult: The testing of the question, whether the mutation rate is different in various parts of the world population, and whether there is any increase in time, by estimation of some selected mutation rates of human hereditary diseases, which would provide theoretically the best source of information.

Some time ago we examined the problem:[73, 84] How large a population has to be screened for mutations, if one wishes to show a trend of the mutation rate in time with a certain degree of precision? Assuming the frequency of new mutations suited for this type of comparison to be between 1 : 5000 and 1 : 10,000, we arrived at the result that for a comparison of two successive periods of ten years, the whole population of the German Federal Republic with about 8 millions of newborns within 10 years would have to be screened in order to show a mutation rate trend with a precision of ± 10 per cent. A research project of this order of magnitude has turned out to be practically impossible to date. Maybe a comparison of this type will be possible somewhere in the future.

Still more important than these practical difficulties are the theoretical ones which are connected with the assumptions required for a proper use of the estimation methods.

First it is assumed that the trait examined is a genetic entity. However, almost all traits which were analysed more exactly have proved to be more or less heterogeneous. This is not so very important, if we are concerned with two mutations, instead of one mutation, with the same mode of inheritance. Here, one mutation rate has only to be split up in two. Haemophilia which was split up about ten years ago into two biochemically-different sex-linked types, A and B, may be mentioned as an example.

The situation becomes more serious if sporadic cases of an anomaly, which had been looked upon as dominant mutations, turn out to be a heterogeneous group of mutations and phenocopies. Retinoblastoma, a malignant tumour of the eye, which is seen in little children, may be mentioned as an example.[36, 74, 77, 80, 81] Here, upon more exact analysis, only between 10 and 20 per cent of the unilateral, and 50-100 per cent of bilateral sporadic cases turned out to be new mutations, whereas the cause of the remaining sporadic cases is unknown.

Other examples of traits, which turned out to be heterogeneous upon a more exact analysis, are achondroplasia[20, 47] and aniridia.[6]

Now, let us proceed to the second important assumption: This is the genetic equilibrium under the influence of selection and mutation only. This assumption, which is especially important for the indirect method, should be tested in all special cases. It has already turned out to be wrong in some traits.

Huntington's chorea might be mentioned as an example.[52, 54] In their comprehensive survey of Huntington's chorea in Michigan, Reed and Neel (1959)[54] showed that an indirect mutation rate estimate gave a much higher value than would be compatible with the very rare mutations directly observed. If they succeeded in ascertaining sporadic cases only fairly exhaustively, the result makes one conclusion very likely: In the Michigan population, more genes for chorea are lost by selection, than can be replaced by mutations. Huntington's chorea is becoming less frequent—at least in Michigan. However, fertility is not only reduced in choreics, but in their healthy sibs as well. Hence, the threat of the disease seems to induce all family members to limit propagation voluntarily. Investigations of Wendt[89] in Germany, which were carried out with much larger material, seem to confirm this result also for Germany. In addition they show that the differences in reproduction rates between choreatics and their healthy sibs on the one side and the population average on the other side have only emerged during the last decades. This may be taken as evidence that the changes in our lives and behaviour due to modern civilization need not exclusively lead to a deterioration of our genetic make-up.

In dominant traits like Huntington's chorea, the assumption of genetic equilibrium can at least be tested. This is not the case in autosomal recessive traits, because here, direct counting of new mutations and comparing with the indirect estimates is impossible. We know on the other side that for recessive anomalies, breaking up of isolates during the last century must have disturbed any genetic equilibrium, which might have existed before. A further source of uncertainty in recessive traits is the possible selective advantage or disadvantage of the heterozygotes in the present as well as in earlier generations. About this point, which must have influenced equilibrium very much, we are left with no direct information whatsoever. In spite of arguments to the contrary, which have been brought forward during the last years,[13] these two reasons are still

TABLE 1. SELECTED MUTATION RATES FOR HUMAN GENES

a) *Dominant mutations:*

No	Trait	Population examined	Mutation rate	Number of mutants/ 1 million gametes	Authors
	more than one estimate				
1	Achondroplasia	Denmark	1×10^{-5}	10	Mørch (1940) corrected by Slatis (1955)
		Northern Ireland	1.3×10^{-5}	13	Stevenson (1957)
2	Aniridia	Denmark	$2.9\,(-5) \times 10^{-6}$	$2.9\,(-5)$	Møllenbach (1947) corrected by Penrose (1956)
		Michigan (USA)	2.6×10^{-6}	2.6	Shaw, Falls, and Neel (1960)
3	Dystrophia myotonica	Northern Ireland	8×10^{-6}	8	Lynas (1957)
		Switzerland	1.6×10^{-5}	16	Klein (1958)
4	Retinoblastoma	England, Michigan (USA), Switzerland,	$6\text{-}7 \times 10^{-6}$	6-7	Vogel (1957)
		Germany, Japan	8×10^{-6}	8	Matsunaga (1961)
	One estimate only				
5	Neurofibromatose	Michigan (USA)	1×10^{-4}	100	Crowe, Schull and Neel (1956)
6	Polyposis intestini	Michigan (USA)	$[1\text{-}3 \times 10^{-5}]$	10-30	Reed and Neel (1955)

No.	Disease	Location	Mutation rate	Rate (× 10⁻⁵ equiv.)	Reference
7	Marfan's syndrome	Northern Ireland	$[4.2\text{-}5.8 \times 10^{-6}]$	4.2-5.8	Lynas (1958)
8	Polycystic disease of the kidney	Denmark	$6.5\text{-}12 \times 10^{-5}$	65-120	Dalgaard (1957)
4	Acrocephalosyndactyly	England	$[3 \times 10^{-6}]$	3	Blank (1960)
10	Osteogenesis imperfecta	Sweden	$0.7\text{-}1.3 \times 10^{-5}$	7-13	Smårs (1961)
11	Diaphyseal aclasis (multiple exostosis)	Germany (Reg.-Bez. Münster)	$6.3\text{-}9.1 \times 10^{-6}$	6.3-9.1	Murken (1963)

b) Sex-linked recessive mutations:

No.	Disease	Location	Mutation rate	Rate (× 10⁻⁵ equiv.)	Reference
12	Haemophilia (A + B)	Denmark	3.2×10^{-5}	32	Andreassen (1943) corrected by Haldane (1947)
		Switzerland	2.2×10^{-5}	22	Vogel (1955)
	Haemophilia A	Germany	5.7×10^{-5}	57	Bitter and Lenz (1963)
	Haemophilia B	Hamburg	3×10^{-6}	3	
	Haemophilia A	Finland	3.2×10^{-5}	32	Ikkala (1960)
	Haemophilia B		$[2.0 \times 10^{-6}]$	2	
13	Duchenne type muscular dystrophy	Utah (USA)	9.5×10^{-5}	95	Stephens and Tyler (1951)
		Northern Ireland	6.0×10^{-5}	60	Stevenson (1958)
		England	4.3×10^{-5}	43	Walton (1955)
		Germany (Südbaden)	4.8×10^{-5}	48	Becker and F. Lenz (1955-56)
		Wisconsin (USA)	9.2×10^{-5}	92	Morton and Chung (1959)
		Leeds (England)	5.1×10^{-5}	51	Blyth and Pugh, (1959)

[Brackets enclose mutation rates corrected for this publication—L. N. M.]

sufficient for me to believe that mutation rate estimates for autosomal recessive traits have little value; hereafter they will not be mentioned.

Having looked at some limitations of the methods, we now proceed to
the results. Table 1 contains a selection of estimates, of which at least the
order of magnitude seems to be fairly reliable. I know very well that against
everyone of these estimates methodological arguments could be raised.
On the other side, the comforting result emerges that in all cases in which
estimations for the same trait were carried out in different populations,
there is a good agreement of the results. However, with only one exception, values have been compared from populations which show a similar genetic composition, and are living under more or less similar environmental conditions. The single exception in which frequency and
mutation rate of a trait have been compared between populations of a
different race, is retinoblastoma. Matsunaga (1959)[35] reported a frequency
of 1 : 22,326 survivors of the first year, which corresponds exactly with
European and Northern American data (5 samples between 1 : 20,288
and 1 : 30,497). A follow-up study of children of sporadic retinoblastoma
patients revealed the same relationship between mutations and phenocopies as had been shown before in European and American populations
(Vogel, 1957a; Tucker et al. 1957 [36, 74, 80]). Hence, also, the mutation rate
might be about the same in the two race groups. Results from populations
which are living in completely different environmental conditions are not
available to date. Investigation might prove to be extremely difficult, as
for complete ascertainment of all cases of a rare hereditary disease, a
highly developed system of medical care is necessary.

Haemophilia has to be mentioned specially. The old estimates[1, 23, 78] are
based on material which had been collected, before a differentiation
between haemophilia A and B became possible. When I estimated the
mutation rate for the material collected in Switzerland by Fonio[18] I discussed the hypothesis that we might be concerned with two closely linked
loci with related functions.[78] Meanwhile, evidence has been brought forward that haemophilia A is fairly closely linked with the deuteranopia
locus, whereas haemophilia B shows only a loose linkage with the
protanopia locus.[90] Under the assumption that the loci for protanopia
and deuteranopia are closely linked, which seemed to be plausible, this
result would contradict close linkage between haemophilia A and B.
Meanwhile, however, a Belgian family with crossing over between the
two colour vision loci has been observed.[28] If this observation is confirmed,
close linkage between these loci is excluded and the genetic relationship
between the two haemophilia loci remains open.

Recently, separate mutation rate estimates have been carried out for
haemophilia A and B in Hamburg. The results which are not yet published will be mentioned again.* All values in Table 1 range from 10^{-4}

---

* I am very grateful to Drs. W. Lenz and Bitter for the permission to use their results.

to 10^{-6}. This does not mean, however, that the mutation rates of dominant and sex-linked recessive visible mutations in man are generally of this order of magnitude. There are many hereditary anomalies with lower mutation rates. Huntington's chorea may be mentioned as one example. A certain frequency is required to make estimation possible. Hence, the mutation rates cited may belong to the highest ones for mutations which lead to clearly discernible and disadvantageous phenotypical deviations.

On the other side, we do not know anything about the frequencies of mutations which do not lead to anomalies, but to smaller deviations in the normal range only, for example blood groups.

Whereas some authors expressed the opinion that for blood group genes an especially low mutation rate has to be assumed, Fleischhacker[17] has shown for the ABO groups where the most parent-child groups have been published that nothing whatsoever can be concluded about mutation rates on the basis of the material available.

3.† *Dependence of the Mutation Rate on the Parental Age and Possible Sex Differences*

In order to learn more about the dynamics of the mutation process some workers have tried to analyse the mutation rates from different points of view. Investigations about dependence of the mutation rates on the age of the parents, especially the father, and about the relative mutation rates in the germ cells of both sexes have primarily to be mentioned. On the basis of statistical results of this type, it even seemed to be possible to draw preliminary conclusions about the molecular nature of human mutations. Examination of the relationship between mutation rate and parental, especially paternal age is made difficult by the fact that most population statistics only contain distribution of births according to mothers', but not to fathers', ages. Hence, control figures are usually insufficient, and only relatively strong deviations can be recognized easily.

Figure 1 shows the increase of mutation rates with paternal age in different traits. A strong increase is found in achondroplasia.[39, 48, 51, 70, 79, 82] Clinical analysis of the cases concerned shows that this increase is confined to a special, but frequent type. It is the dominantly inherited type with change of the facial skeleton, but without decrease in viability, the pug-dog type according to Grebe.[20] The other relatively rare dominant type without changes in the facial skeleton, which was called dachshund-type does not show this strong correlation with paternal age.

There is one other mutation, which was shown to display a similar age correlation. This is the classical type of acrocephalosyndactyly (Apert's syndrome)[7] (Fig. 2). In both cases it has been shown by use of the partial correlation coefficient that the age of the father is indeed, the deci-

†[So numbered in original publication.]

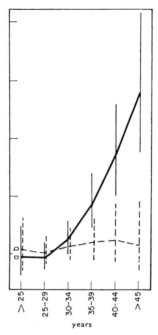

FiG. 1. Relative incidence of dominant mutations in different age groups of the fathers (a) in achondroplasia (175 cases of Mørch, Grebe and Stevenson), (b) in neurofibromatosis, tuberous sclerosis, and osteogenesis imperfecta (109 cases of Borberg and Seedorff. Confidence limits for 95 per cent are included in the figure. Relative incidences are compared with the Danish age distribution of fathers in 1930; cf. Mørch).

sive factor whereas the mothers' age is increased only insofar as older men are, on average, married to older women.[7, 51] It is possible that Marfan's syndrome shows a similar age dependence, but the number of mutations analysed so far is too small for a definite conclusion.[33]

In contrast to this group of dominant traits, there is a different one which shows only a much smaller increase with paternal age. Neurofibromatosis, tuberous sclerosis, and osteogenesis imperfecta might be mentioned here[82]*.... A different indirect method of analysis has shown that we are not concerned with a chance result. This is analysis of birth order, which shows a high correlation with paternal and maternal age.[79, 82] An analysis of birth order for the sporadic cases of a number of dominant traits is contained in Table 2.† All dominant mutations examined show an increase of incidence with birth rank. This result has to be accepted with caution: It is well known that birth order is affected by a number of statistical biases. One possible bias, by which a birth order effect might be coun-

*Figure 3 and discussion relevant to it have been excised.
†The Haldane-Smith-method was used, in which birth order of patients is multiplied by 6 and compared with its expected value.[26]

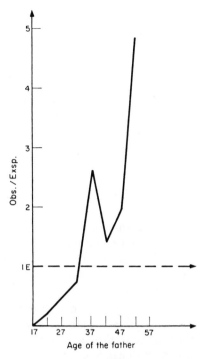

FIG. 2. Relative incidence of Apert's syndrome in different age groups of the fathers (37 cases; data of patients and normal controls quoted from Blank (1960); E: expected value, if no age dependence would exist).

terfeited, is intentional limitation of reproduction after birth of a heavily malformed child. This leads to a cumulation of patients in the last position.

This bias can be avoided, if all sibships are analysed without the last sibling, and all in which the patient is the last child are completely excluded.

TABLE 2. BIRTH ORDER IN DOMINANT MUTATIONS

Trait	Expected	Observed	V	
Aniridia	702	762	2892	
Tuberous sclerosis	222	258	846	
Neurofibromatosis	1992	2052	11,058	
Osteogenesis imperfecta	552	660	2904	$\sqrt{V} = 152.75$
Polyposis intestini	159	210	741	$t = 3.54$
Retinoblastoma (bilateral)	855	1038	4341	$P < 0.0027$
Diaphyseal aclasis	108	150	546	
	4590	5130	23,328	
Achondroplasia	3225	3858	21,519	$\sqrt{V} = 152$
				$t = 5.01$
Apert's syndrome	441	570	1593	$P < 10^{-6}$
	3666	4428	23,112	

Increased birth order is shown also with this limitation. Hence, the effect is not counterfeited by such a type of bias.

But now let us look at birth order in sporadic cases of sex-linked recessive traits. Here, no increase is to be seen. If there would only be no increase, the result would not be too important, as among the sporadic cases of sex-linked recessive traits, only a fraction ($\frac{1}{3}$–$\frac{1}{2}$) is due to mutations in germ cells of their mothers. The others are sons of heterozygous mothers, and here, no birth order effect can be expected.

But in Duchenne type muscular dystrophy, for which much material is available, birth order turned out to be definitely if not significantly decreased. This decrease has to be regarded as a chance result; but it would hardly be compatible with an increase in a substantial part of the material.

Summing up, there is an increase of birth order in the dominant mutations examined, which come from mutations in the germ cells of both parents. No increase is found in sex-linked recessive mutations, which must have originated in germ cells of the mothers only. If the assumption holds true, that a similar mechanism works in both cases, the supposition seems to be justified, that the effect in dominant mutations might be due to mutations in germ cells of the fathers. As we have seen, this supposition is supported by the fact that in the Danish part of the material, where a direct analysis is possible, a direct effect can be shown.

Much more controversial are the results of studies about whether the mutation rate is different in both sexes. Up to now, this problem has only been examined in sex-linked recessive mutations. According to Haldane (1935)[22] the expected number of new mutations (affected sons of homozygous normal mothers) can be calculated under the assumption of equal mutation rates in both sexes, if an estimate of the relative fertility of the affected males is available. First, the analysis seemed to show a much higher mutation rate in male than in female germ cells for both sex-linked traits examined.[23, 25, 79] Later on it was shown in Duchenne type muscular dystrophy with much larger material and refined methods that this result had been counterfeited by incomplete ascertainment of sporadic cases in one of the series investigated.[11, 40, 63] At present, it seems likely that mutation rates for this trait are similar in both sexes. In haemophilia, the material analysed so far contains many inconsistencies. However, in this trait too, the hypothesis of a higher mutation rate in male

TABLE 3. SPONTANEOUS MUTATION RATE IN MALE AND FEMALE MICE
IN 7 SPECIFIC RECESSIVE LOCI (AFTER RUSSELL, 1962)

	Number of gametes tested	Number of mutations	Mutation rate per locus
♂	544,897	32	8.4×10^{-6}
♀	98,828	1	1.4×10^{-6}

P (one-tailed test) = 0.034

germ cells seemed to be very dubious.[31] The material of Bitter and W. Lenz (unpublished),[9] which has recently been collected in Hamburg, and in which a modern and reliable biochemical test for heterozygosity was used, gives a result in favor of a higher mutation rate in males.

There is still a different approach to the problem of a sex difference of mutation rates, which has not been used up to now: The above-mentioned data about increase of mutation rates with paternal age in achondroplasia (Fig. 1) can also be used for examination of a possible sex difference, if the following plausible assumptions are made:

(1) That the whole increase of the mutation rate is due to the fathers, while the mutation rate in mothers is not age-dependent.

(2) That the slope of the curve observed represents the real increase of the mutation rate more or less closely; this means, that chance deviations especially in the lower age groups do not play any significant role.

(3) Let the mutation rate up to the paternal age of about 29, in which group no increase with paternal age is to be seen, be equal in both sexes, while the increase in higher age groups of fathers is due to additional mutations in their germ cells. In this case, the sex difference is estimated as $\delta/\mathcal{Q} = 333/100$. In my opinion, the assumptions on which this estimate is based, are reasonable and even somewhat conservative. But this third assumption can be replaced by the extremely and unrealistically conservative assumption that all mutations found up to the paternal age of 29 appear in the mothers' germ cells, and only the increase which begins around the 30th year is due to mutations in fathers. Even in this case, $\delta/\mathcal{Q} = 117/100$ and this means that a higher mutation rate in males would emerge.

This result strongly favours the view that under the distribution of paternal age which is usual in Western European populations, a much greater part of mutations for this type of achondroplasia does come from male germ cells. I do not claim that the data available prove this conclusion, as statistical confirmation of the result presents difficult problems.*

Here it might be useful to compare the results in man with investigations carried out using seven autosomal recessive loci of mice.[57] Spontaneous mutation rates for the pooled loci turned out to be much higher in males than in females (Table 3). If a one-tailed test is used, which seems to be justified, because a higher mutation rate in females is out of question, the result is statistically significant on the 5 per cent level.

---

* The choice of the normal control population could give an argument against our calculations: We used the distribution of all births in Denmark in the year of 1930 (after Mørch)[39]. The result could theoretically be counterfeited by special properties (unusually many young fathers) in this material. But a comparison with samples from a number of other Western European populations (Germany, Northern Ireland and Australia) has shown that the mean age of fathers is relatively high in the material used. Hence, by use of these data, the effect tends to be somewhat underestimated.

Taken at their face value, the data show a six times higher mutation rate in males than in females.

As is well known, there are no principal differences in germ cell development between mouse and man.

Some authors, including myself, have tried to give an interpretation of the results mentioned, especially the correlation with paternal age, in terms of the conceptions based mainly on experiments in micro-organisms about the molecular basis of the mutation process.[48,79,82]

Comparison of the absolute order of magnitude of mutation rates for single loci shows a surprising correspondence with the average of spontaneous values for the seven test loci in mice[57] (8.4×10^{-6} in ♂♂; 1.4×10^{-6} in ♀♀). Single locus mutation rates in *Drosophila* fit very well, too. In micro-organisms, on the other side, mutation rates are usually some orders of magnitudes smaller.[4,58] This, however, is not surprising, as the time distance between generations is much greater in man, and besides, in micro-organisms generations are only separated by one cell division, while in man many cell generations are interposed. A second variable to be taken into account is the length of the single gene, or the number of mutable sites (nucleotide pairs). Assuming that three nucleotide pairs determine one amino acid, the number of mutable sites for the human hemoglobin genes can be calculated.[27] The order of magnitude is about the same as is usually given for micro-organisms.

The increase of mutation rates with paternal age raised some considerations about the possible mechanism of the mutations concerned. Figure 4 gives the very much simplified diagram of germ cell development in both sexes.[83,55,56] The most important difference is that the 400,-000–500,000 oogonia in the ovaria are formed up to birth or to the first

TABLE 4. SIMPLE MODELS FOR MUTATION AND THEIR
STATISTICAL CONSEQUENCES

	1	2	3	4	5
	Mutation depending on time only	Mutation depending on cell divisions only	Mutation during a certain time before puberty	Mutation after ceasing of divisions	Mutation in mature germ cells
♂ germ cells	Linear increase of mutations with age; no sex difference	Increase of mutations with age; higher mutation rate in ♂	No increase with age; no sex difference	No increase with age; lower mutation rate in ♂	No increase with age; maybe somewhat higher mutation rate in ♂
♀ germ cells	Linear increase of mutations with age	No increase with age; lower mutation rate in ♀	No increase with age	Increase with age; higher mutation rate in ♀	No increase with age; maybe somewhat lower rate in ♀

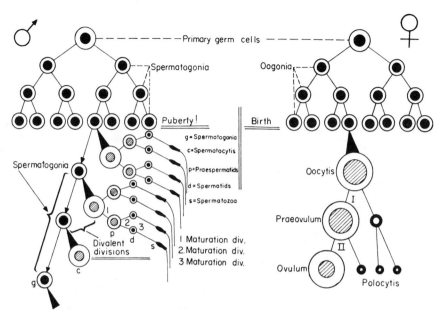

FIG. 4. Simplified diagram of germ cell development in the two sexes (according to Rous-hoven (1940); for ref. see Ref. 79, for a more realistic description see reference in Ref. 85).

months of life, while later on only the maturation divisions follow. Beginning with the impregnation, all oogonia can be formed by about 19 steps of division. In males, on the other side, the stem cells show a proliferation up to sexual maturity. Later on, the germ cells are formed by bivalent divisions of spermatogonia in successive cycles. Hence, the spermatozoon has undergone many more steps of division than the egg cell, and their number grows with the age of the man. Table 4 contains some possible models for the mutation process together with the consequences for human mutation rates. At the first glance, model No. 2 (mutations primarily during divisions of the genetic material as copy errors) seems to be most attractive.[48,79] An increase with paternal age was demonstrated, and some data suggest a higher mutation rate in male than in female germ cells. However, the last result does not seem to be generally valid, and if the increase with paternal age is analysed more exactly, the results turn out to be controversial, too.[82] If we assume that most mutations are due to a copy error, the curve of increase with paternal age can be predicted, taking into account the intensity of spermatogenesis during the reproductive life span[29,34,82,85] (Fig. 5). The highest intensity is observed during the first decade of sexual maturity (about 15–25 years), while reproduction is relatively rare. Hence, the starting point of the curve must be relatively high. During the fourth and especially during the fifth decade, on the other hand, intensity diminishes and the curve is expected to flatten. Comparing the slopes of the observed curves in achondro-

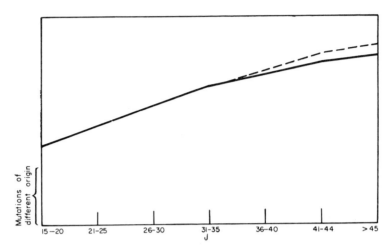

FIG. 5. Cumulative distribution of the number of ejaculations according to age groups and (dotted line) expected increase of spermatogonial divisions and hence, expected increase of mutation frequency with the fathers' age according to the copy error hypothesis. Data from Kinsey and co-workers (1948); MacLeod (1953), ref. in Ref. 82.

plasia as well as in acrocephalosyndactyly with the theoretically deduced curve, we realize that there is no correspondence at all: The mutation rate is very low in younger groups, but the slope of the curves becomes steeper and steeper in every age group.* Taking even into account possible chance variations this result is not compatible with the copy error hypothesis and equal probability of mutation in all divisions. At least auxiliary hypotheses are necessary, which could be constructed on the basis of selection of sexually active men, spermatogonia with an especially high intensity of division, and so on. I personally tend to the opinion that a different age-dependent factor might be involved but I have no idea of what kind this factor might be. For example, no chromosomal aberration has been shown conclusively in any of the traits concerned.

Now, let us turn to the second group of dominant mutations. Here, the starting point and the slope fit much better with the curve expected. But of course, other hypotheses are possible as well and it is not absolutely sure in these mutations that the effect is really due to the fathers only.

One conclusion can be drawn from this very complicated chapter: Human mutations show definite differences in the extent of their dependence on age of the father as well as in relative mutation frequencies in the germ cells of the two sexes. Furthermore, there is no single model for a molecular mechanism of the mutation process, which could explain the actual data in a really adequate manner.

---

* Apart from chance fluctuations in acrocephalosyndactyly, which are apparently due to the smaller number of observations.

4. Investigation on Larger Groups of Mutations

The investigations mentioned so far are concerned with single mutations or with very small groups.

Additionally, it was tried in a more indirect manner to get information about larger groups of mutations.

a. Dominant Lethals

It is a matter of general experience that the relative number of still-born children increases with the age of the parents. The unsophisticated physician will incline to the opinion that this increase might be due to unfavorable conditions in utero and during delivery. But some statistical results of Yerushalmy[91] and Sonneborn[65] seem to show that stillbirth frequency depends primarily on paternal and not so much on maternal age. If this result could be generalized, it could provide a strong argument in favor of the hypothesis that a substantial fraction of stillbirths are due to dominant lethals. Unfortunately, most population statistics do not contain any data about stillbirth frequency and paternal age.

b. Sex-linked recessive lethals

As is well known, the secondary sex ratio at birth is variable with an average of about $\delta/\varphi = 106/100$. It had been suspected for a long time, that the primary sex ratio at impregation was much higher and that during the first months of pregnancy many more male than female foetuses were dying. When the method of determining the nuclear sex at very early stages of development became available, these results were confirmed.[71,85] It is plausible to assume that much of the higher mortality of male zygotes might be due to sex-linked recessive lethals. If one accepts this assumption, [and] knows the number of spontaneous abortions in relation to the birth number, and the sex ratio among the abortions, the total loss of zygotes due to sex-linked lethals can easily be calculated. Frota-Pessoa and Saldanha (1960)[19] analysing population figures given by Stevenson and co-workers (1959),[71,72] arrive at an estimate of 0.0068 sex-linked lethals per germ cell. But this estimation requires numerous assumptions which cannot be taken for granted at all. For example, sex-linked dominant lethals as well as lethals on the Y-chromosome are neglected, and we do not know whether the assumption of a genetic equilibrium between mutation and selection holds true. Hence, the result is a very hypothetical one.

Recently, these considerations have been supplemented by very interesting investigations, which were carried out by Cavalli-Sforza and co-workers in Italy.[10] They examined 58,563 families with about 180,-000 children for dependence of the sex ratio among live- and stillbirth from the age of the grandfather at the birth of the mother. Theoretically, this method permits [one] to trace sex-linked recessive lethals which or-

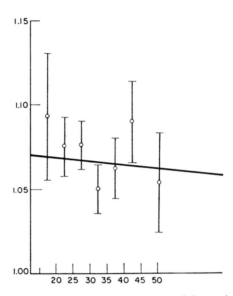

FIG. 6. Sex ratio (σ/φ) in live births in relation to the age of the mother's father at birth of the mother (according to Cavalli-Sforza[10]).

iginated in the germ cells of the grandfather and led to the death of male zygotes in the generation of grandchildren. The investigation rendered two results which are compatible with the underlying genetic hypothesis: First, a negative correlation between the age of the grandfathers and the sex-ratio among the living children was found (Fig. 6). This correlation is not very strong and not significant statistically. Taken at their face value, the data would give a mutation rate estimate of about 2×10^{-5} per locus during the reproductive age in male germ cells, assuming 850 loci at the X-chromosome which are able to produce lethals. The second result is that the sex ratio among the stillborn shows a positive correlation with the age of the grandfather, which is significant on the 5 per cent level, if a one-tailed test is used (Fig. 7). It would be premature to conclude that the data prove the originating of sex-linked lethals in male germ cells and their increase with age, but they contain a strong suggestion that the hypothesis could be correct. Unfortunately, as Dr. Cavalli-Sforza was so kind to inform me, a similar examination of all births of 1960 in Italy (about 800,000) has failed to show the same tendency, teaching us once more that data on sex ratio are a tricky material to work with. A definite decision can be expected from the Italian census of 1962.

c. Autosomal Recessive Lethals

Finally have to be mentioned, the attempts to get information about the mutation rate of autosomal-recessive lethals or detrimentals. They

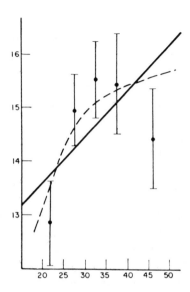

FIG. 7. Sex ratio among stillbirths in relation to the age of the mother's father at birth of the mother (Cavalli-Sforza[10]).

are based on the well-known estimation of lethal and detrimental equivalents from the higher frequency of stillbirths and early deaths in children from consanguineous matings.[41, 63, 60] For a reasonable estimation of total selection against these genes, assumptions about the mean selective disadvantage of heterozygotes are required, for which there are no data available. Hence, *Drosophila* results have to be used; a very dubious procedure. A second assumption, which is almost certainly wrong in these cases, is the existence of a genetic equilibrium. Hence the values calculated are still less reliable than the above-mentioned results about sex-linked recessive lethals.

5. New Approaches to Attack the Problem on a Cellular Basis

Looking at all these statistical investigations together, one cannot escape from the impression that these approaches are more or less exhausted theoretically. Practical progress in limited areas can be expected from the refinements of general population statistics, primarily a better consideration of paternal age on the one hand, and by organization of large-scale investigations on the other. The surveys on frequencies of hereditary diseases carried out in Germany[16, 75, 76, 88] as well as the above-mentioned statistical approaches to the problem of sex ratio and sex-linked recessive lethals, which are under way in Italy,[10] have to be mentioned in this connection.

But there seems to be a strong feeling among human geneticists that completely new methods are required to get deeper insight into the mutation process in man. Hence, approaches to get more information by examining single cells were greeted with great expectations. In this connection the modern cytogenetic methods have to be mentioned.

Here, answers can be expected to the question, whether in a fraction of mutations which lead to well-known hereditary diseases, visible deletions, insertions, translocations etc. are found. Interpretation of results published so far is made difficult by the fact, that nobody knows whether the results published in some cases are specific or not.

A different approach is observation of single cells with serologically or biochemically deviating phenotypes, which are assumed to be due to somatic mutations. Here investigations by Atwood[2, 3] using ingeniously isolated erythrocytes which are free from A- or B-antigen, have raised great hopes. Unfortunately later experiments have shown that the interpretation of the phenotypical changes as being due to mutation is at least controversial. (For similar approaches see Ref. 46.)

Additional information about the mutation process in man can be deduced by analogy from experiments in mammals, especially in mice.[57] The sex ratio of spontaneous mutation rates of seven specific loci has already been mentioned. Besides, experiments in mammals are very important for the problem of physically or chemically induced mutations. But as this field of research will be covered fully in other sessions I shall not enlarge on it.

In conclusion we have to confess that up to now we know something about mutations in man, but in spite of many ingenious approaches we do not know very much. The great number of unsolved problems will be, it is hoped, a continuous challenge for human geneticists.

REFERENCES

1. ANDREASSEN, M., *Haemofili i Danmark*, Munksgaard, Copenhagen, 1943.
2. ATWOOD, K. C., Problems of measurement of mutation rates. Mutations, *Second Conference on Genetics*, 1962, 1–77.
3. ATWOOD, K. C. and SCHEINBERG, L., Somatic variations in human erythrocyte antigens. *J. Cell. Comp. Physiol.* **52**, Suppl. 1958.
4. BECKER, P. E. and LENZ, F., Zur Schätzung der Mutationsrate der Muskeldystrophien. *Z. menschl. Vererb.-u. Konst.-Lehre* **33**, 42, 1955.
5. BECKER, P. E. and LENZ, F., Nachtrag zu der Arbeit "Zur Schätzung der Mutationsrate der Muskeldystrophien". *Z. menschl. Vererb.-u. Konstit.-Lehre* **33**, 463, 1956.
6. BEHNKE, H. and HOLTERMANN, W., Häufigkeit, Vererbung und klinische Ausprägung der Aniridie in Schleswig-Holstein. II. *Internationale Konferenz uber menschliche Genetik*, Rom 1961 Excerpta Medica Foundation.
7. BLANK, C. E., Apert's syndrome (a type of acrocephalosyndactyly), Observations on a British series of 39 cases. *Ann. Hum. Genet.* **24**, 151, 1960.

8. BORBERG, A., *Clinical and Genetic Investigations into Tuberous Sclerosis and Recklinghausen's Neurofibromatosis*. Munksgaard, Copenhagen, 1951.

9. BITTER, and LENZ, W. Unpublished; personal communication, 1963.

9a. BLYTH, H., R. J. PUGH. Muscular dystrophy in childhood. The genetic aspect. *Ann. Hum. Genet.* **23**, 127–163 (1959).

10. CAVALLI-SFORZA, L. L., Indagine speciale su alcune caratteristiche genetiche della popolazione italiana. Note e relazioni, Istituto Centrale de Statistica, Roma, No. 17, 1962.

11. CHEESEMAN, E. A., KILKPATRICK, S., STEVENSON, A. C. and SMITH, C. A. B., The sex ratio of mutation rates of the sex-linked recessive genes in man with particular reference to DUCHENNE type muscular dystrophy. *Ann. Hum. Genet.* **22**, 235, 1958.

12. CHUNG, C. S., ROBINSON, O. W. and MORTON, N. E., A note on deaf mutism. *Ann. Hum. Genet.* **23**, 357–366, 1959.

13. CROW, J. F., *Mutation in Man. Progress in Medical Genetics*, I, 1961.

14. CROWE, F. W., SCHULL, W. J. and NEEL, J. V., *A Clinical, Pathological, and Genetic Study of Multiple Neurofibromatosis*, Springfield, Ill. 1956.

15. DALGAARD, O. Z., *Bilateral Polycystic Disease of the Kidneys*, Munksgaard, Copenhagen, 1957.

16. EBBING, H. C., Die Mutationsrate des Menschen, III, Über Möglichkeiten einer Auswertung der ärztlichen Befunddokumentation für ein Genetik-Register. *Z. menschl. Vererb.-u. Konstit.-Lehre* **35**, 405–419, 1960.

17. FLEISCHHACKER, H., Mutationen im ABO-System? *Anthrop. Anz.* **20**, 271, 1957.

18. FONIO, A., *Die erblichen und die sporadischen Bluterstämme in der Schweiz. Basel*, 1954.

19. FROTA-PESSOA, O. and SALDANHA, P. H., The rate of spontaneous sex-linked mutations and the doubling dose in man. *Ann. Hum. Genet. Lond.* **24**, 367, 1960.

20. GREBE, H., Chondrodysplasie. *Analecta Genet. Rome*, 1955.

21. GUNTHER, M. and PENROSE, L. S., The genetics of epiloia. *J. Genet.* **31**, 413, 1935.

22. HALDANE, J. B. S., The rate of spontaneous mutation of a human gene. *J. Genet.* **31**, 317, 1935.

23. HALDANE, J. B. S., The mutation rate of the gene for haemophilia, and its segregation ratios in males and females. *Ann. Eugen.* **13**, 262, 1947.

24. HALDANE, J. B. S., The rate of mutations of human genes. *Proc. VIII. Int. Congr. Genetics, Stockholm*, 267, 1948.

25. HALDANE, J. B. S., Mutation in sex-linked recessive type of muscular dystrophy. A possible sex difference. *Ann. Hum. Genet.* **20**, 344, 1956.

26. HALDANE, J. B. S. and SMITH, C. A. B., A simple exact test for birth order effect. *Ann. Eugen.* **14**, 117 (1947–49).

27. HILSCHMANN, N., Die Globine. *Blut. Z. f. Blutforschung*, VII, Suppl.-Heft, 333–343, 1961.

27a. IKKALA, E. *Haemophilia*, Helsinki 1960.

28. KALMUS, H., Distance and sequence of the loci for protan and deutan defects and for glucose-6-phosphate dehydrogenase deficiency. *Nature* **194**, 251, 1962.

29. KINSEY, A. C., POMEROY, W. B. and MARTIN, C. E., *Sexual Behaviour in the Human Male*. Philadelphia and London, 1948.

30. KLEIN, D., La dystrophie myotonique (Steinert) et la myotonie congénitale (Thomsen) en Suisse. *J. Génét. Hum.* **7,** Suppl. 1958.

31. KOSOWER, N., CHRISTIANSEN, R. and MORTON, N. E., Sporadic cases of hemophilia and the question of a possible sex difference in mutation rates. *Am. J. Hum. Genetics* **14,** 159–169, 1962.

32. LYNAS, M. A., Dystrophia myotonica with special reference to Northern Ireland. *Ann. Hum. Genet.* **21,** 318, 1957.

33. LYNAS, M. A., Marfan's syndrome in Northern Ireland: an account of thirteen families. *Ann. Hum. Genet.* **22,** 289, 1958.

34. MACLEOD, J. and GOLD, R. Z., The male factor in fertility and infertility. VII. Semen quality in relation to age and sexual activity. *Fertil. and Steril.* **4,** 194, 1953.

35. MATSUNAGA, E. and OGYU, H., Genetic study of retinoblastoma in a Japanese population. *Jap. J. Hum. Genet.* **4,** 136, 1959.

36. MATSUNAGA, E., Genetic study on sporadic retinoblastoma in Japan. *Annual Rep. of the Nat. Inst. of Genetics* **12,** 1961.

37. MEDICAL RESEARCH COUNCIL of GREAT BRITAIN, *The Hazards to Man of Nuclear and Allied Radiations*. London, 1956.

38. MØLLENBACH, C. J., *Medfodte defekter i ojets indre hinder, Klinik og arvelighedsforhold*. Munksgaard, Copenhagen, 1947.

39. MØRCH, E. T., *Chondrodystrophic Dwarfs in Denmark*. Munksgaard, Copenhagen, 1941.

40. MORTON, N. E. and CHUNG, C. S., Formal genetics of muscular dystrophy. *Amer. J. Hum. Genet.* **11,** 360, 1959.

41. MORTON, N. E., CROW, J. F. and MULLER, H. J., An estimate of the mutational damage in man from data on consanguineous marriages. *Proc. Nat. Acad. Sci.* **42,** 855–863, 1956.

42. MURKEN, J. D., Über multiple cartilaginäre Exostosen. *Z. menschl. Vererb.-u. Konst.-Lehre* **36,** 469–505, 1963.

43. NACHTSHEIM, H., Die Mutationsrate menschlicher Gene. Naturwissenschaften **41,** 385, 1954.

44. NEEL, J. V., The study of human mutation rates. *Amer. Naturalist* **86,** 129, 1952.

45. NEEL, J. V., Some problems in the estimation of spontaneous mutation rates in animals and man. *Wld. Hlth. Organization*, Geneva, 1957.

46. NEEL, J. V., Mutations in the human population. *Methodology in Hum. Genet.* 203–244, 1962.

47. NEEL, J. V., SCHULL, W. J. and TAKESHIMA, K., A note on achondroplasia in Japan. *Jap. J. Hum. Gen.* **4,** 165–172, 1959.

48. PENROSE, L. S., Parental age an mutation. *Lancet* II, 312, 1955.

49. PENROSE, L. S., In: *The Hazards to Man of Nuclear and Allied Radiations* London, 1956.

50. PENROSE, L. S., Mutation in man. *Acta Genet. Basel*, **6,** 169, 1956.

51. PENROSE, L. S., Parental age in achondroplasia and mongolism. *Amer. J. Hum. Genet.* **9,** 167, 1957.

52. REED, T. E., The definition of relative fitness of individuals with specific genetic traits. *Amer. J. Hum. Genet.* **11,** 137, 1959.

53. REED, T. E. and NEEL, J. V., A genetic study of multiple polyposis of the colon. (With an appendix deriving a method of estimating relative fitness). *Amer. J. Hum. Genet.* **7**, 236, 1955.

54. REED, T. E. and NEEL, J. V., Huntington's chorea in Michigan. 2. Selection and Mutation. *Amer. J. Hum. Genet.* **11**, 107, 1959.

55. ROOSEN-RUNGE, E. C., Kinetics of spermatogenesis in mammals. *Ann. N.Y. Acad. Sci.* **55**, 574, 1952.

56. ROOSEN-RUNGE, E. C., Quantitative studies on spermatogenesis in the albino rat. III. Volume changes in the cells of the seminiferous tubules. *Anat. Rec.* **123**, 385, 1955.

57. RUSSELL, W. L., In: Report of the United Nations Scientific Committee on the Effects of Atomic Radiation S. 106, Tab. X.

58. SCHULL, W. J. (Ed.), *Mutations. Second Conference on Genetics of the Macy-Foundation.* The University of Michigan Press, Ann Arbor, 1960.

59. SCHULL, W. J. and NEEL, J. V., Radiation and the sex ratio in man. *Science* **128**, 343, 1958.

60. SCHULL, W. J., Inbreeding effects on man. *Eugenics Quart.* **6**, 2, 1959.

61. SHAW, M. W., FALLS, H. F. and NEEL, J. V., Congenital aniridia. *Amer. J. Hum. Genet.* **12**, 4, 1960.

62. SLATIS, H. M., Comments on the rate of mutation to chondrodystrophy in man. *Amer. J. Hum. Genet.* **7**, 76, 1955.

63. SMITH, C. A. B. and KILKPATRICK, S., Estimates of the sex ratio of mutation rates in sex-linked conditions by the method of maximum likelihood. *Amer. J. Hum. Genet.* **22**, 244, 1959.

64. SMÅRS, G., *Clinical, Genetics Epidemiological and Socio-Medical Aspects.* Svenska Bokförlaget, 1961.

65. SONNEBORN, T. M., Paternal age and stillbirth in man. *Rec. Gen. Soc. Amer.* **25**, 661, 1956.

66. STEPHENS, F. E. and TYLER, F. H., Studies in disorders of muscle. V. The inheritance of childhood progressive muscular dystrophy in 33 kindreds. Amer. J. Hum. Genet. **3**, 111, 1951.

67. STEVENSON, A. C., Muscular dystrophy in Northern Ireland I. An account of the condition in fifty-one families. *Ann. Eugen.* **18**, 50, 1953.

68. STEVENSON, A. C., Muscular dystrophy in Northern Ireland. II. An account of nine additional families. *Ann. Hum. Genet.* **19**, 159, 1955.

69. STEVENSON, A. C., Muscular dystrophy in Northern Ireland IV. Some additional data. *Ann. Hum. Genet.* **22**, 231, 1958.

70. STEVENSON, A. C., Achondroplasia: an account of the condition in Northern Ireland. *Amer. J. Hum. Genet.* **9**, 81–91, 1957.

71. STEVENSON, A. C., DUDGEON, M. Y. and MCCLURE, H. O., Observations on the result of pregnancies in women resident in Belfast II. Abortions, hydatidiform moles and ectopic pregnancies. *Ann. Hum. Genet. Lond.* **23**, 305, 1959.

72. STEVENSON, A. C. and WARNOCK, H. A., Observations on the result of pregnancies ending in 1957. *Ann. Hum. Genet. Lond.* **23**, 382, 1959.

73. STROBEL, D. and VOGEL, F., Ein statistischer Gesichtspunkt für des Planen von Untersuchungen über Änderungen der Mutationsrate beim Menschen. *Acta Genet.* (Basel), **8**, 274, 1958.

74. TUCKER, D. P., STEINBERG, A. G. and COGAN, C. G., Frequency of genetic transmission of sporadic retinoblastoma. *A. M. A. Archives of Ophthalmology* **57**, 532–535, 1957.

75. V. VERSCHUUR, O. and EBBING, H. C., Die Mutationsrate des Menschen. Forschungen zu ihrer Bestimmung, I. *Z. menschl. Vererb.-u. Konstit,-Lehre* **35**, 93–99, 1959.

76. V. VERSCHUUR, O., Die Mutationsrate des Menschen, Forschungen zu ihrer Bestimmung, IV. Die Häufigkeit krankhafter Erbmerkmale im Bezirk Münster. *Z. menschl. Vererb.-u. Konstit.-Lehre* **36**, 383–412, 1962.

77. VOGEL, F. Über Genetik und Mutationsrate des Retinoblastoms. *Z. menschl. Vererb.-u. Konstit.-Lehre* **32**, 308, 1954.

78. VOGEL, F., *Vergleichende Betrachtungen über die Mutationsrate der geschlechtsgebunden—rezessiven Hämophilieformen in der Schweiz und in Dänemark*. Blut I, 91–109, 1955.

79. VOGEL, F., Über die Prüfung von Modellvorstellungen zur spontanen Mutabilität an menschlichem Material *Z. menschl. Vererb.-u. Konstit.-Lehre* **33**, 470, 1956.

80. VOGEL, F. Neue Untersuchungen zur Genetik des Retinoblastoms (Glioma retinae) *Z. menschl. Vererb.-u. Konstit.-Lehre* **34**, 205, 1957.

81. VOGEL, F., Die eugenische Beratung beim Retinoblastom (Glioma retinae). *Acta Genet. (Basel)* **7**, 344, 1957 b.

82. VOGEL, F., Gedanken über den Mechanismus einiger spontaner Mutationen beim Menschen. *Z. menschl. Vererb.-u. Konstit.-Lehre* **34**, 389, 1958.

83. VOGEL, F., Die spontane Mutabilität menschlicher Gene. *Arch. der Julius Klaus-Stift. f. Vererb.-Forsch., Sozialanthropologie u. Rassenhygiene* XXXVI, 1/4, 1961.

84. VOGEL, F., Grundsätzliche Erwägungen zu Untersuchungen über den Ansteig der Mutationsrate beim Menschen. i.: *Schriftenreihe d. Bundesministers f. Atomkernenergie u. Wasserwirtschaft,* 17, 1960.

85. VOGEL, F., *Lehrbuch der allgemeinen Humangenetik*. Berlin-Götingen-Heidelberg, 1961.

86. WALTON, J. N., On the inheritance of muscular dystrophy. *Ann. Hum. Genet.* **20**, I, 1955.

87. WATSON, J. D. and CRICK, H. C., The structure of DNA. *Cold Spring Harb. Symp. quant. Biol.* **18**, 123, 1953.

88. WENDT, G. G., Praktische Erfahrungen bei der Sammlung aller Fälle von Huntington'scher Chorea aus dem Bundesgebiet. in: *Schriftenreihe d. Bundesminister f. Atomkernenergie u. Wasserwirtschaft* **17**, 1960.

89. WENDT, G. G., Die Fruchtbarkeit van Kranken mit Huntington'scher Chorea. *Second International Conference of Human Genetics,* Rome, 1961 (Excerpta Medica).

90. WHITTAKER, D. L., COPELAND, D. L. and GRAHAM, J. B., Linkage of color blindness to Hemophilias A and B. *Amer. J. Hum. Genet.* **2**, 149–158, 1962.

91. YERUSHALMY, J., Infant and maternal mortality in the modern world. *Ann. Amer. Acad. Polit. Soc. Sci.* **237**, 134, 1945.

Human Population Cytogenetics

W. M. COURT-BROWN and P. G. SMITH

Human population cytogenetics is concerned with the frequency of chromosome aberrations, their causation and the selective forces that may be exerted on their carriers. Its pursuit entails the determination of the incidence of different aberrations in defined populations, the examination of the families of abnormal propositi and the clinical scrutiny of affected individuals. So far much of human cytogenetics has been concerned with identifying and examining persons with abnormal karyotypes, and our knowledge of incidences is limited. We do know that aberrations, detectable at mitosis, are by no means uncommon in the population and that they are remarkably frequent in certain special groups. A little is known about the crude incidences of some forms of sex-chromosome aneuploidy and we are beginning to have some appreciation of the incidence of the most commonly detectable type of structural heterozygosity. It is not yet possible, however, to estimate with confidence the frequency of origin of the latter. In general much more information is necessary and this paper discusses some of the problems of obtaining and using it.

1. Retrospective Studies

The factors that underlie the origin of an abnormal karyotype may be sought for by recording such data as birth order and parental ages at birth and comparing them with those from a suitable control population. The same general approach can be used to search for possible environmental agents which, acting on one or other parent, may enhance the risk of their conceiving a karyotypically abnormal child—e.g., exposure to ionizing radiations or to infective agents. The technique of retrospective study can also be used to investigate whether individuals with a particular karyotypic aberration show undue susceptibility to physical or mental disease, by comparing their histories of morbidity with those of matched controls. Some of these forms of study are open to the general criticism that the recall of past events may be influenced by the knowledge of the presence of an abnormality, and the validation of such data requires the most accurate cross-checking against sickness records. This criticism, however, is obviously not relevant to the type of study that depends on recording such data as parental ages at birth, but even here difficulties have been encountered in obtaining control information.

Reproduced by permission of the publisher and P. G. Smith from British Medical Bulletin, *25; 74–80 (1969)*

William M. Court-Brown, *a physician and population geneticist, received his M.D. degree in 1942 from Andrews University. He worked for several years in radiotherapy and in 1950 turned to the study of the harmful effects of ionizing radiations. He published several papers dealing with the factors governing the length of the latent period between exposure to radiation and the onset of gastro-intestinal symptoms. His research interests were later directed toward human cytogenetics, and until his death, in December, 1968, he was Director of the Medical Research Council's Clinical and Population Cytogenetics Research Unit, Edinburgh. Some of his publications in this field include* Abnormalities of the Sex Chromosome Complement in Man, *with others (London, 1964); and* Chromosome Studies in Adults *(Cambridge, 1966).*

A classical example of the use of the retrospective technique is provided by the work of R. L. Jenkins and L. S. Penrose of more than thirty years ago, which emphasized the significance of advancing maternal age to the risk of the conception of mongols (see Penrose & Smith, 1966). Recent evidence points to the similar influence of maternal age on the conception of other forms of autosomal trisomy, and also of 47,XXY males and 47,XXX females. However, this influence does not hold for 45,X females (Lindsten, 1963) or apparently for 49,XXXXY males (W. M. Court-Brown and P. Law, in preparation). Sigler, Lilienfeld, Cohen & Westlake (1965) undertook an extensive study of the radiation-exposure histories of the mothers of mongols and matched controls. When all forms of exposure were considered, there was a marginally significant increase of exposure to the mothers of the mongols ($P < 0.05$), but, when fluoroscopic and therapeutic exposure were considered alone, then the level of significance was more impressive ($P < 0.01$). However, the evidence suggests that radiation exposure may explain only a small number of instances of mongolism.

A good deal of work has been done on infection, following the report by Collmann & Stoller (1962) which appeared to indict infectious hepatitis. The theory is that, if infections are relevant, then the births of the abnormal individuals will be clustered in time and the clusters themselves will have a definitive relationship to the antecedent epidemics under suspicion. So far the suspected role of infectious hepatitis has not been confirmed by other studies (Baird & Miller, 1968). Virtually no work has been done on the problem of examining the histories of morbidity of karyotypically abnormal individuals and matched controls. An attempt was made to examine the past psychiatric history of 47,XXX females in this way (Kidd, Knox & Mantle, 1963), but the technique poses considerable difficulties, especially in the field of mental disturbance.

2. Incidence Studies

Much epidemiological work in cytogenetics occupies an intermediate position between retrospective and prospective studies. It is based on the

P. G. Smith, an applied mathematician, was born in England in 1942. He received his B.S. from City University, London, in 1963. From 1965 to 1967 he worked in the Medical Research Council's Statistical Research Unit, London, on problems relating to the epidemiology of leukemia. His current research interests include the epidemiology of chromosomal abnormalities and the study of the production of chromosomal aberrations by radiation.

premise that if the survey of some specialized group—e.g., high-grade mental defectives—shows an aberration frequency significantly in excess of that found in the newborn, then this points to a special risk linked with the particular aberration.

Unfortunately the populations studied are often not sufficiently defined, and two major caveats stem from the implications inherent in comparisons of the frequency of an abnormality in the newborn with that in an older group. First, such comparisons assume that there has been no temporal change in incidence, and this could lead to misleading or false conclusions. As an example of such a change, Matsunaga (1967) has drawn attention to the effect in Japan of the passage of the Eugenic Protection Law in 1948, which permitted voluntary sterilization and induced abortions for economic reasons. By 1960 the live birth-rate had dropped to about 17/1,-000 from about 34/1,000 in 1947, while over the same period the proportion of live births to mothers aged over 35 years had fallen from 20% to 6%. Matsunaga estimates that this remarkable change in the maternal age distribution may have reduced the frequency of mongolism in the newborn by about 40%, and presumably it has also reduced the numbers of 47,XXY males and 47,XXX females.

The second assumption is that the abnormality does not materially affect mortality risks. We can say unequivocally that the mortality risks of the live-born autosomal trisomies are substantially increased. The risks for live-born 45,X females are almost certainly increased, particularly from death due to cardiovascular malformations, and, furthermore, it is likely that the mortality risks for children who show an unbalanced form of autosomal structural heterozygosity are considerably greater than for normal children. However, most of the comparative work between the newborn and other groups has been done on the frequency of males with an abnormal nuclear sex and to a lesser extent of doubly chromatin-positive females.

The work of Maclean, Harnden, Court-Brown, Bond & Mantle (1964) and Bochkov (1965) suggests that sex-chromosome aneuploidy may be associated with some increased perinatal and infantile mortality risks. If this is so, then incidence comparisons between the newborn and older groups will lead to underestimates of the total risks associated with these abnormal sex-chromosome complements. It must also be remembered that there may be marked temporal changes in the expectation of life of affected individuals. This effect has been clearly shown for mongols, whose mean survival age has approximately doubled between 1932 and 1963 from 9 to 18 years (see Penrose & Smith, 1966).

An example of the type of comparison that has been made, and one which illustrates the snags inherent in imprecise definitions of study populations, is that of the incidences of chromatin-positive males in the newborn and among the mentally subnormal treated in hospital (for details see Court-Brown, 1969). Information is available on more than 41,000 consecutive live male births from nuclear-sexing surveys yielding an incidence of chromatin-positive males of about 1.7/1,000. The nuclear sexing of at least 13,000 male defectives treated in hospital shows the incidence of chromatin-positive males to be raised to about 9.4/1,000. Unfortunately, most of the latter surveys provide no information either on the age structure of the study populations or on the distribution of the patients by intelligence quotient (I.Q.). There are, however, three studies of ordinary hospitals for the mentally subnormal, based on over 3,500 patients, which permit allocation to the ranges of I.Q. < 20, 20–49, and 50 or more. The studies are mutually compatible and show the frequency of chromatin-positive males/1,000 examined to be 2.2, 9.0 and 15.2 in the three I.Q. groups respectively. The grossly defective show, as might have been expected, a frequency not obviously different from the newborn, while the frequency is greatest among high-grade defectives.

The intelligence quotient becomes important in assessing the significance of the findings on nuclear sexing in specialized hospitals—e.g., those for the hard-to-manage mentally retarded males in Sweden (Hambert, 1966) and the English maximum-security hospitals of Rampton and Moss Side (Casey, Segall, Street & Blank, 1966; Casey, Street, Segall & Blank, 1968). These hospitals are for patients who require special security on account of violent or persistent antisocial behaviour. Both the Swedish and English studies yielded a frequency of chromatin-positive males of about 20/1,000, significantly in excess of that from the ordinary hospitals for the mentally subnormal, and for some time this led to the idea that criminal behaviour per se accounted for the excess of abnormal males. However, it is now clear that much of, and possibly all, the excess is accounted for by the different distributions, according to I.Q., of the patients in the ordinary hospitals for the mentally subnormal and those in the special hospitals. All this does not detract from the possibility that criminal acts may be a common reason for the presence of the chromatin-positive males, but information on this can come only from more comprehensive studies.

The problem of the accurate definition of populations studied is again relevant to the frequency of 47,XYY males in certain selected groups. There is no doubt that these males are unusually common in certain populations, for example, those of the British maximum-security hospitals (Jacobs, Brunton, Melville, Brittain & McClemont, 1965; Casey, Blank, Street, Segall, McDougall, McGrath & Skinner, 1966; Jacobs, Price, Court-Brown, Brittain & Whatmore, 1968). Since their discovery, these males have been searched for in ordinary hospitals for the mentally subnormal, but here their distribution between hospitals appears non-random. The telling point appears to be whether the hospital makes a practice of admitting patients referred by a Court of Justice, even though it is in no way formally a maximum-security hospital. Thus, Close, Goonetilleke, Jacobs & Price (1968) found two of 19 adult males of 183 cm. or more in height to have a 47,XYY complement, in a hospital for the mentally subnormal in which a high proportion of patients were detained because of behavioural disturbances. P. A. Jacobs and colleagues (unpublished data) sought for males of 183 cm. or more in a number of Scottish hospitals for the mentally subnormal. Only 31 were found, including two with a 47,-XYY complement and one with 47,XYY/48,XXYY complement. Significantly, all three came from the same hospital, and the XYY males from the same ward, a security ward. However, the hospital could not be designated other than an ordinary hospital for the mentally subnormal. In contrast the same authors examined 606 males of all ages, unselected for height, from an English hospital for the mentally subnormal, without finding a single example of a 47,XYY male. The important points are that this hospital does not especially cater for those with antisocial conduct and it is run on more open lines than usual. All this indicates the necessity for a close appraisal of the admission policies of hospitals, together with details of the age and sex structure of the patients, and in general a more rigorous definition of study populations than has been the case so far.

Often a proportion of adults asked to participate in a study decline to do so, and the bias thus introduced is likely to be most important in incidence studies involving karyotypic aberrations that can be associated with behavioural disorders. A good example was encountered by Jacobs, Price, Court-Brown, Brittain & Whatmore (1968) in their chromosome survey of patients in a maximum-security hospital. There were 203 men in one wing of the hospital, of whom 187 agreed at the first request to provide a blood sample, and among these there were five 47,XYY males. Nine of the remaining 16 agreed on a second request and among these were two 47,XYY males ($P < 0.05$). The final seven non-co-operators included a 48,-XXYY male who had been found in an earlier study (Maclean *et al.* 1962). In effect, it is impossible to know to what extent bias is introduced by refusals. If the refusal rate is low, then not too much harm may be done. However, the results of studies which fail to include a substantial proportion of the originally chosen population need to be treated with some circumspection. For instance, Kjessler (1966) selected 178 men for study who had attended a subfertility clinic over a period of 12 months, but for one reason or another, including non-co-operation, could report on only 135 men, or about 76%.

It is appropriate to discuss the incidence of autosomal structural heterozygosity in the population, and the point has to be made, and will be elaborated later, that the data refer to structural aberrations detectable in somatic cells by current techniques. The discussion is confined to the results of studies in Edinburgh. Table I shows the results from three surveys of male babies in three Edinburgh hospitals, those in hospitals A and B being continuing surveys of all live-born males, whereas the survey in hospital C was of randomly selected babies and completed some time ago. Of the 1,788 babies studied, two were recognized to have an autosomal structural abnormality. It should be noted in Tables I–IV that the abnormal karyotypes are divided into sex-chromosome aneuploidy, autosomal aneuploidy, structural aberrations of a sex chromosome or of autosomes and other abberrations. The last-named column in these tables contains a total of three abnormal subjects, two 46,XX males and a prisoner with a complement 47,XY,mar+. The extra chromosome was a small one, morphologically unlike any of the normal karyotype. The man was without physical phenotype abnormality and there is doubt as to how to classify this type of aberration. The complements of the abnormal cases are given at the foot of each table.

Tables II and III list male and female groups respectively, examined for various reasons but with no cause to suspect that any group would show an unusual frequency of abnormal individuals. In the absence of better data these groups may be considered as representative of the general adult population, although understandably the purist could criticize this assumption. The findings between the groups in each table are mutually compatible, as are the findings in aggregate for the two sexes. In all, 998 individuals have been studied and five found to be structural heterozygotes, or one in about 200. The information in Table IV is on males and is from six groups, all adults except Group I, and all selected from penal institutions with the exception of Group VI. Furthermore, Groups IV and V were selected by height to facilitate the identification of 47,XYY males, whereas the employees of an industrial organization (Group VI) were selected solely on the basis of height. In these groups one would expect a raised incidence of sex-chromosome aneuploidy because of the manner of selection, and this was found; but there were no prior reasons for suspecting a raised incidence of autosomal structural heterozygosity. In the event, six such heterozygotes were found in 2,047 males, or one in about 340.

The aggregate data on structural heterozygosity from Table I, from Tables II and III combined, and from Table IV have been tested for homogeneity. In spite of the incidence in babies being rather low, the findings are compatible ($\chi_2^2 = 3.70$, $P < 0.10$). Combining all the information from

TABLE I. CHROMOSOME STUDIES ON LIVE-BORN MALE BABIES

Group	Number studied	Number of cells*	Aneuploidy		Structural aberrations		Other aberrations	Total abnormalities†
			Sex chromosomes	Autosomes	Sex chromosomes	Autosomes		
Hospital A	1,006	2	3	4	0	0	1	9
Hospital B	516	2	0	0	0	0	0	0
Hospital C	266	10	0	0	0	1	0	1
All groups	1,788	—	3	4	0	2	1	10

*The basic number counted and analysed per individual.

†Abnormalities:
Hospital A: 47, XXY (2); 47, XYY; 47, XY,G+ (4); 45, XY,D—, D—,t(DqDq)+; 46, XX
Hospital C: 45, XY, D—, E—, mar+ /46, XY, E—, mar+

TABLE II. CHROMOSOME STUDIES ON ADULT MALES WITH NO KNOWN BIAS FOR INCREASED ABERRATION FREQUENCY

Group number	Group	Number studied	Number of cells*	Aneuploidy		Structural aberrations		Other aberrations	Total abnormalities§
				Sex chromosomes	Autosomes	Sex chromosomes	Autosomes		
I	General-practice sample†	207	30	0	0	1	0	0	1
II	Spouses‡	60	2	0	0	0	1	0	1
III	Radiation workers	68	100	0	0	0	1	0	1
IV	Benzene workers	68	100	0	0	0	0	0	0
V	Thorotrast patients	27	100	1	0	0	0	0	1
VI	Ankylosing spondylitics	206	50–100	0	0	0	1	0	1
All groups		636	—	1	0	1	3	0	5

*The basic number counted and analysed per individual
†See Court-Brown, Buckton, Jacobs, Tough, Kuenssberg & Knox (1966)
‡Males marrying into families in which a structural aberration is segregating
§Abnormalities:

Group I: 46, Xinv(Yp + q−)
Group II: 45, XY, D−,D−,t(Dq,Dq)+
Group III: 45,XY,D−D−,t(Dq,Dq)+
Group V: 45,X/47,XYY
Group VI: 45,X,D−,D−,t(Dq,Dq)+

TABLE III. CHROMOSOME STUDIES ON ADULT FEMALE GROUPS WITH NO KNOWN BIAS FOR INCREASED ABERRATION FREQUENCY

Group number	Group	Number studied	Number of cells*	Aneuploidy		Structural aberrations		Other aberrations	Total abnormalities§
				Sex chromosomes	Autosomes	Sex chromosomes	Autosomes		
I	General-practice sample†	231	30	0	0	0	1	0	1
II	Spouses‡	43	2	0	0	0	1	0	1
III	Luminous dial painters	63	100	0	0	0	0	0	0
IV	Thorotrast patients	25	100	0	0	0	0	0	0
	All groups	362	—	0	0	0	2	0	2

*Basic number counted and analysed per individual
†See Court-Brown et al. (1966)
‡Females marrying into families in which a structural aberration is segregating
§Abnormalities:
Group I: 46, XX/46, XX, t(Dq+; Dq−)
Group II: 46, XX, t(I?−; Bq+)

161

TABLE IV. VARIOUS YOUNG AND ADULT MALE GROUPS

Group number	Group	Number studied	Number of cells*	Aneuploidy		Structural aberrations		Other aberrations	Total abnormalities†
				Sex chromosomes	Autosomes	Sex chromosomes	Autosomes		
I	Approved schools (all heights)	340	2	3	0	0	0	0	3
II	Scottish Borstals‡ (all heights). New entrants	607	10	3	0	0	2	1	6
III	Prison allocation centres. Sentences 1 year or more (all heights)	302	2	0	0	0	2	0	2
IV	Scottish prisons (178 cm. or more in height)	419	2	3	0	0	1	1	5
V	Young Offenders Institutions§ (178 cm. or more in height)	91	2	1	0	0	0	0	1
VI	Industrial organization (employees, 183 cm. or more in height)	288	2	2	0	0	1	0	3
	All groups	2,047	—	12	0	0	6	2	20

*The basic number counted and analysed per individual
†Abnormalities:
  Group I: 46, XY/47, XXY; 47, XYY (2)
  Group II: 47, XYY; 47, XXY (2); 46, XX; 46, XY, inv(Cp–q+); 46, XY, lh+, t(l?+; Cq–)
  Group III: 45, XY, D–, D–, t(Dq, Dq)+; 45, XY, D–, t(Dq, Gq)+
  Group IV: 47, XYY; 47, XXY (2); 46, XY, t(l?–; 16q+); 47, XY, mar+
  Group V: 47, XYY
  Group VI: 47, XXY (2); 45, XY, D–, D–, t(Dq, Dq)+

‡Penal institutions for young persons between 16 and 21 years of age. The period of Borstal training depends on the response of the inmate, the maximum length being two years

§Penal institutions for young persons aged between 17 and 21 years who the courts do not think will benefit from any other form of detention. Most of the inmates will have already spent some time in a Borstal, to which they cannot be sent twice

162

Tables I–IV shows that 14 individuals with autosomal structural heterozygosity have been found among 4,833 examined, or about 0.3%. This is somewhat lower than the estimate made by Court-Brown, Buckton, Jacobs, Tough, Kuenssberg & Knox in 1966, of about 0.5%. If the mosaic female in Group I of Table III is excluded, then there are 13 structural heterozygotes whose abnormality may be presumed to have followed a prezygotic event. Another point is that these 13 heterozygotes contain six with the Robertsonian translocation of the type t(DqDq), or approximately one in 800 individuals examined. This is not significantly different from an estimate by Court-Brown (1967) that about one in 1,000 of the population would be carriers of this, the most commonly detectable form of structural heterozygosity.

3. Prospective Surveys

Incidence studies have raised many questions both for the clinician and the geneticist, and for the former many of the questions relate to genetic counselling. Some of the effects of the common forms of aneuploidy are predictable, for example the sterility of 47,XXY males and of 45,X females (although the possibility of mosaicism can never be discounted), but we are far from being sufficiently informed about the risks of other effects, especially those on intelligence or behaviour. We cannot yet advise with any authority on the risk of mental subnormality for a male baby with a 47,XXY complement and we are almost completely in the dark about the risks contingent on a 47,XYY karyotype. In a 47,XXX female, we do not yet appreciate the extent of the risk of secondary amenorrhoea nor are we able to predict the likelihood of her ending her days in a hospital for the mentally subnormal. Our acquaintance with the effects on reproductive fitness of randomly ascertained forms of structural heterozygosity is only minimal, and probably too often has discouraging advice been offered to families who have been identified through congenitally malformed propositi or because of gross impairment of reproductive ability.

The phenotypic effects of karyotypic abnormalities will be put into realistic perspective for the clinician by the examination of the results of prospective surveys, particularly those of unselected live-born babies. What is required is the identification of individuals with an abnormal karyotype, the choice of suitably matched controls and the long-term follow-up of the abnormal individuals and the controls.

The clinician will not be the sole beneficiary of prospective studies, for these are essential for the geneticist faced with the necessity to understand the dynamic situation created by the perpetual influx into the population of karyotypically abnormal individuals. Through the examination of their parents it will be possible to estimate the frequency of origin of new structural aberrations, and the examination of the selective forces exerted on such aberrations must occupy geneticists for decades to come.

We already know that, by comparison with other detectable forms of structural heterozygosity, translocations of the type t(DqDq) are maintained in the population at a relatively high frequency. Hamerton (1968) has adduced evidence from a study of the offspring of male carriers of this translocation as opposed to female carriers, which favours positive prezygotic selection among the former. Our understanding of the significance of this important form of chromosome polymorphism will be hastened by the formal prospective study.

Fortunately aberrant karyotypes are sufficiently frequent among the live born to justify routine chromosome analysis. Blood from a skin puncture is simply obtained and so easily cultured that the failure rate should be appreciably less than one per cent, while there is no evidence that the few failures are other than random. Mass studies of the newborn will start to become a feasible proposition in about two to three years' time, with the introduction at an operational level of microdensitometric direct-scanning techniques linked with computer analysis of the digitized information.

4. Cytological Problems

No account has yet been taken of the various snags inherent in the cytological techniques for examining populations, and only a few can be mentioned here. The first problem is the efficiency of detection of structural heterozygotes from the analysis of mitotic cells, and there are important limitations to this. Paracentric inversions cannot be seen at all, and the only detectable pericentric inversions are those following breaks non-equidistant from the centromere and sufficiently so for the visual recognition of the change in arm ratio. Reciprocal translocations will go unrecognized where nearly equal amounts of material are exchanged, while presumptive simple deletions can be suspected only when relatively large amounts of material are involved. This problem of the inefficiency of recognition has long worried those concerned with the study of chromosome-damaging agents, particularly in radiation cytogenetics. In fact H. J. Evans (personal communication, 1968) reckons that the efficiency of scoring symmetrical re-arrangements in cultured lymphocytes following radiation exposure may be as low as about 20%. It is not permissible to use the experience so gained to correct the findings from the examination of somatic cells in population studies in order to provide a more realistic estimate of the true frequency of structural heterozygotes, but it cannot be denied that the real frequency must be appreciably greater than that based on the examination of mitotic cells.

A better estimate would be obtained from the study of meiotic chromosomes, and the experience of McIlree, Price, Court-Brown, Tulloch, Newsam & Maclean (1966) is relevant. These authors did meiotic and mitotic studies in parallel on 50 subfertile males with sperm counts lower than 40×10^6/ml. and ranging down to zero. Two men were found whose mitotic chromosomes were indistinguishable from normal, but each of whom showed a multivalent association at diakinesis compatible with a reciprocal translocation. These observations demonstrate, therefore, that there must be, as expected, structural heterozygotes in the population not recognizable in mitotic cells. Meiotic studies are possible on males, but not as a formal population study except where there is the clinical justification of subfertility, whereas the technique for the study of female meiotic cells is being developed (Henderson & Edwards, 1968; Jagiello, Karnicki & Ryan, 1968; Yuncken, 1968) and is obviously more difficult than that for males. To obtain an indication of the findings in ordinary men and women, perhaps the best that can be hoped for will be studies on gonadal tissue exposed during the course of surgical operations, such as, for instance, the repair of inguinal herniae in males. Given that such examinations are ethically justifiable, then the results will have to be assumed as representative of the general population.

The value of nuclear sexing is limited to the identification of individuals with an abnormal number of X chromosomes, or, in the instance of females with two X chromosomes, of those with a sufficiently gross structural aberration of one X chromosome to be reflected in the size of the sex-chromatin body. An important disadvantage of nuclear sexing as the primary technique of ascertainment in examining male populations is the failure to ascertain those with an abnormal sex-chromosome complement where there is but a single X chromosome. It is essential that such abnormal males should not be overlooked in the study of some groups—and the 47,XYY male is one example—while it is important to ascertain among subfertile males such complements as 45,X/46,XY (Kjessler, 1966) or 45,X/46,Xinv(Yp + q −) or 45,X/46,XYdic (McIlree *et al.* 1966). A good argument can be made, therefore, for the use of chromosome analysis and nuclear sexing for ascertaining the karyotypically abnormal, certainly in specialized groups and indeed for all population studies, on the grounds that this procedure will minimize the error of failure to ascertain certain forms of mosaicism where the number of lymphocytes counted per individual is low (see below). These forms are those containing a normal cell line appropriate for the phenotypic sex of the individual, and an abnormal one in which the number of X chromosomes is at variance with the phenotypic sex, such as mosaics of the types 46,XY/47,XXY,45,X/46,XX and 46,XX/47,XXX. The argument is based on the knowledge that there can be considerable variation in the proportions of the constituent cell lines between tissues in a mosaic. For example, males are occasionally found whose buccal mucosal cells are chromatin-positive but whose lymphocytes are consistently 46,XY, and presumably the reverse can be true.

Nuclear sexing has its own problems. In experienced hands about 50% of cells on average are scored as chromatin-positive in buccal smears from individuals in whom there are two apparently normal X chromosomes, and it is quite possible, using this technique alone, to overlook a 45,X/ 46,XX female. There are two techniques where over 90% of cells from normal females are scored as chromatin-positive, but one technique, the examination of a monolayer of cultured fibroblasts, is useless for survey work, whereas the other, the examination of amniotic cells (Klinger, 1957), can be used only for studies of babies (Robinson & Puck, 1967). It may appear a counsel of perfection to suggest the joint use of chromosome analysis and nuclear sexing for ascertainment, but this may become possible in the future with the application of pattern-recognition techniques. In the meantime, not too much will be lost if the primary method of ascertainment is restricted to chromosome analysis, and this raises the question of how many cells to study per individual in a survey. The heart of the problem is again that of the detection of mosaicism, whether of mosaics in which one or more lines are aneuploid or those in which one or more lines are heterozygous for a structural aberration.

Ford writes elsewhere on mosaicism (see page 104 of this Bulletin), and here it is sufficient to say that the pronouncement that an individual is not a mosaic needs to be qualified by a statement on the types of cells studied and the numbers of each type counted and analysed. It is not usually feasible, however, to do large-scale surveys on other than cultured lymphocytes, and statements on incidence will relate only to this technique, even though individuals identified as abnormal are then further studied through the examination of other types of cell.

Dr Park Gerald of Boston (personal communication, 1966) initiated the study of only two cells in surveying newborn babies, and it is a valid criticism that this will lead to an underestimate of the incidence of mosaicism. It is equally valid, however, to point out that such a technique will establish reliable incidences for karyotypic aberrations presumed to be present in every cell. At present the number of cells examined per individual in survey work is determined by the type of abnormality being sought and by the resources of the laboratory involved, and it may range from two to upwards of 30–50, or even more. What seems important is that the description of a survey might include not only as precise a definition as possible of the population being studied, but also a statement on technical procedure which covers the basic number of cells examined per individual for ascertainment purposes, and the procedure then adopted when an abnormal individual is discovered. If ascertainment is based on two cells, it is important to know the train of events following either the finding that two cells differ or that two show the same abnormality. Once an abnormal cell has been found, then the question of how many more cells are examined will rest on local practice governing the recognition of mosaicism, and the full report of a survey might include a statement on this. In the future there is bound to be an increasing interest in searching for differences in the results from different centres that may be due to racial or demographic factors, and for this purpose the provision of the sort of information noted above seems desirable.

These thoughts about standardizing the technical features of surveys post-date much of the work reported in Tables I–IV, and it has to be admitted that the listed groups are not homogeneous in terms of the basic number of cells studied. The full descriptions of these surveys and others will be presented elsewhere.

5. *The Direct Study of Environmental Agents*

Earlier we noted how the retrospective survey can be used as an indirect form of inquiry to examine environmental factors that may adversely influence the risk of conceiving a child with an abnormal karyotype. Chromosome studies, however, may be used to determine directly whether a population has been or is being exposed to a chromosome-damaging agent, although at present the scope of this inquiry is limited to agents producing relatively large effects. Most studies have been on ionizing radiations, and it is now established that, given a great enough dose in vivo, an increased frequency of aberrations is detectable in culture of lymphocytes from blood samples taken immediately on the cessation of exposure. This was shown for extensive fluoroscopic procedures in diagnostic radiology by Bloom & Tjio (1964), for therapeutic doses of medium kilovoltage *x* rays to part of the body (Buckton, Langlands, Smith & McLelland, 1967b), and for a single dose of 2 Mev *x* rays to the whole body and limited to 50 rads or less (Langlands, Smith, Buckton, Woodcock & McLelland, 1968). It has also been shown that an increased frequency of lymphocytes with aberrations is detectable for up to 10 years or more after extensive exposure of the body to large doses of radiation (Buckton, Jacobs, Court-Brown & Doll, 1962; Norman, Sasaki, Ottoman & Fingerhut, 1966; Buckton, Court-Brown & Smith, 1967a).

The aberrations are of two general classes, unstable and stable: the former include dicentric and ring chromosomes and acentric fragments, whereas the latter include inversions and reciprocal translocations. By their nature, unstable aberrations increase the probability of the affected cell's failing to survive division, and the level of cells with unstable aberrations progressively falls with time after exposure to radiation; this does not appear to be the case for cells with stable aberrations. The efficiency of detection of unstable aberrations is high, but, as already noted, that of stable aberrations may be as low as about 20%. For this reason the effects of ionizing radiations and other chromosome-damaging agents are usually assessed in terms of the production of unstable aberrations, e.g., the frequency of dicentric chromosomes/1,000 cells. In man, practically all studies have been on cultured lymphocytes, and, in fact, the rate of disappearance from the peripheral blood of cells with unstable aberrations has been used to estimate the mean life-span of the lymphocyte. It now seems clear that some lymphocytes survive for prolonged periods in vivo, one estimate of the mean life-span being nearly 1,600 days (Buckton *et al.* 1967a). It is likely that unstable aberrations will quickly disappear from cells with a high rate of turnover, e.g., erythocyte precursors.

As the mean survival time of some lymphocytes is long, and as there is no evidence so far that the extent of contained damage modifies the life expectancy of the cell, the examination of cultured lymphocytes has a potential for indicating exposure to a chromosome-damaging agent and for providing evidence that a population is being exposed to an undue amount of such an agent in its environment. An example of this approach is provided by the men in Group III (Table II), all of whom worked at an atomic energy establishment. Some were exposed only to the general background of the establishment, but the majority were workers exposed to raised background levels of radiation. In no instance did exposure exceed the internationally agreed permissible limit. The object of the study was to explore the possible effects of occupational exposure, and the finding of one man to be a structural heterozygote was incidental. The effect of this exposure is judged by examining the number of dicentric chromosomes/1,000 cells for the men who were grouped into four different levels of exposure. Those with no special occupational exposure had a dicentric frequency of 1.3/1,000 cells, comparable to that of the ordinary population. However, for the groups of men with mean accumulated doses of 4, 27 and 84 rads of γ rays, the relevant dicentric frequencies were 2.2, 5.0 and 6.7/1,000 cells, differences not explicable by age (Court-Brown, Buckton & McLean, 1965).

It is an over-simplification to say that the observed differences reflected just the total accumulated exposure, for the question of the rate of accumulation has been ignored. Among other things the level of aberrations found in lymphocytes from an individual working or living within an increased radiation background must be influenced by both the rate of aberration production and the rate of disappearance of lymphocytes with aberrations. The dynamics of the effects of chronic exposure are not simple and evidence of an increased aberration frequency, judged from unstable aberrations, cannot be transformed at present into mutagenic risks to gonadal cells. Nevertheless this form of inquiry can be argued to represent a valid extension of population cytogenetics.

ACKNOWLEDGMENT

The authors are indebted to Mrs Pamela Law for her help.

REFERENCES

BAIRD, P. A. & MILLER, J. R. (1968) *Br. J. prev. soc. Med.* **22** 81
BLOOM, A. D. & TJIO, J. H. (1964) *New Engl. J. Med.* **270** 1341

BOCHKOV, N. P. (1965) In: *Proceedings of the Symposium on the Mutational Process, Prague, 9–11 August 1965*, p. 121. Academia, Prague

BUCKTON, K. E., COURT-BROWN, W. M. & SMITH, P. G. (1967a) *Nature, Lond.* **214,** 470

BUCKTON, K. E., JACOBS, P. A., COURT-BROWN, W. M. & DOLL, R. (1962) *Lancet,* **2,** 676

BUCKTON, K. E., LANGLANDS, A. O., SMITH, P. G. & McLELLAND, J. (1967b) In: Evans, H. J., Court-Brown, W. M. & McLean, A.S., ed. *Human radiation cytogenetics (Proceedings of an International Symposium held in Edinburgh, 12–15 October 1966)*, p. 122. North-Holland, Amsterdam

CASEY, M. D., BLANK, C. E., STREET, D. R. K., SEGALL, L. J., McDOUGALL, J. H., McGRATH, P. J. & SKINNER, J. L. (1966) *Lancet,* **2,**859 [Letter]

CASEY, M. D., SEGALL, L. J., STREET, D. R. K. & BLANK, C. E. (1966) *Nature, Lond.* **209,** 641

CASEY, M. D., STREET, D. R. K., SEGALL, L. J. & BLANK, E. C. (1968) *Ann. hum. Genet.* **32,** 53

CLOSE, H. G., GOONETILLEKE, A. S. R., JACOBS, P. A. & PRICE, W. H. (1968) *Cytogenetics,* **7,** 277

COLLMANN, R. D. & STOLLER, A. (1962) *Am. J. publ. Hlth,* **52,** 813

COURT-BROWN, W. M. (1967) *Human population cytogenetics.* North-Holland, Amsterdam

COURT-BROWN, W. M. (1969) *Int. Rev. exp. Path.* **7,** 31

COURT-BROWN, W. M., BUCKTON, K. E., JACOBS, P. A., TOUGH, I. M., KUENSSBERG, E. V. & KNOX, J. D. E. (1966) *Chromosome studies on adults.* (Eugenics Laboratory Memoirs, No. 42.) Cambridge University Press, London

COURT-BROWN, W. M., BUCKTON, K. E. & McLEAN, A. S. (1965) *Lancet,* **1,** 1239

HAMBERT, G. (1966) *Males with positive sex chromatin.* Akademiförlaget-Gumperts, Göteborg

HAMERTON, J. L. (1968) *Cytogenetics,* **7,** 260

HENDERSON, S. A. & EDWARDS, R. G. (1968) *Nature, Lond.* **218,** 22

JACOBS, P. A., BRUNTON, M., MELVILLE, M. M., BRITTAIN, R. P. & McCLEMONT, W. F. (1965) *Nature, Lond.* **208,** 1351

JACOBS, P. A., PRICE, W. H., COURT-BROWN, W. M., BRITTAIN, R. P. & WHATMORE, P. B. (1968) *Ann. hum. Genet.* **31,** 339

JAGIELLO, G., KARNICKI, J. & RYAN, R. J. (1968) *Lancet,* **1,** 178

KIDD, C. B., KNOX, R. S. & MANTLE, D. J. (1963) *Br. J. Psychiat.* **109,** 90

KJESSLER, B. (1966) *Karyotype, meiosis and spermatogenesis in a sample of men attending an infertility clinic.* Karger, Basel

KLINGER, H. P. (1957) *Acta anat.* **30,** 371

LANGLANDS, A. O., SMITH, P. G., BUCKTON, K. E., WOODCOCK, G. E. & McLELLAND, J. (1968) *Nature, Lond.* **218,** 1133

LINDSTEN, J. (1963) *The nature and origin of X chromosome aberrations in Turner's syndrome.* Almqvist & Wiksell, Stockholm

McILREE, M. E., PRICE, W. H., COURT-BROWN, W. M., TULLOCH, W. S., NEWSAM, J. E. & MACLEAN, N. (1966) *Lancet,* **2,** 69

MACLEAN, N., HARNDEN, D. G., COURT-BROWN, W. M., BOND, J. & MANTLE, D. J. (1964) *Lancet*, **1**, 286

MACLEAN, N., MITCHELL, J. M., HARNDEN, D. G., WILLIAMS, J., JACOBS, P. A., BUCKTON, K. E., BAIKIE, A. G., COURT-BROWN, W. M., MCBRIDE, J. A., STRONG, J. A., CLOSE, H. G. & JONES, D. C. (1962) *Lancet*, **1**, 293

MATSUNAGA E. (1967) In: *Proceedings of the World Population Conference, Belgrade, 30 August–10 September 1965*, vol. 2, p. 481. United Nations, New York

NORMAN, A., SASAKI, M. S., OTTOMAN, R. E. & FINGERHUT, A. G. (1966) *Blood*, **27**, 706

PENROSE, L. S. & SMITH, G. F. (1966) *Down's anomaly*. Churchill, London

ROBINSON, A. & PUCK, T. T. (1967) *Am. J. hum. Genet.* **19**, 112

SIGLER, A. T., LILIENFELD, A. M., COHEN, B. H. & WESTLAKE, J. E. (1965) *Bull. Johns Hopkins Hosp.* **117**, 374

YUNCKEN, C. (1968) *Cytogenetics*, **7**, 234

Some Observations on the Effect of Inbreeding on Mortality in Kure, Japan

WILLIAM J. SCHULL and JAMES V. NEEL

In 1958, Schull presented data on the relationship between inbreeding and the frequency of stillbirths and death during the first month of life in three Japanese cities, Hiroshima, Kure, and Nagasaki. Expressed as the ratio of regression coefficient to intercept, i.e., B/A, the values for the three cities were 11.52, 1.22, and 2.65, respectively. Morton (1961) attempted to dismiss the small effects of inbreeding on stillbirths and early mortality in Nagasaki and Kure as due to "some environmental disturbance," nature unspecified. Heterogeneity of the population was also suggested as a factor in the findings. Subsequently, we (Neel and Schull, 1962; Schull and Neel, 1965) have shown that in Nagasaki the relatively small effect of inbreeding on mortality persists throughout the first ten years of life and, in addition, through a careful consideration of the distributions with respect to consanguineous marriages of a number of socio-economic variables, have been unable to find any evidence for the alleged environmental disturbance. It has now been possible to conduct a similar follow-up in Kure. The average surviving child in Kure at the time of the study was 15 years of age. Inasmuch as 90% of the deaths prior to age 20 occur during the first 15 years of life *(Ninth Japanese Life Tables)*, these figures should be almost synonymous with pre-reproductive mortality. It will be shown that in Kure, also, the low initial mortality persists, with no evidence that the finding is any less valid than that for Hiroshima and Nagasaki.

The Cohort Studied

The cohort to be described arose as an outgrowth of a comprehensive attempt in the years 1948–1953 to determine whether there could be observed any difference between children born to parents one or both of whom were exposed to the atomic bombings of Hiroshima and Nagasaki and children born to suitable "control" parents. At the initiation of this attempt, it was uncertain whether an adequate number of nonexposed parents from whence to select a comparison group would exist in the aforementioned cities. To cope with this possibility, Kure, a city some

William J. Schull, *a human-population geneticist, was born in Missouri in 1922. He received his B.S. in 1946 and his M.S. in 1947 from Marquette University, and his Ph.D. in 1949 from Ohio State University. Presently he is Professor of Human Genetics and Anthropology at the University of Michigan where he is associated with James V. Neel. His major field of interest and research has been in population genetics. He worked intermittently in Japan from 1949 to 1964 where, as a member of the Atomic Bomb Casualty Commission, he conducted extensive research into the effects of atomic radiation on an exposed human population. Since then he has conducted population-genetics studies in Chile. He is the current (1970) president of the American Society of Human Genetics.* Author of numerous publications in human-population genetics, he has edited or co-authored Human Heredity, with J. V. Neel (Chicago, 1954); The Effect of Exposure to the Atomic Bombs on Pregnancy Termination in Hiroshima and Nagasaki, with R. C. Anderson, et al. (National Academy of Sciences National Research Council Publication 461: Washington, D.C., 1956); Genetics and the Epidemiology of Chronic Diseases, edited with J. V. Neel and M. W. Shaw (1956); Mutations, Ed. (Ann Arbor, 1962); Genetic Selection in Man, Ed. (Ann Arbor, 1963); and The Effect of Inbreeding on Japanese Children, with J. V. Neel (New York, 1965).*

twenty miles to the east and south of Hiroshima, was studied also until it became apparent that the possibility cited would not eventuate. Thus during the years 1948–1950, with the co-operation of the municipal authorities, an attempt was made to register all pregnant women in Kure sometime between the twenty-first week of gestation and parturition. Registration occurred when these women or their representatives presented themselves to register for certain rations to which their status entitled them. The economic stringencies of the immediate postwar period in Japan were such that virtually all women who were eligible availed themselves of these special rations.

Elsewhere we have described in detail the completeness of this system of registration and the nature of the information which was collected (Neel and Schull, 1956; Schull, 1958; Schull and Neel, 1965). Briefly, the following obtained: At the time of her registration for ration purposes, each pregnant woman completed the first two-thirds of a questionnaire which included such items as identifying information, consanguinity, a short summary of past reproductive performance, and pertinent details concerning her present pregnancy. This questionnaire was filled out in duplicate. The original was given to the registrant, who in turn presented it to the attendant at her delivery; the duplicate was retained by the Atomic Bomb Casualty Commission (ABCC). Upon termination of the pregnancy, the midwife or physician in attendance completed the questionnaire by answering certain questions pertaining to the characteristics of the child and delivery and notified the ABCC of the termination of the pregnancy. Subsequently, a physician in the employ of the Commission called to examine the child and to verify certain observations reported by the attendant. In this manner, it was possible to define a cohort of some 8,211 pregnancies of known gestation (at least twenty-one weeks) which terminated in single births of known sex. Some of the individuals residing in Kure during the years of this study were exposed to the atomic bombings of Hiroshima and Nagasaki. Pregnancies occurring to such individuals are not included among the 8,211 if the exposed individual(s) received an exposure which exceeded five roentgens on the average. Pregnancies which were stated to be induced and which terminated in an infant weighing less than 2500 grams as well as pregnancies which terminated in a child of unknown birthweight also are not included among the 8,211.

Of the total pregnancies, in 575 instances the parents were related as first cousins, first cousins once removed, or second cousins. The reliability of these assertions of relationship has been considered elsewhere in connection with the presentation of similar data obtained in Hiroshima and Nagasaki (Schull and Neel, 1965). In brief, there are undoubtedly errors in ascertainment; however, these are not apt to be of such a number and kind that they significantly alter the nature of the results to be presented.

We are here concerned with mortality among the 565 of these 575 children who were liveborn and a suitable comparison group selected from among the 7,636 conceptuses where the parents were unrelated. The latter sample was chosen by the simple expedient of selecting every pregnancy registration involving unrelated parents where the terminal digit of the number assigned at registration was zero. To bring the group so chosen more nearly in size with the sample of pregnancies from related parents, those "zero" infants whose subterminal registration digits were 2 or 5 were also excluded. These procedures resulted in a sample of 606 pregnancy terminations where the parents had reported they were unrelated. Of these, 597 terminated in a liveborn infant.

Ascertainment of Fact and Cause of Death

There has existed in Japan since the last quarter of the 19th century a system of compulsory family registration. Vital events affecting the composition of a given family or the status of members of that family must be reported to the office having custody of the family's record, termed the *koseki*. An indispensable part of the system is the legal or permanent address of the family, known as the *honseki*. Changes in the latter must also be reported to the *koseki* office. Knowledge of the last or any recent *honseki* of an individual is a virtual guarantee that the survival status of that individual can be determined indefinitely.

As a rule, births, deaths, and stillbirths are declared to local offices in the place of occurrence within seven days for deaths and stillbirths and fourteen days for births. If the office of declaration is not the office of custody of a given family's *koseki,* the latter office is notified through administrative channels designed to accomplish such transfers of information. The office of custody upon receipt of information which alters a given *koseki* is required to revise the appropriate record immediately. The *koseki* is a public document which may be perused by anyone, and a certified copy or abstract may be obtained for a nominal charge, which may be waived under certain circumstances.

Declarations of death in Japan must be accompanied by a medical certificate from the attending physician. In the event a physician is not in attendance at a particular death or that death is unnatural, a coroner is sent to examine the corpse and to certify the cause of death. These certificates of causes of death are filed at local health centers and are under the jurisdiction of the Ministry of Health and Welfare; the *koseki* are under the jurisdiction of the Ministry of Justice.

An accounting of our success in ascertaining the survival status of the 1,162 children of interest here and of determining the causes of death for those no longer alive follows:

Total number of *koseki* to be located		1,162
Number located	1,153	
No *koseki* exists; family not Japanese	5	
Koseki presumably exists but not located	4	
Number of *koseki* inspected and survival status obtained		1,153
Child alive	1,050	
Child dead	103	
Number of death certificates to be found		103
Cause of death obtained	102	
Cause of death not obtainable	1	

Clearly, failure of follow-up because of lack of *koseki* can scarcely be regarded as an important source of bias; the four children of Japanese parents whose survival status is unknown are too few to influence the data significantly, even if all the cases involved children no longer alive. Moreover, since all four cases are the offspring of unrelated parents, if the probability of death is greater for the child not traced than the child traced, the effect of inbreeding upon mortality to be reported, small as it is, can only be smaller.

The Effect of Inbreeding on Mortality

Some years ago, Morton, Crow, and Muller (1956) advanced an ingenious genetic argument which asserts that the logarithm of the proportion of individuals, S_i, having coefficient of inbreeding F_i that survive some specified period of risk of death is a linear function of a constant α and some multiple β of the coefficient of inbreeding; that is

$$\log S_i = \alpha + \beta F_i$$

Now it can be shown that maximum likelihood estimates of α and β, say

A and *B*, can be obtained from the simultaneous solution of the equations

$$\sum_i (N_i p_i / P_i) = \sum_i N_i$$
$$\sum_i (N_i p_i F_i / P_i) = \sum_i N_i F_i$$

where N_i, p_i, and F_i are, respectively, the number of observations in the *i*th inbred class characterized by the observed proportion of deaths p_i and the inbreeding coefficient F_i, and where

$$P_i = 1 - \exp - (A + BF_i)$$

The above equations may be solved by trial-and-error or a number of iterative procedures, e.g., the generalized Newton-Raphson technique. In our experience, convergence is generally rapid.

Given certain assumptions (an equilibrium population, nonsynergistic gene action, etc.), the Morton, Crow, and Muller argument asserts that the ratio of B/A will be large if at the majority of loci, or at least a substantial minority, genetic variability is maintained through a balance of recurrent mutation and selection, that is, if the genotype of "optimal" fitness in the Darwinian sense is a homozygote(s). On the other hand, this ratio will be small if genetic variability is maintained primarily through a balance of opposing selective forces or if the nongenetic contribution to mortality is appreciably larger than the genetic, irrespective of how genetic variability may be maintained. Although such a large number of objections have been raised to this argument that its utility must be regarded as doubtful, the ratio of B/A remains a convenient way to summarize inbreeding effects.

Table 1 sets forth the mortality in the first fifteen years of life, on the average, of the 1,153 children whose survival status was determinable. Inspection suffices to reveal no conspicuous relationship between mortality and inbreeding for either males or females or the sexes combined. While mortality among liveborn males is somewhat less than expected on the basis of the *Ninth Japanese Life Tables,* it is not significantly so. Mortality among liveborn females is in very close agreement with these tables; we observe that 9.33% of young females fail to attain their fifteenth birthdays, whereas 9.49% are not expected to do so.

To afford a further insight into the accord between these data and the experiences predicted from the *Ninth* and *Tenth Japanese Life Tables,* we present Table 2. Before we examine this table, however, a remark or two about the life tables seems appropriate. The *Ninth Life Tables* were computed on the basis of the 1950 census of Japan and on the vital statistics from October 1, 1950 to September 30, 1952; the *Tenth Life Tables* stem from the 1955 Japanese census and the vital statistics for the one year following October 1, 1955. In view of the basic populations and the periods of observation, it should be apparent that neither of these tables

TABLE 1. MORTALITY AS ASCERTAINED IN 1964 AMONG CHILDREN BORN ALIVE
IN KURE IN THE YEARS 1948–1950 BY SEX OF CHILD
AND PARENTAL RELATIONSHIP

	Males	Females	Total
Unrelated			
Alive	263	275	538
Dead	26	25	51
Total	289	300	589
Proportion	.0900	.0833	.0866
Second cousins			
Alive	66	53	119
Dead	9	8	17
Total	75	61	136
Proportion	.1200	.1311	.1250
First cousins once removed			
Alive	56	51	107
Dead	3	2	5
Total	59	53	112
Proportion	.0508	.0377	.0446
First cousins			
Alive	140	146	286
Dead	11	19	30
Total	151	165	316
Proportion	.0728	.1152	.0949
Total			
Alive	525	525	1,050
Dead	49	54	103
Total	574	579	1,153
Proportion	.0854	.0933	.0893

describe precisely the mortality to be expected in Kure in the years under
discussion. Furthermore, the life tables represent a synthesis of both
urban and rural mortality risks, and, to the extent that the latter differ
from the former, the life tables fail to represent the risks anticipated in
an urban area. And, finally, in a society such as the Japanese where the
recent trend has been toward lessened risks of death at all ages in the pre-
reproductive years, life tables computed as they are upon the age-specific
mortality rates which obtain in a particular year or group of years fail to
reflect this trend adequately. With due regards for these reservations, it is
of interest that the data reveal the risks of death in Kure in the early years
to be approximately those anticipated on the basis of the *Ninth* and, in the
later years, the *Tenth Life Tables*. Certainly these data afford no basis for
the supposition that infantile and childhood mortality in Kure were per-
turbed by some unusual event or events which render suspect the mortality
experiences of these children, nor for that matter is there any evidence of
such events in the medical annals of the city. Since there is no direct evidence that the data set forth in Table 1 are unrepresentative of mortality
experiences in other urban areas of Japan in the years since 1948, we turn

now to the results of fitting the exponit model, that is, the argument of Morton, Crow, and Muller, to these data.

Estimates of A and B and the ratio of the two for males, for females, and for the sexes combined are as follows:

	A	B	B_F	B/A
Males	.0978	−.3879	−.6871	−3.97
Females	.0886	.4359	.2521	4.92
Sexes combined	.0929	.0405	.1940	0.44

In no instance is the effect of inbreeding statistically significant. The differences between the various values of B illustrate once again how sensitive these indicators are to extraneous sources of variation and/or how large the sampling error may be (Schull and Neel, 1965). Moreover, exclusion of the three cases of accidental death does not materially alter the conclusions these data provoke ($A = 0.0881$; $B = 0.1281$).

As one measure of the internal consistency of the observations from Kure, we present a further set of estimates of B, namely B_F, derived from the data when the outbred class is excluded. Comparison of the mortality observed in the outbred groups with that predicted from the inbred fails to disclose evidence of significant internal heterogeneity in these data. However, it must be borne in mind that the power of the test against alternatives which deviate but slightly from the null hypothesis is not great.

TABLE 2. MORTALITY AS ASCERTAINED IN 1964 AMONG CHILDREN BORN ALIVE IN KURE IN THE YEARS 1948–1950 BY AGE AT DEATH

Years of life completed	Deaths observed			Total deaths expected	
	Outbred	Inbred	Total	Ninth*	Tenth†
0	33	30	63	61.33	45.25
1	7	9	16	11.36	6.14
2	4	5	9	8.62	4.94
3	2	3	5	6.89	3.89
4	—	2	2	5.42	2.99
5	1	1	2	4.21	2.22
6	2	—	2	3.20	1.63
7				2.41	1.25
8				1.78	1.00
9	1	1	2	1.35	.84
10				1.05	.76
11				.89	.73
12	1	1	2	.84	.73
13				.90	.73
14				1.04	.76
Total	51	52	103	111.29	73.86
Males	26	23	49	55.29	39.09
Females	25	29	54	56.00	34.77

Ninth Japanese Life Tables, published 1955.
†*Tenth Japanese Life Tables*, published 1960.

Cause of Death

To complete this picture of mortality and its relationship to inbreeding, we now examine the causes of death among the 103 liveborn infants who failed to survive to their fifteenth birthdays. Information pertinent to the cause of death was available from the death certificate on record with the Public Health Department in Kure. Though not without faullt, these certificates appear of sufficient accuracy to permit assigning the cause of death of the deceased children to broad classes of causes. The results of such classification are to be seen in Table 3. For these purposes, a child was said to have died of prematurity if the birthweight was less than 2500 grams and

TABLE 3. MORTALITY AS ASCERTAINED IN 1964 AMONG CHILDREN BORN ALIVE IN KURE IN THE YEARS 1948-1950 BY REPORTED CAUSE OF DEATH AND PARENTAL RELATIONSHIP

Cause of death	Unrelated	Parental relationship		
		Second cousins	1½ cousins	First cousins
Accident	3			
Birth accident	1			1
Congenital defect	1		1	4
Infectious disease	18	9	3	14
Neoplasia	1	1		
Prematurity	4	2		4
Combinations				1
Other and unknown	23	5	1	6
Total	51	17	5	30

death occurred in the first month postpartum. In point of fact, the vast majority of deaths ascribed to prematurity occurred in the first week of life. "Birth accidents" include primarily birth injuries. "Combinations" refer to deaths attributable to two or more causes. The category "Other and Unknown" encompasses two groups, namely, those deaths where the stated cause defies ready classification and those where no data exist with regard to the cause of death. The former constitute the bulk of this category.

Cause of death cannot be assigned, it will be noted, to even these broad categories in more than one-third of all cases. Clearly, therefore, interpretations of these data must be guarded, especially since relatively more unassigned cases occur among the outbred than among the inbred children. Nonetheless, we call particular attention to the more frequent occurrence of deaths among the inbred attributable to congenital defects and infectious disease. With regard to the former, Schull (1958) has reported a significant association between major congenital malformations and inbreeding in Hiroshima, Kure, and Nagasaki, and, of course, many such defects are incompatible with life. The apparent increase with inbreeding of deaths ascribable to infectious disease is noteworthy in view of a similar effect reported for Hiroshima and Nagasaki (Schull and Neel, 1965).

Discussion

Despite the paucity of relevant observations on man, Morton apparently accepts as established without equivocation the hypothesis that genetic variability is maintained in the majority of instances through a balance between mutation and selection and holds, therefore, that when data which deviate from this concept arise they must of necessity be biased. The least imaginative challenge in the present instance is, of course, to assert that the B/A ratio is spuriously low because of some "environmental disturbance," with one implication being that the intercept, the value associated with the outbred children, has been inflated as a consequence of an unusual degree of nongenetic death. In the earlier data as well as those presented here, it should be noted that those infants whose parents were not related do not have a conspicuously high mortality experience, as judged by the *Ninth Japanese Life Tables*. It would seem, then, that one must conclude that this "environmental disturbance" has depressed the mortality experience among the inbred children. Such could possibly be the case if, for example, socioeconomic status increased with increasing inbreeding, since mortality is apparently inversely related to social status. However, in both Hiroshima and Nagasaki, socioeconomic status *decreases* with increasing inbreeding; presumably the same may be true in Kure. With respect to Morton's attempt (1961) to impute the small effect to population heterogeneity, we have great difficulty understanding why a Japanese population should be stigmatized as heterogeneous when a sample of migrants from a triracial population in Brazil apparently is not (Morton, 1964). The later assertion "that all studies of inbreeding effects in man, even the most careful and expensive, were carried out in heterogeneous populations with little control over sociological variables that might be confounded with consanguinity" (Morton, 1964) is at best obfuscatory and at worst a calculated effort to misrepresent. While we would be the first to admit that in this area of population genetics *no* study, past or present, has controlled *all* sources of extraneous variation nor, for that matter, is any future study apt to do so, some investigators have made more than a little effort to delineate and control possible sociological concomitants (see Schull and Neel, 1962).

In spite of the polarization which seems to have characterized many of the arguments over the role of "classical" and "balanced" loci in the maintenance of genetic variability, the concealed genetic load of a population, i.e. that revealed by inbreeding, patently cannot be wholly a consequence of one or the other kind of loci but must, in fact, be some function of the two. Certain conjectures about the relative contributions of these two types of loci as well as others which can be envisaged can be and have been made from data derived from a study of the effects of inbreeding, but precise estimates are as yet impossible. When taken at face value, the present data in the context of the Morton-Crow-Muller argument may be interpreted as evidence (1) that the contribution of "classical" loci to the maintenance of the genetic load of the Japanese is small, (2) is not small but is obscured by a relatively large nongenetic component, or (3) that the average overdominance, h, is relatively large and hence the ratio criterion fails to discriminate effectively between the contributions of "classical" and "heterotic" loci. In short, within the context of present formulations of the nature of inbreeding effects, the data lead to no firm conclusions. Since we do not feel the B/A observed in Kure differs "significantly" from that observed in Hiroshima and since elsewhere we have argued that the apparent differences in the magnitude of the inbreeding effects observed in all the various studies in Japan to date are not significant and that "there is no convincing evidence that the effects of inbreeding on mortality differ in the major racial groups nor that these effects are large" (Schull and Neel, 1965, pp. 90–113), the utility of the Morton-Crow-Muller argument is, at best, dubious.

Summary

Analysis of the relationship of mortality in the first fifteen years of life to inbreeding in a cohort of 1,153 children born in Kure, Japan, fails to disclose a significant effect of inbreeding on mortality. When taken at face value, a small increase in pre-reproductive mortality is noted with increased inbreeding. These data can be interpreted as evidence contravening the notion that a substantial fraction of the genetic load of the Japanese manifested in death prior to maturity is maintained by "classical" loci, but other interpretations are possible.

ACKNOWLEDGMENTS

This work was conducted under the aegis of the Atomic Bomb Casualty Commission, a co-operative research agency of the U.S. National Academy of Sciences-National Research Council and the Japanese National Institute of Health of the Ministry of Health and Welfare, with funds provided by the U.S. Atomic Energy Commission, the Japanese National Institute of Health, and the U.S. Public Health Service.

REFERENCES

Japanese Life Tables, The Ninth. 1956. Tokyo: Welfare Minister's Secretariat of Japan, Division of Health and Welfare Statistics, pp. 1–13.

Japanese Life Tables, The Tenth. 1960. Tokyo: Welfare Minister's Secretariat of Japan, Division of Health and Welfare Statistics, pp. 1–27.

MORTON, N. E. 1961. Morbidity of children from consanguineous marriages. *Prog. Med. Genet.* 1: 261–291.

MORTON, N. E. 1964. Genetic studies of northeastern Brazil. *Cold Spring Harbor Symposium Quant. Biol.* 29: 69–79.

MORTON, N.E., CROW, J. F., AND MULLER, H. J. 1956. An estimate of the mutational damage in man from data on consanguineous marriages. *Proc. Nat. Acad. Sci.* (U.S.) 42: 855–863.

NEEL, J. B., AND SCHULL, W. J. 1956. *The Effect of Exposure to the Atomic Bombs on Pregnancy Terminations in Hiroshima and Nagasaki.* Washington, D. C.: National Academy of Sciences-National Research Council Publ. 461.

NEEL, J. B., AND SCHULL, W. J. 1962. The effect of inbreeding on mortality and morbidity in two Japanese cities. *Proc. Nat. Acad. Sci.* (U.S.) 48: 573–582.

SCHULL, W. J. 1958. Empirical risks in consanguineous marriages: Sex ratio, malformation, and viability, *Amer. J. Hum. Genet.* 10: 294–343.

SCHULL, W. J., AND NEEL, J. V. 1962. The child health survey: A genetic study in Japan. In *The Use of Vital and Health Statistics for Genetic and Radiation Studies.* New York: United Nations, pp. 171–194.

SCHULL, W. J., AND NEEL, J. V. 1965. *The Effect of Inbreeding on Japanese Children.* New York: Harper and Row.

Extraneous Variation in Inbreeding Studies

NEWTON FREIRE-MAIA and J. B. C. AZEVEDO

Studies on inbreeding effects in man, including those with data on variables not directly related to the problem, such as age distribution, socioeconomic status, illiteracy rates, etc., may fail to disclose the action of some extraneous agents capable of disturbing the results.

Studies on Japanese populations are generally concordant in showing relatively low inbreeding effect and suggesting the unimportance of the mutational model to explain the inbred load. In spite of the careful, detailed, and ample investigations of Schull and Neel (1965), a doubt remains that some environmental disturbance may be responsible for the situation (Morton, 1966).

Since, in each set of data, the inbred subsamples apparently more similar socioeconomically to the control seem to be those showing the lowest inbreeding coefficients, we calculated the values of A and B by using only these two groups of data for some Japanese populations. For estimating B, the simplified formula based on the method of Freire-Maia and Freire-Maia (1964) was used. (See formula 10 in Freire-Maia, 1964.) A is the natural logarithm of the frequency of survivors in the control subsample.

Table 1 shows that, for the data obtained in Japan, the new estimates of B are 3, 12, and 69 times larger than the original values based on the regression technique of Morton et al. (1956). The corresponding values of B/A range from 13 to 31. The estimates of B obtained for Japanese and their descendants living in Brazil range from 0.9 to 2.6 and those of B/A from 6 to 19. One of the new estimates is negative while the other two are substantially larger than those based on the regression technique, namely 6 and 10. The corresponding values of B/A are 44 and 73. For the total, B equals 6 and B/A, 46. Our data on Japanese populations in Brazil include abortions, miscarriages, stillbirths, and infant-juvenile mortality (up to the age of 20 years), whereas only segments of this range are represented in the other data.

On the basis of these new estimates, the inbred load among Japanese may be larger than is now generally admitted. As a matter of fact, it is interesting to recall that one of the highest estimates of B known at present is that based on Japanese data from Kuroshima (Schull et al., 1962). Accord-

Newton Freire-Maia, *a population geneticist, was born in Minas Gerais, Brazil, in 1918. He received his D.Sc. from the University of Rio de Janiero in 1960. Currently he is professor in and head of the Department of Genetics, Federal University of Parana, Brazil. His publications in population genetics deal particularly with consanguineous marriages and the concept of the genetic load, as does his article included here. Much of his data is drawn from extensive field research among various populations in Brazil. He has also conducted genetic investigations in Central Portugal. He is president of the Council of the Brazilian Genetics Society, an editor of numerous genetics journals, and in 1968 he received the Brazilian National Genetics Prize. He is author of many works on population genetics including* Genetica Medica, *with A. Freire-Maia (São Paulo, 1965);* Populações Brasileiras, Aspectos Demográficos Genéticos e Antropológicos, *with F. M. Salzano (São Paulo, 1967); and* Problems in Human Biology: The Study of Brazilian Populations *(Detroit, forthcoming).*

ing to Krieger (1966), these data gave $A = 0.097$, $B = 4.733$, and $B/A = 48.7$.

We are aware of the pitfalls inherent in investigations on inbreeding effect in man not only through the analysis of data obtained among Japanese but also among Caucasians and Negroes. The low inbreeding effect on mortality, which was verified among Caucasians in the south of the State of Minas Gerais in Brazil (Freire-Maia, Freire-Maia, and Quelce-Salgado, 1963; Freire-Maia, 1963), for instance, seems to be due to bias by socioeconomic differentials. A new investigation in the same area, but using siblings as controls, revealed an inbreeding load only a little lower than that verified among non-Caucasians (i.e., Mulattoes and Negroes), both showing the same order of magnitude ($B \sim 3$–4); B/A is roughly 10 in both samples (J. B. C. Azevedo and N. Freire-Maia, unpublished observations).

Parenthetically, it seems important to mention that Krieger (1966) found a high homogeneity of load estimates for postnatal deaths among five

TABLE 1. EFFECTS OF INBREEDING ON JAPANESE POPULATIONS
IN JAPAN AND BRAZIL

Population	Regression technique			Present suggestion		
	A	B	B/A	A	B	B/A
Kure, Japan (1)	0.09	0.04	0.4	0.09	2.75	30.6
Hiroshima, Japan (2)	0.09	0.53	5.9	0.09	1.84	20.4
Nagasaki, Japan (2)	0.10	0.11	1.1	0.10	1.28	12.8
Bauru, SP, Brazil (3)	0.15	0.87	5.8	0.14	6.19	44.2
Tupã, SP, Brazil (4)	0.12	1.43	11.9	0.14	−0.46	−3.3
Assai, Pr, Brazil (4)	0.14	2.59	18.5	0.14	10.27	73.4
BRAZIL, TOTAL	0.14	1.61	11.5	0.13	6.00	46.2

(1) Schull and Neel, 1966; (2) Schull and Neel, 1965; (3) Freire-Maia, Guaraciaba, and Quelce-Salgado, 1964; (4) N. Freire-Maia and N. Takehara (unpublished data).

João Bosco da Costa Azevedo (Federal University of Parano, B.S. and Lic. Sci., 1965), human geneticist, was born in Guidoval, Brazil, in 1941. He is presently affiliated with the Laboratory of Human Genetics, Faculty of Philosophy, Marília, Brazil. To date he has done population-genetics field research on white and non-white populations in Brazil. His publications based on these studies include "Efeitos do endocruzamento sôbre a mortalidade e a morbidade em populações sul-brasileira. Segunda investigação," with F. A. Marçallo, Ciência e Cultura, 15:192 (1963); "Uma reanálise do problema da carga genética em brancos e negros do sul de Minas," with N. Freire-Maia, Ciência e Cultura, 17:156 (1965); "Consanguinidade e solteirismo," Ciência e Cultura, 20:184 (1968); and "Inbreeding and celibacy," Proceedings of the XII International Congress of Genetics (Tokyo), 1:300 (1968).

investigations with sib controls (including our own data), compared with a highly significant heterogeneity among estimates from 15 studies with controls taken at random from the general population. The values of B in the former studies (sib controls) ranged from 1.0 to 2.7 and those of B/A from 8 to 22 (data from Brazil, S. Helena, and the United States). These parameters in the second group ranged, respectively, from -0.5 to 4.7 and from -1 to 49 (data from Brazil, Sweden, Tanganyika, Japan, Germany, the United States, and France).

Analysis of the total data from five surveys in the south of Minas Gerais, including the results of the first two already published, showed Caucasians with low and non-Caucasians with high inbreeding load, but the estimates based only on the control group and on sibships with coefficients of inbreeding equal to or lower than $1/64$, as suggested above, disclosed a different situation. According to this procedure, $A = 0.34$, $B = 4.54$, and $B/A = 13$ for Caucasians and $A = 0.39$, $B = 5.56$, and $B/A = 14$ for non-Caucasians. Although these estimates of B appear to be somewhat inflated on the basis of the majority of the data in the literature, there is no reason to suspect that Caucasians and non-Caucasians have inbred loads of different magnitude. This is a confirmation of the results obtained with the use of siblings as controls, as shown above.

We also would like to point out that it is impossible to obtain a good estimate of the ratio of the genetic load in an inbred population to that in a panmictic one. A rough estimate of this ratio is obtained by dividing B by A, which includes a certain fraction of deaths due to nongenetic factors. As this value is not known, A may be used as a rough approximation in *populations with high standards of living*, where prenatal and infant-juvenile deaths have relatively low incidence and may be largely due to genetic factors. In the Brazilian regions we studied, however, these events are known to be primarily due to the poor conditions of living. The ratio B/A may be highly misleading in such populations, if taken at face value, due to the enormous error introduced in A, which is much larger than the genetic damage in a random mating population. Therefore, the ratio of the inbred load to the outbred load is expected to be much larger than the crude B/A ratio. (See, for instance, Freire-Maia, 1960.)

It is not the intention of this note to conclude that estimates of A and B based only on the two above mentioned subsamples are *always* more reliable than all the others suggested by Morton, Crow and Muller (1956), namely B, B_{FO}, B_{FF}, and B_{CO}. Our purpose is only to call attention to the possibility that, at least in some situations, they may be based on less biased data.

Due to the large errors of the estimates presented here and based on our suggestion, these estimates are not statistically different from those based on the classical MCM method. We suggest, therefore, that efforts be concentrated on the accumulation of data on the lowest inbreeding levels easily detectable in man, namely those with values of F equal to or a little lower than $1/64$, in order to obtain estimates of lethal equivalents with a larger precision than those reported here.

From the above words, the reader will correctly conclude that the interest of this note rests much more on the suggestion of a new approach than on the results already obtained with it.

ACKNOWLEDGMENTS

The Laboratory of Human Genetics is supported by grants from the Brazilian National Research Council, the Rockefeller Foundation, the Research Council of the Federal University of Paraná, and the Gulbenkian Foundation.

REFERENCES

FREIRE-MAIA, N. 1960. Deleterious mutations in man. *Eugen. Quart.* 7: 193–203.

FREIRE-MAIA, N. 1963. The load of lethal mutations in white and Negro Brazilian populations. II. Second survey, *Acta Genet. Stat. Med.* (Basel) 13: 199–225.

FREIRE-MAIA, N. 1964. On the methods available for estimating the load of mutations disclosed by inbreeding. *Cold Spring Harbor Symp. Quant. Biol.* 29: 31–40.

FREIRE-MAIA, N. AND FREIRE-MAIA, A. 1964. Estimate of the genetic load disclosed by inbreeding. *Genetics* 50: 527–529.

FREIRE-MAIA, N., FREIRE-MAIA, A., AND QUELCE-SALGADO, A. 1963. The load of lethal mutations in white and Negro Brazilian populations. I. First survey. *Acta Genet. Stat. Med.* (Basel) 13: 185–198.

FREIRE-MAIA, N., GUARACIABA, M. A., AND QUELCE-SALGADO, A. 1964. The genetical load in the Bauru Japanese isolate in Brazil. *Ann. Hum. Genet.* (Lond.) 27: 329–339.

KRIEGER, H. 1966. *Inbreeding Effects in Northeastern Brazil.* Ph.D. Thesis, University of Hawaii.

MORTON, N. E. 1966. Review of *The Effects of Inbreeding on Japanese Children*, by W. J. Schull and J. V. Neel. *Eugen. Quart.* 13: 276–278.

MORTON, N. E., CROW, J. F., AND MULLER, H. J. 1956. An estimate of the mutational damage in man from data on consanguineous marriages. *Proc. Nat. Acad. Sci.* (U.S.) 42: 855–863.

SCHULL, W. J., AND NEEL, J. V. 1965. *The Effects of Inbreeding on Japanese Children*. New York: Harper and Row.

SCHULL, W. J., AND NEEL, J. V. 1966. Some further observations on the effect of inbreeding on mortality in Kure, Japan. *Amer. J. Hum. Genet.* 18: 144–152.

SCHULL, W. J., YANASE, T., AND NEMOTO, H. 1962. Kuroshima: The impact of religion on an island's genetic heritage. *Hum. Biol.* 34: 271–298.

Natural Selection

Overview

If then, animals and plants do vary, let it be ever so slightly or slowly, why should not variations or individual differences, which are in any way beneficial, be preserved and accumulated through natural selection, or the survival of the fittest? (Charles Darwin, *The Origin of the Species*)

Thus, in 1859, Darwin set forth his revolutionary theory that all species contain individuals who differ, and that the *variation* within any species is the material upon which *natural selection* operates. In his classic work he suggested that the "natural selection of numerous successive, slight, favorable variations" produces evolutionary changes within a species which in turn result in "intermediate gradations" between species rather than radical differences. Actually, the naturalist Alfred Russel Wallace, then working in Malaya, had independently proposed the identical theory in an essay sent to Darwin in 1858. At a meeting of the Linnean Society in 1858, the work of both men—Wallace's 1858 essay and Darwin's 1844 memoir in which he originally proposed his ideas concerning evolution—was presented jointly.

Darwin named the agent by which selection operates "survival of the fittest," referring by the phrase to the *differential mortality* of individuals within a species. His concept has since been broadened to include the selective agent of *differential fertility*. Differential mortality operates on individuals prior to reproductive age, and therefore determines that group of individuals who survive and may potentially produce the offspring who will constitute the next generation of a population. Differential fertility operates within this surviving group

187

of individuals who have reached reproductive age. It refers to the differences in the contribution made to the next generation by individuals due to their inequalities in reproductive performance. Although differential mortality and differential fertility operate through separate means and on different age groups, their ultimate genetic consequences for a population are much the same. For example, a person who survives to the age of 100 but who never reproduces contributes no more genetically to the next generation than the infant who dies in the first year of life. In this example, selection operates against the infant through differential mortality and against the centenarian through differential fertility, and each is "genetically dead" for the population.

Determining Darwinian Fitness

Selection is based on the differences in fitness among individuals. The so-called *Darwinian fitness* of an individual is measured both by his capacity to survive and his capacity to leave descendents. Ideally, a particular fitness value (or adaptive value) of an individual should refer to his whole genotype, but this ideal measurement remains highly theoretical. In practice, quantification of total genotypic fitness is impossible, and most estimations of selection in human populations are done on the basis of simple individual-gene models involving the so-called "marginal genotype" and changes in allele frequencies at a single locus in a gene pool. Marginal genotype is defined as "the genetic endowment of one or more individuals or gametes at the locus under consideration" (Mettler and Gregg, 1969).

In any population, the various genotypes, with respect to a single locus, may differ in their Darwinian fitness; that is, one genotype may have an advantage over any other, either due to high fertility or lower mortality. Whatever the cause, the advantage may be measured in terms of the relative number of offspring produced by all the genotypes at that locus born in a single generation. The numerical ratio based on the number of offspring produced by the different genotypes is the relative "Darwinian fitness," "survival value," or "adaptive value" of the class of genotype being considered. This value is often represented by the symbol W. The complementary factor of W is the *selective coefficient* (s) which is a measurement of the intensity of selection operating against the genotype in question. Thus, $W = 1 - s$.[1]

---

[1]Using data from the classic study done by Ernst Mørch (1941) in Denmark on the growth anomaly called chondrodystrophic dwarfism, the relative Darwinian fitness (W) of the dwarfs may be calculated as follows:

The total number of chondrodystrophic dwarfs, living and dead, for whom records were available at the time of Mørch's study was 108. These 108 dwarfs produced 27 children. Because the anomaly is due to the presence of a dominant gene, one-half of the 27 offspring, or 13.5, could be expected to be dwarfs. (The expected and observed number of dwarf offspring were not significantly different.) Of the 108 abnormal alleles which could

The average fitness ($\overline{W}$) of a population is the sum over all genotypes of the relative fitness at each locus times the frequency of each. This average fitness is actually a population's reproductive rate, and $\overline{W}$ has to be close to 1 over any extended period of time. If it were not, the population would become either extinct from loss of numbers or infinite from unlimited increase.

Interaction of Selection and Mutation

Unlike Darwin, we now know that in any population there are genotypes that differ from one another and that the differences are due to mutation. Briefly stated, mutation provides the raw material of evolution in the form of alleles, and selection then determines the fate of these alleles in a gene pool. The interaction of mutation and natural selection in populations was originally formulated mathematically in the work of Ronald Fisher (1930), and was subsequently elaborated by the formal geneticists Sewall Wright (1931) and J. B. S. Haldane (1932). Fisher's original theorem states, "The rate of increase in fitness of an organism at any time is equal to its genetic variance in fitness at that time." Assuming that all genetic variance originates with mutation, Fisher calculated the mathematical probabilities that a mutation might be established in the gene pool of a population. Under his formulation, harmful mutants are constantly eliminated or maintained at low frequencies by natural selection, but advantageous mutants, often effecting imperceptible changes in the organism, are incorporated into a population's gene pool at a regular rate which can be mathematically defined. Over time an advantageous mutation may eventually either *displace* the former allele(s) or achieve a state of *equilibrium* with its allele(s) through the action of opposing, balanced selective forces acting on the locus in question.

Selection may therefore operate as both a *stabilizing* and a *dynamic* force in a population. In either role it exerts systematic pressure on the

---

have been passed on, approximately 13.5 were transmitted; therefore, the proportion of alleles still in existence after one generation was 13.5:108 or 0.125. The reproductive performance of the dwarfs was compared to that of their normal sibs to obtain the dwarfs' relative fitness (W). The 108 chondrodystrophics had 457 normal sibs who themselves had a total of 582 children, living and dead. The potential number of normal alleles, that is, $2 \times 457 = 914$, had therefore, decreased in the ratio of 582:914, or 0.6368. Thus, the relative Darwinian fitness of the allele for chondrodystrophic dwarfism is:

$$W = \frac{0.125 \text{ (dwarfs)}}{0.6368 \text{ (normals)}} = 0.1963.$$

In summary, relative to the normal alleles, only approximately 20 percent of all the chondrodystrophic alleles present in one generation were transmitted to the next, approximately 80 percent were eliminated. The 80 percent is the selective coefficient (s), and is a measure of the intensity of selection operating against the dwarfs.

(Because the original figures, as presented by Mǿrch, were subject to error for various reasons [see Slatis, 1955], the calculated selective coefficient and Darwinian fitness are not exact. The data are given merely to illustrate the method for calculating s and W.)

gene pool so that the direction and magnitude of genetic changes are determinate in principle (see Chapter I). As a stabilizing agent it both eliminates deleterious mutations each generation (*normalizing selection*) and maintains a gene pool's existing allele frequencies through time in equilibrium with the environment (*stabilizing selection*). Simultaneously, selection is also an important dynamic agent of evolution, changing a population's genetic composition as the environment changes (*directional selection*). A disadvantageous allele may be maintained at a very low frequency in a gene pool (generally 1 percent or less) by the balanced opposing forces of normalizing natural selection and recurrent mutation. An advantageous allele may increase in a gene pool and attain a frequency greater than that which can be maintained by recurrent mutation alone. If the latter situation occurs, the result is *genetic polymorphism*, which is defined as "the occurrence together in the same habitat at the same time of two or more distinct forms of a species in such proportions that the rarest of them cannot be maintained merely by mutation" (Ford, 1953).

Genetic Polymorphism

Genetic polymorphism may be *transient* or *balanced*. In the case of transient polymorphism natural selection is operating as a dynamic agent of evolutionary change. An advantageous mutant gradually displaces the normal allele(s) in the gene pool until the latter is reduced to a low frequency maintained by recurrent mutation alone. At this point, of course, by definition a polymorphism no longer exists at that particular locus, and selection is of a normalizing type. Until this point is reached, however, a polymorphism does exist.

Balanced polymorphism, on the other hand, is the result of natural selection operating as a stabilizing agent. In certain situations, an advantageous mutation may increase in the gene pool to a specific frequency (greater than 1 percent and less than 99 percent) where it stabilizes, neither increasing nor decreasing in frequency through subsequent generations (see Table IV-1). With respect to that locus, the gene

TABLE IV-1. SICKLE-CELL MODEL FOR SELECTION
IN A HYPOTHETICAL POPULATION

| | Phenotypes | | | |
	Normal	Sickle-cell trait	Sickle-cell anemia	Totals
Genotypes	Hb-A/Hb-A	Hb-A/Hb-S	Hb-S/Hb-S	—
Initial proportions	0.64	0.32	0.04	1.00
Selective coefficient (s)	0.25	0	1.00	—
Relative fitness ($1 - s$)	0.75	1.00	0	—
After selection	0.48/0.80	0.32/0.80	0	0.80
New proportions	0.60	0.40	0	1.00
Expected offspring after random mating	0.64	0.32	0.04	1.00

pool is in a state of stable equilibrium called balanced polymorphism, and the specific allele frequencies tend to be restored to equilibrium whenever they are disturbed. This state is reached because at specific gene frequencies, whatever they may be, the opposing selective forces acting on the locus in question are balanced. As long as the selective forces do not change, the gene frequencies will not change. In this way genetic diversity in a population which does not depend on recurrent mutation is maintained.

The simplest case of balanced polymorphism is that which is maintained by the selective advantage of the heterozygotes over either of the homozygotes. Table IV-1 depicts the much cited situation in which the heterozygous sickle-cell trait has a selective advantage over either of the homozygotes, normal or sickle-cell anemia, at the hemoglobin locus. Since fitness is relative, the fitness of the most favored genotype (here the heterozygote Hb-A/Hb-S) is set at 1.00. In this hypothetical case, all the abnormal homozygotes (Hb-S/Hb-S) die of severe anemia ($s = 1.00$), and, relative to the heterozygotes, 25 percent of the normal homozygotes (Hb-A/Hb-A) die of malaria ($s = 0.25$). Under the specific conditions postulated here, the gene pool is in equilibrium at the hemoglobin locus, but in practice the fitness of the normal homozygote is very difficult to estimate. Undoubtedly it differs from population to population, and the gene pools in question may or may not be in equilibrium at this locus.

Genetic polymorphisms are very common phenomena in all human populations. The modern techniques of biochemistry are now revealing how widely individual men and populations differ in the various enzymes and proteins of the body. In his article included in this chapter, the biochemical geneticist Harry Harris reviews some of the numerous biochemical variants present in human populations. At some loci there is one allele that can be regarded as a normal form which is almost universally present and of which variants are extremely rare. At other loci a universal normal gene is also present, but variant alleles are present in some populations at sufficient frequencies to constitute genetic polymorphisms. Harris has electrophoretically screened European and Negro populations for 18 different enzyme systems, and has found that about one-quarter of the enzymes in his arbitrarily selected series exhibit polymorphism in each population group examined.

Natural selection, especially as it operates through the advantage of the heterozygote, is generally postulated as being responsible for the existence of the majority of human polymorphisms. But for the most part it has not been proven operative in human populations, the above case of the abnormal hemoglobins being the notable exception. Rather, one is led to conclude that the nature of the forces operating to maintain the numerous genetic polymorphisms in man must be more complex than heterozygosity alone. For example, Harris is of the opinion that the incidence and distribution of alleles in populations depend not only on the rate of mutation and differential selection, but also on *chance factors* ("drift") which may fortuitously lead to the elimination or the spread of particular alleles within a gene pool. However, the idea that drift could be of any great importance in evolution per se has been deprecated primarily on the assumption that almost no mutations are neutral. But recently the Japanese geneticist M. Kimura (1968) has calculated the rate of evolution in terms of nucleotide substitutions. His estimated rate is much higher than has been traditionally accepted, and Kimura concludes that many of the mutations involved must be neutral or nearly neutral ones which have been established in finite populations by chance or drift. The idea of neutral mutations contradicts classic evolutionary theory, which stipulates that all mutations are either lethal, deleterious, or adaptive to an organism. Yet as the biochemical evidence accumulates, this chance mode of evolution, so-called "non-Darwinian evolution," is finding more acceptance in scientific circles (cf. King and Jukes, 1969).

However, the classic concept of mutation and natural selection as elaborated and synthesized by Fisher, Wright, and Haldane continues to provide the framework within which genetic change in populations may be theoretically predicted. But a large gap exists between this mathematical-deductive theory of natural selection and the documentation of its actual operation in natural populations. "Evolution in process," that is, the observed displacement through time of a typical (wild) form by a mutant form within a population, has been recorded in laboratory populations, particularly those of rapidly reproducing organisms such as bacteria, viruses, and *Drosophila.* Yet, few evolutionary changes in natural populations have actually been observed and recorded. The most famous case of "evolution observed in process" is Ford's (1953) documentation of the dramatic change in the moth populations living in the industrial areas of England from a typical white form to a melanic (*carbonaria*) form. Field studies showed that in an environment of increasing smoke and soot the melanic moths were protected by their coloring from predatory birds, and in some areas the better adapted mutant *carbonaria* form attained a frequency of over 90 percent in the population. It might be predicted that these populations will eventually approach 100 percent melanic, and the polymorphism for coloration will have been lost. The documented spread of the mutant form through the moth populations provides a unique example of natural selection observed to be operating dynamically, as directional selection, to produce a transient polymorphism as the populations genetically change in response to a changing environment.

Evolution is difficult to observe and record in any natural population, and demonstration of its actual operation in human populations is practically impossible. The time span of a human generation, approximately thirty years, and the consequent lack of vital information covering more than two or three generations in any population, combine to make the conclusions regarding the actual operation of natural selection in human populations strictly inferential. The lack of time depth in human studies accounts for insurmountable difficulties in the interpretation of genetic data pertinent to natural selection. For example, the population geneticist is never certain whether the polymorphism he observes is a balanced or a transient one; only observations over many generations would provide the information necessary to remove the uncertainty.

The study of natural selection in human populations is further stymied by the almost total lack of knowledge about the actual nature of the selective pressures acting on man. In 1798, Thomas Malthus proposed that the "vices of mankind are active and able ministers of depopulation" (referring to "discouragements to marriage," the "depopulation of large towns," and the like); that "sickly seasons, epidemics, pestilence, and plague advance in terrific array, and sweep off their thousands and ten thousands [of people]"; and that famine is "the last, most dreadful resource of nature." Today, our enumeration of the actual selective forces operating on mankind has not extended much past Malthus's original ones, which may be rephrased as those aspects of human behavior which influence reproduction, disease, and malnutrition and starvation. Along these lines, Arno Motulsky discusses in the article included here the probable important selective role disease may have played historically in the establishment of some of the human polymorphisms and the evolution of human populations. For several reasons, however, any conclusions regarding the exact genetic impact disease may have had must remain speculative. First, there is a lack of information on past populations in the form of vital records, genetic data, and so on. Second, disease acts on the total organism, and the specific underlying genetic factors that make some people susceptible and others immune to specific diseases have not been determined (see Harris article below).

The Malarial Hypothesis

The case described by Motulsky of the various red-cell polymorphisms which are present in numerous human populations is a unique one, because the probable nature of the selective pressure maintaining these polymorphisms is partially understood. In 1954, the English geneticist Anthony Allison first examined the *malarial hypothesis;* that is, that the red-cell polymorphism of normal Hemoglobin A and abnormal Hemoglobin S, present in many populations of Africa, is a balanced polymorphism maintained by the selective advantage of the heterozygote (sickle-cell trait) in the presence of *hyperendemic* or *holoendemic* falciparum malaria. Since Allison's original publication, numerous studies on the population distribution of other red-cell abnormalities such as Hemoglobins C and E, glucose-6-phosphate dehydrogenase (G-6-PD) deficiency and thalassemia have provided further evidence that these polymorphisms also are linked to hyperendemic or holoendemic malaria (Motulsky, 1964).

Yet even for this widely accepted case, the evidence supporting the malarial hypothesis, however convincing, is indirect. The evidence is of two types: (1) the overlapping geographical distribution of the abnormal red-cell polymorphisms with past or present endemic malaria, and (2) the advantage of the heterozygote in a malarial environment. Each of the three articles reproduced here, by Motulsky, Siniscalco, *et al.*, and Wiesenfeld, deals with the evidence supporting the malarial hypothesis.

With reference to the first type of evidence, Motulsky presents data and distribution maps showing the overlapping world distribution of the various red-cell polymorphisms and endemic malaria. Siniscalco and his colleagues cite similar data, but as they relate to a very small circumscribed area, the island of Sardinia. Their study is an excellent example of the type of field work that must be done to test the malarial hypothesis in human populations. They were able to show a negative correlation in Sardinia between altitude and frequencies of G-6-PD deficiency and the thalassemia trait. They conclude that "malignant malaria is the obvious ultimate factor" in that the correlation is positive when population gene frequencies are compared with both the former incidence of malaria mortality in an area and malarial parasite rates in school children now living at various altitudes.

To the overlapping geographical distribution of malaria and the sickle-cell trait in Africa, Wiesenfeld adds a third element; that is, the distribution of the Malaysian agricultural complex. His hypothesis is that the spread of slash-and-burn agriculture throughout parts of Africa created the conditions for development of endemic malaria, which in turn put more intense selective pressure on populations, thereby increasing the frequencies of the sickle-cell trait. Wiesenfeld concludes that the dynamic interactions of disease, genetics, and agriculture account for their widespread association in African populations. In summary, with respect to the first type of evidence, as Allison (1963) has stated, "The only convincing argument from gene distribution against the malaria protection hypothesis would be the finding of high frequencies of abnormal hemoglobin among populations known to have lived for long periods in malaria-free environments." This situation has not yet been reported, and as more genetic data is collected from tropical and subtropical populations, the distribution evidence continues to support the malarial hypothesis.

The second type of evidence, the advantage of the heterozygote, derives from two kinds of data. First, the clinical picture of a lower parasite density in sickle-cell-trait carriers, which in turn is associated with lower morbidity and mortality from malaria, supports the thesis that the heterozygote is afforded protection against malaria (Allison, 1961). Similar evidence has also been reported for G-6-PD deficiency where the malaria-parasite rate in normal homozygous females is 2 to 80 times higher than that in heterozygous females (Luzzatto, Usanga, and Reddy, 1969). Second, the hypothesis is further supported by the differential age distribution of the hemoglobin genotypes in a population exposed to endemic malaria (see Table IV-2). In the case of Hemoglobin

TABLE IV–2. FREQUENCIES OF HEMOGLOBIN GENOTYPES WHERE GENETIC EQUILIBRIUM IS DETERMINED BY DIFFERENTIAL MORTALITY FROM MALARIA*

Genotype	Offspring frequency	Adult frequency	Fitness
	Expected	Expected	
Hb-A/Hb-A	0.6529	0.616	0.943
Hb-A/Hb-S	0.3103	0.384	1.238
Hb-S/Hb-S	0.0367	0	0
	Observed	Observed	
Hb-A/Hb-A	0.659	0.616	0.935
Hb-A/Hb-S	0.310	0.384	1.239
Hb-S/Hb-S	0.031	0	0

*Expected frequencies of normal sickle-cell heterozygous and sickle-cell homozygous subjects in newborn and adult populations, assuming that the Hb-S gene frequency is 0.19, and that equilibrium is maintained through differential mortality alone. The observed values which are from the Musoma district of Tanganyika do not differ significantly from the expected.

Source: Modified from Allison, 1965.

S, age-stratified studies have shown that the proportion of sickle-cell-trait carriers is progressively greater in groups of advancing age (Motulsky, 1964). This is the picture to be expected if the selective agent is differential mortality in childhood rather than differential fertility of adults. Differential childhood morbidity and mortality is the vital pattern characteristic of a population affected by hyperendemic or holoendemic disease. In such a population the surviving adults would have already been exposed and acquired some degree of disease immunity. In contrast, an epidemic disease affects all age groups if the population has not been previously exposed to the disease.

Yet even these data, although externally convincing, elicit some uncertainty with regard to explanation and interpretation. For example, the exact mechanism by which heterogzygous possession of an abnormal red cell or red-cell enzyme provides protection against malaria is unknown. Motulsky's suggestion that an abnormal cell might in some way provide a less satisfactory environment for the growth and survival of the malarial parasite remains a plausible but purely speculative general explanation for the biological advantage of the heterozygote. Furthermore, the observed differential age distribution of genotypes would occur if the sickle-cell trait conferred an advantage in the presence of *any* constant selective pressure, not necessarily malaria. Thus, in the absence of other substantiating data, the evidence relating to possible heterozygote advantage cannot be used exclusively to prove the malarial hypothesis.

Even in the simplest case of a balanced polymorphism maintained by the selective advantage of the heterozygote, the underlying factors which create the total situation in which the heterozygote becomes and remains the most "adapted" genotype may be very complex. For example, the general malarial hypothesis leaves unanswered several questions regarding the totality of the ecological situation within which a red-cell mutation such as *Hb-S* can reach so high a frequency (in some tribes of West Africa 40 percent of the population are sickle-cell-trait carriers). Actually, the history of the human red-cell polymorphisms is the history of endemic malaria, which in turn is the history of the "cultural evolution" of human populations.

In his article in this chapter, Stephen Wiesenfeld develops a series of mathematical models to reconstruct the probable historical dynamics of the three interdependent variables—genetics, desease, and culture—which have interacted to produce the situation we observe today in West Africa. His hypothesis is "that the development and differentiation of the Malaysian agricultural system is intimately bound to changes in the gene pool of populations using this agricultural system." In part, his conclusions are similar to those originally set forth by the physical anthropologist Frank Livingston (1958); that is, that the development and spread of slash-and-burn agriculture in Africa created the environmental conditions necessary for the presence and spread of hyperendemic malaria. Mortality from malaria then put increased selective pressure on the gene pools of those agricultural populations. Under this pressure the mutant allele Hb-S increased in frequency as a result of the advantage it conferred to the heterozygous carrier. Wiesenfeld carries Livingston's argument one step further. Using a negative-feedback mathematical model, he demonstrates that with increasing frequencies of the sickle-cell trait a partial reduction in malaria is achieved by the reduction of the proportion of people capable of being fatally infected. The presence of more immune people means the release of more human energy into the futher development and maintenance of agriculture to meet the demands of an increasing population. Selection in this case appears to be dependent on the two variables: the frequency of the sickle-cell-trait heterozygotes and the size and density of the population.

As previously discussed in Chapter II, the size and structure of a population constitute an integral part of the total environment within which a gene pool exists and evolves. In order to understand the genetic evolution of populations, it is necessary to understand the various factors influencing the size and structure of population. From our knowledge of past and present human populations we know that agriculture is a factor which has had a tremendous impact on population size and density. It has been estimated that a hunting-gathering subsistence economy, such as that practiced by South African Bushmen or Australian Aborigines, could support an average population of 8.0 persons per 100 square miles. By contrast, agriculture, as practiced in a village-farming economy, provides for an average density of 2,500 persons per 100 square miles (Braidwood and Reed, 1957). Such population expansion, made possible by agriculture, is important in determining the fate of mutations in a gene pool. In expanding populations, larger numbers of people provide increased chances for mutation and the ultimate survival of the mutant (Kojima and Kelleher, 1962). The introduction and spread of agriculture in tropical and subtropical areas provided ideal population conditions for the occurrence and survival of a random mutation, such as *Hb-S*.

Importantly, however, other selective pressures derive from the increased population density of agricultural populations, which make the situation of the abnormal hemoglobin polymorphism more complex than that maintained by infectious disease alone. For example, populations which depend on a staple-carbohydrate crop also experience severe nutritional stresses in the form of protein-food deprivation. The deprivation is especially important during early childhood when the nutritional disease of protein-calorie malnutrition (PCM) may be responsible for approximately 40 percent of all mortality in the first four years of life (Wills and Waterlow, 1958). Where malaria and malnutrition are found together, they may combine to cause the death of a child when neither one acting alone would have been a fatal agent. It has been estimated that in Gambia, for example, the combined effects of malaria and malnutrition result in a mortality of about 50 percent during the first four years (Wills and Waterlow, 1958). It is quite possible that under such circumstances the heterozygous sickler with his lower malarial-parasite load is better able to survive the nutritional stresses than the normal homozygote who suffers additionally from a dense parasite load and therefore a more severe case of malaria. In this manner, disease and nutritional stresses, both the result of agriculture and increased population density, may act in conjunction as extremely effective selective agents, killing millions of children in the tropical and subtropical populations of the world each year.

The assumed relationship between natural selection and the other human polymorphisms is even more difficult to prove than the situation just described, primarily because the selective forces involved have not been identified. For example, the numerous extensive investigations undertaken on the human blood-group polymorphisms in attempts to demonstrate the relevant selective forces at work only serve to illustrate the extreme complexity of the situation facing the investigator. There is no doubt that selection operates in the form of immunological incompatibility in the cases of the Rh, ABO, and possibly the MN blood-group systems, but in some cases it is the heterozygote that appears to be at an advantage, while in others it is the homozygote (Newcombe, 1963; 1965). Furthermore, associations between the blood groups and various chronic and internal diseases have been shown to exist, but for the most part these diseases cause morbidity and mortality after reproductive age. Therefore, these proven associations have been of little value in identifying the role of such diseases as effective selective agents (see Mourant, 1959). However, other data showing a possible relationship between the ABO blood-group polymorphism and infectious disease is more promising. For example, in a smallpox epidemic in India, individuals of blood groups A or AB experienced a significantly more severe course of the disease and higher mortality than individuals of blood groups O and B (Chakravartti, *et al.*, 1966). In this case it seems that epidemic-disease mortality may be related to the allele frequencies at the ABO locus.

Differential Mortality and Fertility

It is probable that natural selection operating through differential mortality is less important among modern populations where differential fertility appears to be the more effective agent. Crow (1958) has devised a formula for calculating an index of the total selection intensity (I_t) of any population. Crow's total index has two components: I_m, due to mortality prior to reproductive age; and I_f, due to differences in fertility among women who have reached reproductive age.[2] Application of Crow's formula to the demographic data of a population provides a rough measure of the relative selective importance of differential mortality versus differential fertility in that population. Following Crow's method, Spuhler (1962) measured selection in three mutually exclusive population categories, namely, tribal, state, and national (see Table IV–3). Of the ten populations included in each category, five in the tribal, two in the state, and one in the national showed an index of mortality *higher* than the index of fertility. The data appear to confirm that differential fertility is the more important

---

[2] The formula used for calculation of I_t is

$$I_t = I_m + (1/p_s)I_f,$$

where $I_m = p_d/p_s$; and p_d = number of individuals counted at birth who die before reproductive age; and $p_s = \bar{x}/\bar{x}_s$, number of individuals who survive reproductive age and have varying numbers of progeny; $\bar{x}$ = over-all mean number of births and $\bar{x}_s$ = mean number of births per surviving parent; and where $I_f = V_f/\bar{x}_s^2$; and V_f = variance due to fertility.

TABLE IV-3. SELECTION INTENSITY IN SELECTED POPULATIONS SHOWING MAXIMUM AND MINIMUM I_m AND I_f

Tribal Populations	I_m	I_f	I_f/p_s	I_t
1. Bosogo Bantu*	1.778	0.141	0.392	2.170
2. Yao, Nagasaland	0.190	0.349	0.415	0.605
3. Ramah Navaho	0.374	1.572	2.159	2.533
4. Yakö	0.377	0.211	0.290	0.667
State Populations				
1. Bengali Villages	0.456	0.217	0.316	0.772
2. Trinidad and Tabago	0.127	0.875	0.987	1.114
3. Malta and Gozo	0.202	1.028	1.236	1.438
4. Hutterites	0.218	0.136	0.166	0.384
National Populations				
1 & 4. Mexico†	0.490	0.613	0.915	1.405
2. England and Wales	0.036	1.210	1.254	1.290
3. Switzerland	0.062	1.496	1.588	1.650

*The populations are listed in the following order: (1) Maximum I_m, (2) Minimum I_m, (3) Maximum I_f, and (4) Minimum I_f.
†Mexico has both the maximum I_m and the minimum I_f.

Source: Modified from Spuhler, 1962.

agent of natural selection, but definitive conclusions based on such data must be made cautiously for several reasons.

Reliable demographic data are lacking for most populations of the world, and the few figures available are usually very rough approximations. Furthermore, strict genetic interpretation of the indices is impossible in that selection is effective only to the extent that the differential mortality or fertility is genetically determined, whereas the indices, I_m and I_f, are derived from data in which both genetic and nongenetic elements are represented. Crow's index therefore does not provide an accurate estimate of the intensity of natural selection operating on a population. Yet, if they were based on sound demographic data, the two indices should give an indication of the relative *opportunity* provided by the mortality and fertility of a population for the operation of natural selection. In the included article, the demographer Dudley Kirk uses the United States vital statistics in this manner to assess the possible genetic consequences of the changes in mortality and fertility which have occurred within the last century in this country. He concludes that the dramatic reduction of mortality and fertility has resulted in a relaxation of selection intensity never before experienced by man, one that should have far-reaching, albeit unknown, effects on the genetic structure of future generations.

Kirk's data and conclusions are relevant to a question that always arises: Is man still evolving under the influence of natural selection? The answer is *affirmative* and derives from the obvious fact that man continues to modify his environment, often radically, and in so doing continually changes the selective forces operating on his gene pools. Migration to and settlement in new environments; control and elimination of disease; effective control of fertility; introduction of probable new agents of morbidity and mortality, such as air pollution, insecticides, and chemical food additives, are all hallmarks of modern man. Inasmuch as man's actions can and do affect the selective pressures operating on his gene pool, the predictable result is the continuing evolution of *Homo sapiens* now and in the future.

Human Polymorphism

See the following reading: "Enzyme and Protein Polymorphism in Human Populations," by Harry Harris, p. 202.

Polymorphism and Differential Mortality

See the following readings: "Metabolic Polymorphisms and the Role of Infectious Diseases in Human Evolution," by Arno G. Motulsky, p. 222; "Population Genetics of Haemoglobin Variants, Thalassaemia and Glucose-6-Phosphate Dehydrogenase Deficiency, with Particular Reference to the Malaria Hypothesis," by M. Siniscalco, L. Bernini, G. Filippi, B. Latte, P. Meera Khan, S. Pomelli, and M. Rattazzi, p. 253; and "Sickle-Cell Trait in Human Biological and Cultural Evolution," by Stephen L. Wiesenfeld, p. 273.

Differential Mortality and Fertility

See the following reading: "Patterns of Survival and Reproduction in the United States: Implications for Selection," by Dudley Kirk.

REFERENCES TO THE LITERATURE CITED IN THE OVERVIEW

1. ALLISON, A. C., 1961. Genetic factors in resistance to malaria. *Ann. N.Y. Acad. Sci.*, 91:710–729.
2. ———, 1963. Malaria and G-6-PD deficiency. *Nature*, 197:609.
3. ———, 1964. Protection afforded by sickle-cell trait against subtertian malarial infection. *Brit. Med. J.*, 1:290–294.
4. ———, 1965. Population genetics of abnormal haemoglobins and glucose-6-phosphate dehydrogenase deficiency, p. 365–391. In G. H. P. Jonxis (ed.), *Abnormal Haemoglobins in Africa*. Oxford, Blackwell Scientific Publications.
5. BRAIDWOOD, R. J. AND C. A. REED, 1957. The achievement and early consequences of food production: A consideration of the archeological and natural-historical evidence. *Cold Spr. Har. Sympos. Quant. Biol.*, 22: 19–31.
6. CHAKRAVARTTI, M. R., B. K. VERMA, T. V. HANURAV, AND F. VOGEL, 1966. Relation between smallpox and the ABO blood groups in a rural population of West Bengal. *Humangenetik*, 2: 78–80.

7. CROW, J. F., 1958. Some possibilities for measuring selection intensities in man. *Hum. Biol.*, 30: 1–13.

8. DARWIN, C., 1960. *The Origin of Species.* New York, Mentor Books.

9. FISHER, R. A., 1930. *Genetical Theory of Natural Selection.* Oxford, Clarendon Press.

10. FORD, E. B., 1953. The genetics of polymorphism in the Lepidoptera. *Advanc. Genet.*, 5: 43–87.

11. HALDANE, J. B., 1932. *The Causes of Evolution.* London, Longmans Green.

12. KIMURA, M., 1968. Evolutionary rate at the molecular level. *Nature*, 217: 624–626.

13. KOJIMA, K. AND T. KELLEHER, 1962. Survival of mutant genes. *Amer. Nat.*, 96: 329–346.

14. KING, J. L. AND T. H. JUKES, 1969. Non-Darwinian evolution. Science, 164: 788–798.

15. LIVINGSTON, F. B., 1958. Anthropological implications of sickle cell gene distribution in West Africa. *Amer. Anthrop.*, 60: 533–562.

16. LUZZATTO, L., E. A. USANGA, AND S. REDDY, 1969. Glucose-6-phosphate dehydrogenase deficient red cells: Resistance to infection by malarial parasites. *Science*, 164: 839–841.

17. MALTHUS, T. H., 1798. An essay on the principle of population. In G. Hardin (ed.), *Population, Evolution, Birth Control.* W. H. Freeman, San Francisco.

18. METTLER, L. E. AND T. G. GREGG, 1969. *Population Genetics and Evolution.* Prentice Hall, New Jersey.

19. MØRCH, E. T., 1941. *Chondroplastic Dwarfs in Denmark.* Ejnar Munksgaard, Copenhagen.

20. MOTULSKY, A. G., 1964. Hereditary red cell traits and malaria. *Amer. J. Trop. Med. Hygen.*, 13: 147–158.

21. MOURANT, A. E., 1959. Human blood groups and natural selection. *Cold Spr. Har. Sympos. Quant. Biol.*, 24: 57–63.

22. NEWCOMBE, H. B., 1963. Risk of fetal death to mothers of different ABO and Rh blood types. *Amer. J. Hum. Genet.*, 15: 449–464.

23. ———, 1965. Risks in ABO and Rh incompatibility. *Amer. J. Hum. Genet.*, 17: 97–98.

24. SLATIS, H. M., 1955. Comments on the rate of mutation to chondrodystrophy in man. *Amer. J. Hum. Genet.*, 7: 76–79.

25. SPUHLER, J. N., 1962. Empirical studies on quantitative human genetics, p. 241–251. In United Nations World Health Organization *Seminar on the Use of Vital and Health Statistics for Genetic and Radiation Studies*, 1960. World Health Organization, Geneva.

26. WILLS, V. G. AND J. C. WATERLOW, 1958. The death-rate in the age group 1–4 years as an index of malnutrition. *J. Trop. Ped.*, 4: 167–170.

27. WRIGHT, S. W., 1931. Evolution in Mendelian populations. *Genetics* 16: 97–159.

Enzyme and Protein Polymorphism in Human Populations

HARRY HARRIS

A very large number of different enzymes and proteins are made in the human organism and there are now good reasons to believe that the amino-acid sequence of each of their polypeptide chains is coded in the DNA of a separate gene locus. So there is a vast array of so-called "structural" gene loci in the genetic constitution of each individual. Furthermore it has been shown that, at certain loci, many different alleles determining structurally distinct versions of the corresponding polypeptide may exist in human populations. Most of these are quite rare. But in some cases certain alleles at a particular locus may be sufficiently frequent as to give rise to what is often referred to as genetically determined polymorphism. That is a situation in which individual members of the population can be categorized into two or more separate types, each relatively common and each characterized by the distinctive manner in which they synthesize the particular enzyme or protein.

In the present paper we will be mainly concerned with considering the incidence and distribution of such alleles at different loci, and with the general question of how they may have come to assume the frequencies that we observe. An idea of the general nature and dimensions of the problem can perhaps be best obtained by considering briefly some well-studied examples.

1. Some Examples of Multiple Allelism and Polymorphism

a. The Haemoglobin Variants

During the past twenty years a considerable number of genetically determined variants of haemoglobin have been discovered by the investigation of patients with various haematological abnormalities and also by random surveys of individuals in different populations (Lehmann & Huntsman, 1966). The great majority of these appear to differ from so-called normal haemoglobin by a single amino-acid substitution in one or other of the constituent polypeptide chains, and they can be attributed to mutational events which have resulted in only a single base change in the corresponding gene (Crick, 1967). In fact more than 40 different β-chain variants in which the specific amino-acid substitution has been defined are now known, and also at least 20 different α-chain variants. No doubt many more occur.

Reproduced by permission of the publisher and Harry Harris from the British Medical Bulletin, 25:5–13 (1959).

Harry Harris, a physician and biochemical geneticist, was born in England in 1919. He received his M.D. from Cambridge University in 1943. Between 1947 and 1953 he did research work in human genetics in the Galton Laboratory and the Department of Biochemistry, University College, London. Since 1965 he has been Galton Professor of Human Genetics and Biochemistry, University College. He is also Honorary Director of the Medical Research Council's Human Biochem-, ical Genetics Research Unit. In 1966 he was elected a Fellow of the Royal Society, and in 1968 was awarded the William Allan Memorial Award of the American Society of Genetics. Presently he is editor of the English genetic journal Annals of Human Genetics. His books on human biochemical genetics include An Introduction to Human Biochemical Genetics (Cambridge, 1953); Human Biochemical Genetics (Cambridge, 1959); Garrod's Inborn Errors of Metabolism (Oxford, 1963); The Principles of Human Biochemical Genetics (Amsterdam, 1970).

Many of these mutant alleles can be regarded as abnormal or deleterious because they produce some degree of haematological disease (sometimes very severe) in either the heterozygous or homogygous state. But some are relatively harmless even in homozygotes, and it is likely that these may be somewhat under-represented in the sample because of a bias against their selection in favour of those leading to obvious haematological disease.

One can readily classify the various alleles into two main groups in terms of their incidence. The first and much the smallest group comprises those alleles which are found relatively frequently in one or another human population. In the case of the β locus, it includes the allele that determines Hb S (or sickle-cell haemoglobin), which is very common in tropical Africa, where in different populations it may occur in 5–40% of all individuals; the allele determining Hb C, which is more localized in West Africa; the allele determining Hb E, which is common in many populations in South-East Asia; and the allele determining Hb D Punjab, which occurs in appreciable frequency in certain populations in India. (For detailed references see Livingstone (1967).)

The second group of alleles are all rare. They have turned up in an irregular manner in a wide variety of populations and the majority have been seen in members of only a single family. It is obviously difficult to get estimates of their incidence, but it appears from population surveys carried out in Europe that perhaps 1 in 1,000 people may be heterozygous for one or other of the many alleles which give rise to electrophoretically detectable variant of adult haemoglobin (see Lehman & Carrell, p. 14 of this Bulletin). They may represent mutants of either the α or β locus. In the case of the α-chain locus, all the structural variants so far defined appear to belong to this group of rare alleles.

The incidence of these various haemoglobin alleles has generally been accounted for in terms of classical theory. It is supposed that the primary amino-acid sequences of the polypeptide chains of so-called normal haemoglobin have been evolved by natural selection and are now more or less optimal for the species. But fresh mutations resulting in structural alterations of proteins are always occurring at a finite rate and essentially at random. In the vast majority of cases, the altered protein structure is expected to be in some degree functionally inferior to the normal type and so is at a selective disadvantage. Consequently the incidence of the rare variants is assumed to be mainly determined by a balance between the rate of recurrence of fresh mutations and their elimination by natural selection. Very occasionally, however, a variant occurs which, at least in the particular environment obtaining at the time, confers some advantages on people carrying it, and so it tends to spread. Usually the peculiar advantage is confined to heterozygotes, and the homozygotes are less fit. Under these circumstances the frequency of the allele tends to approach an equilibrium situation, a so-called balanced polymorphism, in which the relatively increased contribution of the allele to the next generation by heterozygotes is offset by the relatively reduced contribution from homozygotes. The classic example of this is of course the sickle-cell allele in Central Africa. Homozygotes with sickle-cell anaemia generally die in early life and contribute virtually nothing to the next generation. Yet the allele is very common. This is evidently owing to the fact that the heterozygotes have a better chance of surviving to adult life and so of leaving more children than normal homozygotes, because they are less likely to die from malaria, which is a major cause of mortality in this part of the world (Allison, 1964). Presumably, as malaria is eradicated the incidence of the sickle-cell allele will progressively decline and eventually become very rare, like most other β-chain alleles.

The special features of the sickle-cell polymorphism are a very severe pressure of selection against the homozygote, and yet a very high frequency of the allele. This means that the selective advantage favouring the heterozygotes has to be quite considerable to maintain the polymorphism. But even in these circumstances it proved to be quite difficult to demonstrate the effect and, although the evidence is now very convincing, it is still in certain aspects incomplete. In other common polymorphisms, selection against the homozygote appears to be much less marked and so the increased fitness of the heterozygote required to maintain the allele frequency need be only small, and may be virtually undetectable by available methods. Indeed although the polymorphisms of the other haemoglobin variants, such as Hb C in West Africa, and Hb E in South-East Asia, are often also assumed to have been due to selective survival of the heterozygotes with respect to malaria, which is endemic also in these areas, substantial direct evidence in support of this has not yet been obtained.

b. Glucose-6-Phosphate Dehydrogenase (G6PD)

The structure of G6PD is coded at a gene locus on the X chromosome, and more than 20 different variant forms, each apparently determined by a different allele at this locus, have been identified (for references, see World Health Organization (1967)). They have been shown to differ from the normal or standard form of the enzyme and from one another in such properties as electrophoretic mobility, Michaelis constants, thermostability and pH optima and, although the nature of the structural differences has so far been defined in only one case (Yoshida, 1967), it seems very probable that most or all of them are due to single amino-acid substitutions in the protein, similar to those found in the haemoglobins. An important point is that these structural differences often result in quite marked differences in the level of activity of the enzyme in red cells and other tissues of individuals carrying the various alleles.

Mostly these alleles are very rare, but certain of them have an unusually high incidence in particular populations and give rise to characteristic polymorphisms. For instance, besides the allele Gd^B which determines the so-called normal or standard form of the enzyme, two other alleles Gd^{A-} and Gd^{A+} both occur in many African populations with gene frequencies of between about 0.1 and 0.2, though they are rare or absent elsewhere. The variant protein determined by Gd^{A-} causes the well-known Negro form of G6PD deficiency which is the basis of primaquine sensitivity and certain other adverse drug reactions. However, apart from this drug idiosyncrasy, individuals carrying this allele appear to be in other respects quite healthy. The other common variant in Negroes that is determined by the allele Gd^{A+} is associated with only a very slight reduction in enzyme level, and this is apparently harmless.

In many populations living in Southern Europe and the Middle East, a different sort of G6PD polymorphism occurs because of the high incidence of the allele $Gd^{Mediterranean}$. This determines another striking form of G6PD deficiency, and it predisposes to the haemolytic disease known as favism, which may occur when affected individuals eat fava beans, a common feature of the diet in this part of the world. There are probably also other G6PD alleles which occur commonly in particular areas, for example Gd^{Canton} in South-East Asia and Gd^{Athens} in Greece, though their distributions have not yet been worked out in detail.

Because populations which have a high incidence of one or another form of G6PD deficiency come from areas in which malaria is or has been in the past a major cause of mortality, it has been suggested (Motulsky, 1964) that here, as in the case of the sickle-cell gene, malaria may have been an important selective agent in determining the prevalence of particular G6PD alleles (e.g., Gd^{A-} in Negro populations, and $Gd^{Mediterranean}$ in Southern European and Middle Eastern populations). The malaria parasite might proliferate less well in individuals whose red cells were G6PD-deficient and therefore metabolically abnormal. There is some, though as yet not very extensive, evidence (Gilles, Fletcher, Hendrickse, Lindner, Reddy & Allan, 1967) to suggest that these alleles may indeed confer some selective advantage in terms of malarial morbidity or mortality. But one must also note that the Gd^{A+} allele, though as prevalent as the Gd^{A-} allele in Africa and similarly rare or absent elsewhere, does not result, like the Gd^{A-}, in a marked enzyme deficiency.

c. The Haptoglobin Variants

Another extensively studied example is the serum protein, haptoglobin (for recent review and references, see Giblett (1968)). There are two sorts of polypeptide chains, α and β, and most of the variations which have been observed can be attributed to multiple alleles at the α-gene locus. The findings here, however, are in striking contrast to those obtained with haemoglobin and G6PD, because there are three alleles (Hp^{1S}, Hp^{1F} and Hp^{2FS}) which are common and wide spread. In European and African populations all three are found, though with differing frequencies, and in Asiatic populations Hp^{1S} and Hp^{2FS} both have a significant incidence but Hp^{1F} is rare or absent. As with haemoglobin and G6PD, a number of very rare alleles at the haptoglobin loci (both α and β) occur.

A special point of interest about the haptoglobin polymorphism is that it is possible to infer from the structural differences in the protein something about the origin of the alleles (Smithies, Connell & Dixon, 1962). The α polypeptides determined by Hp^{1F} and Hp^{1S} each contain 83 amino-acid residues and differ in only a single one (Black & Dixon, 1968). The α polypeptide determined by Hp^{2FS} is nearly twice as long (142 residues) and appears to represent an end-to-end fusion of the hp1Fα and the hp1Sα polypeptides, with a sequence of 24 residues missing at the site of fusion. It presumably originated as the result of a mutational event involving a chromosomal re-arrangement in an individual who happened to be heterozygous for Hp^{1F} and Hp^{1S}. In other words, the new allele probably arose in a population already polymorphic for the Hp^{1F} and Hp^{1S} alleles. Furthermore the peculiar structure of the polypeptide results in a rather characteristic polymerization of the haptoglobin molecule which is readily detected by starch-gel electrophoresis and, since this effect has not been seen in haptoglobins in other species, including higher apes (Parker & Bearn, 1961), it seems quite likely that the mutational event giving rise to the Hp^{2FS} allele occurred only after the separation of the human line. Nevertheless it has spread throughout the species and today is the commonest of the three alleles in most human populations.

So one might imagine that the Hp^{2FS} allele conferred some distinctive selective advantage. Yet it is difficult to see from what is known of the differences in the properties and function of the proteins (Giblett, 1968) what exactly this might be, or for that matter how selective forces affecting the $Hp^{1F} Hp^{1S}$ polymorphism might be operating. Certainly individuals of the different common haptoglobin types do not appear different in any obvious way in fitness. Hence one must assume either that the selective differences, if they occur, are very slight, or that they were for some unknown reason much more significant in the past but have been minimized and rendered trivial by subsequent changes in the environment.

d. Phosphoglucomutase

Many forms of phosphoglucomutase enzyme protein with apparently similar catalytic specificities have been demonstrated in different individuals (Hopkinson & Harris, 1966, 1968). There appear to be at least three different gene loci (designated PGM_1, PGM_2, and PGM_3) which are separate and not closely linked, and each locus determines a distinct set of two or three enzyme proteins (so-called isozymes). Since the electrophoretic properties of the several members of a particular set, but not of the other sets, are all similarly affected by allelic substitutions at the corresponding locus, one presumes that they contain a common polypeptide chain, which is not present in the isozymes of the other sets.

Multiple alleles determining electrophoretically distinct variants of the corresponding isozymes have been shown to occur at each of the three loci, but their incidence and distribution differ considerably from locus to locus. They have been mainly studied in populations of European or of African origin, though a number of other groups have also been examined. At locus PGM_1, two common alleles occur. In Europeans their frequencies are about 0.76 and 0.24, and they are both present with very similar frequencies among Africans and indeed in certain other population groups. So the polymorphism appears to be wide spread and much the same in very different areas. Besides these common alleles, a number of others have also been found. They are, however, individually extremely rare. At locus PGM_3, two common alleles also occur both in Europeans and in Africans, but their relative frequencies differ strikingly. In Europeans the frequencies of these two alleles are about 0.74 and 0.26, whereas in Africans they are about 0.34 and 0.66 respectively. At locus PGM_2 there is one allele which predominates in both Europeans and Africans, and the other alleles that have been detected are all much less common. In Africans, however, there is some degree of polymorphism because, besides this common or standard allele, another occurs with a frequency of about 0.01 and, since this has not been detected in the many thousands of Europeans that have been tested, there must be a considerable difference in its incidence between the two groups.

These phosphoglucomutase variants were discovered in the course of an electrophoretic screening programme deliberately aimed at searching for common polymorphic differences. The individuals studied were, in the main, normal and healthy, and there was no indication that the common variant types of the enzyme were associated with any marked functional differences which might be of selective significance. It seems therefore that, if such differences do occur, they are probably very subtle and relatively small in magnitude.

2. The Extent of Polymorphism

These examples illustrate something of the degree of allelic variation that may occur at different gene loci. At some loci, although multiple alleles may be demonstrable, there is one allele that can be regarded as the standard or normal form and is almost universally present, while all the others are extremely rare. At other loci (e.g., the β-haemoglobin locus and the G6PD locus), although a standard allele occurs and is recognizable as such, there are in some populations, but not in others, alleles which are present in sufficient frequency as to give rise to a common polymorphism. At still other loci (e.g., the α-haptoglobin locus and the PGM_1 and PGM_3 loci), polymorphism is the rule. Two or more alleles occur relatively frequently and are widely distributed in many different populations. Indeed in some cases there appears to be no valid reason for regarding one allele rather than another as the so-called normal or standard form.

A question which obviously arises is: what is the relative incidence of these various situations among gene loci in general, and in particular how often do polymorphisms occur? Do the haemoglobin, G6PD, haptoglobin, phosphoglucomutase and other polymorphisms represent special and rather unusual forms of variation not typical of enzymes or proteins in general, or are they examples of a relatively common phenomenon? It would clearly be of some interest to know what fraction of the very large number of proteins and enzymes which are formed in the human organism exhibits these sorts of variation. Since the structure of each protein is presumed to be determined by at least one gene locus, we are in effect asking: at what proportion of this vast array of gene loci do two or more relatively common alleles occur in different human populations?

In principle, it should be possible to get an approximate answer to this question by examining in detail a series of arbitrarily chosen enzymes and proteins in randomly selected individuals and preferably in several different populations. In each case one would aim to see whether or not common polymorphism is demonstrable, and in what proportion of the enzymes or proteins it occurs. One might also hope to obtain some information about the occurrence of rare variants.

During the past few years the Medical Research Council's Human Biochemical Genetics Research Unit, working in London, has been engaged in this kind of project (Harris, 1966, 1967). We have mainly studied enzymes which occur in red cells, because these provide a convenient source of material if one wishes to examine large numbers of different individuals. But in most cases the enzymes are also found in other tissues and there is no reason to believe that they are unrepresentative of enzymes in general, at least as far as the incidence of inherited variation is concerned. More recently the placenta has been found to be a useful source of material for such investigations.

The enzymes studied were chosen in an essentially arbitrary fashion, the only criteria adopted being whether they happened to occur in blood or placenta, and whether a suitably sensitive method could be devised by which a search for structural variants could be conveniently conducted in a reasonably sized sample of the general population. The work has mainly been done on English people, but samples from other population groups, mainly Africans, have been studied, though less extensively. Where variant forms of an enzyme were found, detailed family studies were carried out to determine their genetic basis.

The general procedure used to search for enzyme differences was starch-gel electrophoresis.[1] This method is particularly powerful in the detection of differences in molecular charge, such as those that might be produced by the substitution of a basic or acidic amino acid for a neutral amino acid or vice versa. It is not, however, very sensitive to other types of molecular difference (e.g., the substitution of one neutral amino acid by another) and so, even if the electrophoretic conditions employed were optimal, this method could be expected to detect only a proportion of all the possible forms of variation in enzyme structure that might occur.

So far 18 different enzymes have been examined in varying degrees of detail in the course of this project, and a considerable number of genetically determined electrophoretic variants have been identified. Many of the alleles involved are relatively infrequent. But, in six of the enzymes, evidence was obtained for two or more alleles, each with a frequency of at least 0.01 in the particular population. The relatively common individual differences so produced can be regarded as polymorphisms in the sense that this term is conventionally used in genetics. In the case of one of these enzymes, phosphoglucomutase, as has been previously mentioned, polymorphism attributable to independent allelic variation at three separate and unlinked loci was identified. Thus altogether eight polymorphic loci were discovered during this investigation.

The frequencies of different alleles found at these loci are summarized in Table I. In five cases (red-cell acid phosphatase, phosphoglucomutase PGM_1 and PGM_3, adenosine deaminase, and peptidase D), the polymorphism occurs in both Europeans and Africans, although the allele frequencies vary somewhat between these contrasting population groups. At one locus (adenylate kinase), polymorphism was found only among the Europeans, and in two (peptidase A and PGM_2) only among the Africans. Thus about one-quarter of the enzymes in this arbitrarily selected series exhibited polymorphism in each population group. Clearly the phenomenon is a relatively common one.

If anything, the results must under-estimate the true incidence of polymorphism simply because the enzymes were scrutinized only for elec-

[1] See McDougall & Synge, *Br. Med. Bull.* 1966, **22**, 115.—ED.

TABLE I. RESULTS OF AN ELECTROPHORETIC SURVEY OF EIGHTEEN ARBITRARILY CHOSEN ENZYMES IN EUROPEAN AND NEGRO POPULATIONS. ONLY COMMON ALLELES (FREQUENCY > 0.01) ARE LISTED. MANY "RARE" ALLELES WERE ALSO DETECTED

Enzymes	Europeans			Negroes			References
	Allele 1	Allele 2	Allele 3	Allele 1	Allele 2	Allele 3	
Red-cell acid phosphatase	0.36	0.60	0.04	0.17	0.83	—	Hopkinson, Spencer & Harris (1964)
Phosphoglucomutase							
Locus PGM_1	0.76	0.24	—	0.79	0.21	—	Spencer, Hopkinson & Harris (1964)
Locus PGM_2	1.00	—	—	0.99	0.01	—	Hopkinson & Harris (1966)
Locus PGM_3	0.74	0.26	—	0.34	0.66	—	Hopkinson & Harris (1968)
Adenylate kinase	0.95	0.05	—	1.00	—	—	Fildes & Harris (1966)
Peptidase A	1.00	—	—	0.90	0.10	—	Lewis & Harris (1967)
Peptidase D (prolidase)	0.99	0.01	—	0.95	0.03	0.02	Lewis & Harris, unpublished work
Adenosine deaminase	0.94	0.06	—	0.97	0.03	—	Spencer, Hopkinson & Harris (1968)

Other enzymes studied were: phosphohexoseisomerase, malate dehydrogenase, isocitrate dehydrogenase, red-cell hexokinase, lactate dehydrogenase, methaemoglobin reductase, red-cell pyrophosphatase, pyruvate kinase, placental acid phosphatase, peptidases B and C, and a red-cell "oxidase". None of these showed common electrophoretic polymorphism, though a number of rare variants were identified.

trophoretic differences. Furthermore the discriminative power of even this technique has been found to vary considerably from enzyme to enzyme because of technical problems, and it is quite possible that in some cases polymorphic variation has been missed.

It is difficult to arrive at a satisfactory estimate of the total number of different enzymes and proteins which occur in the human organism, and which are presumably coded at separate gene loci. But they must certainly number many thousands. If the results of the enzyme survey can be taken as at all representative, then in any given population of individuals there must be a significant fraction of loci at which two or more relatively common alleles occur. So probably thousands of polymorphisms, each involving a different enzyme or protein, exist. It is not without significance that essentially the same conclusion has been reached by Lewontin & Hubby (1966) on the basis of a similar electrophoretic survey of enzymes and proteins in a quite different species, *Drosophila pseudo-obscura*.

An interesting point which emerges concerns the degree of individual diversity which must actually occur in human populations. Some idea of this can be obtained by considering together the several enzyme polymorphisms that have already been demonstrated in a single population. Relevant data on ten loci involving eight different enzymes in the English population are given in Table II. Since each of these polymorphisms appears to occur independently of the others, it follows that a very large number of combinations of enzyme phenotypes may occur among individuals in the general population. By combining the frequencies given in column three of the table, one finds that the most frequent combination of phenotypes will occur in less than 2% of the population. Furthermore, from column four one can show that the chance that two randomly selected people in the population would have exactly the same combination of enzyme types is about 1 in 200. Column five provides another way of looking at the data. It gives the proportion of individuals in the population who are heterozygous for alleles at each of the loci. It appears that approximately 97% of people in this population must be heterozygous at at least one of the ten loci listed in the table. Thus quite a high degree of individual differentiation in enzyme make-up can be demonstrated from this quite limited series of examples. This must surely represent only the tip of the iceberg, and one may plausibly conclude that, in the last analysis, every individual will be found to have a unique enzyme constitution.

An obviously important problem is how far these common enzyme variations are reflected functionally. Not much information is as yet available about this. But, in a few cases, significant differences in activity levels between the several common phenotypes of a particular enzyme have been detected. In the case of red-cell acid phosphatase, for example, there are six electrophoretically distinct phenotypes that can be distinguished, and they represent the homozygous and heterozygous combinations of three common

TABLE II. ENZYME POLYMORPHISM IN THE ENGLISH POPULATION. (DATA ON TEN LOCI)

Enzymes	Number of alleles with frequency greater than 0.01	Frequency of commonest phenotype	Probability of two randomly selected individuals' being of same type	Proportion of population who are heterozygous	References
Red-cell acid phosphatase	3	0.43	0.34	0.51	Hopkinson et al. (1964)
Phosphoglucomutase					
Locus PGM_1	2	0.58	0.47	0.36	Spencer et al. (1964)
Locus PGM_3	2	0.55	0.45	0.38	Hopkinson & Harris (1968)
Placental alkaline phosphatase	3	0.41	0.30	0.50	Robson & Harris (1967)
Liver acetyltransferase	2	0.50	0.50	0.50	Price Evans & White (1964)
Adenylate kinase	2	0.90	0.82	0.10	Fildes & Harris (1966)
Serum cholinesterase					
Locus E_1	2	0.96	0.92	0.04	Kalow & Staron (1957)
Locus E_2	2	0.90	0.82	0.10	Robson & Harris (1966)
Phosphogluconate dehydrogenase	2	0.96	0.92	0.04	Parr (1966)
Adenosine deaminase	2	0.88	0.79	0.11	Spencer et al. (1968)
All enzymes combined	—	0.018	0.005	—	

alleles. These phenotypes, presumably because of the structural differences in the enzyme present, also differ in their average total level of red-cell phosphatase activity (Hopkinson, Spencer & Harris, 1964). For instance, homozygous individuals with the so-called type B form of the enzyme show on average about 50% greater activity than homozygous individuals of type A, and the heterozygote type BA is intermediate in this respect. Similar differences are seen with the other phenotypes. Differences in activity have also been noted in the case of the common serum cholinesterase phenotypes (Kalow & Staron, 1957; Harris, Hopkinson, Robson & Whittaker, 1963) and in the phosphogluconate dehydrogenase phenotypes (Parr, 1966).

Another interesting example is provided by the polymorphism of a liver enzyme which behaves as an acetyltransferase (Price Evans & White, 1964). This polymorphism was discovered when it was found that people differed very markedly in the rate at which they acetylate and therefore inactivate the drug isoniazid, which is used in the chemotherapy of tuberculosis. About 50% of Europeans are so-called rapid inactivators because they have a relatively high level of the transferase whereas, in the other 50%, the level of activity of the enzyme is very low and the drug is inactivated very slowly. This polymorphic difference is due to two common alleles, the slow inactivators being homozygous for one, and the rapid inactivators heterozygous or homozygous for the other. It is of interest that in Europeans the allele resulting in very low enzyme activity is more than twice as common as the allele determining the active enzyme, the gene frequencies being roughly 0.7 and 0.3.

3. Rare Variants

Besides the relatively common alleles which give rise to the so-called polymorphisms, a considerable number of rare alleles determining different variant forms of particular enzymes or proteins have also been found during the course of population surveys. Thus evidence for 5–10 different rare alleles at loci determining the structures of the enzymes phosphohexoseisomerase (Detter, Ways, Giblett, Baughan, Hopkinson, Povey & Harris, 1968), placental alkaline phosphatase (Robson & Harris, 1966), peptidase A (Lewis & Harris, 1967, and unpublished work), lactate dehydrogenase (Kraus & Neely, 1964; Davidson, Fildes, Glen-Bott, Harris, Robson & Cleghorn, 1965), and phosphoglucomutase (Hopkinson & Harris, 1966) has been obtained in the course of electrophoretic studies in which samples from several thousand individuals have been examined. And numerous other examples of rare variants of particular enzymes or proteins picked up in the course of routine investigations could be cited. In fact it is beginning to appear that, if virtually any enzyme or protein is examined by sufficiently sensitive methods in a large enough number of individuals, one or more rare variants are likely to be detected. The multiplicity of rare haemoglobins mentioned earlier illustrates the same phenomenon, and shows how very many different rare alleles at single loci may be demonstrable if a protein is subjected to particularly intensive investigation.

The precise incidence of any particular one of these rare alleles is obviously difficult to determine. Some appear to occur with gene frequencies of between 10^{-2} and 10^{-3} in certain populations, but in the majority of cases the individual gene frequencies are probably much lower. Nevertheless at any single locus so many different rare alleles may exist that an appreciable fraction of the population can be heterozygous for one or another of them. For instance, from the presently available data it seems that the fraction of the population which is heterozygous for one or other of the rare alleles determining electrophoretically detectable variants of such enzymes as phosphohexoseisomerase, phosphoglucomutase (PGM_1 and PGM_2), the red-cell peptidases A and B, and lactate dehydrogenase, is in each case in the order of 1 in 300 to 1 in 700. Since presumably only a proportion of the structural variants of any given enzyme will be detectable electrophoretically, the true fraction is in fact likely to be somewhat higher. If, as is not improbable, there are many other loci at which there is a similar incidence of "rare" alleles, the multiplicity of rare enzyme and protein variants so produced could in toto contribute quite significantly to the diversity among individuals in the population.

Because these "rare" variants have mainly been discovered in the course of random population surveys, they have usually been observed only in heterozygotes, who synthesize the common form of the enzyme or protein (or, where there is polymorphism, one or other of the common forms), as well as the variant. Such individuals are usually normal and healthy. However, in some cases the variant enzyme or protein may be functionally defective, so that in the homozygous state it could result in significant clinical abnormality. In fact most rare "recessive" diseases can probably be attributed to rare alleles of this sort (for further discussion, see Harris (1968)).

It is not yet clear what proportion of the rare alleles at any particular locus may lead to obvious clinical abnormality in the homozygous state. Probably it varies considerably from locus to locus according to the structure and also the function of the enzyme or protein involved. However, judging from the properties of many of the rare enzyme and protein variants as they have been observed in heterozygotes, it seems that the proportion that is likely to be markedly deleterious in the homozygous state is, at many loci, quite small.

4. Some General Considerations

If we extrapolate from these various observations, the picture that is beginning to emerge can perhaps be formulated in the following way. As a result of mutations in the remote or more recent past, there is likely to be present in any sizeable human population more than one allele at virtually every gene locus which codes for a specific enzyme or protein, and at many loci a considerable number of such alleles probably occur. In general they produce structurally distinct forms of the enzyme or protein and these in most cases probably differ from one another by single amino-acid substitutions. The majority of these alleles are very rare. However, at perhaps 25% or more of all such loci, at least two alleles—each present in an appreciable fraction of the population (at least 2%)—may occur.

The incidence and distribution of these alleles in populations must depend essentially on three main factors: (i) the rate at which fresh mutations occur; (ii) differential selection due to the effects, exerted by the enzyme and protein variants that the alleles determine, on the survival and fitness of individuals who carry them; and (iii) chance factors which may lead fortuitously to the elimination of particular alleles from the population or to their spread.

a. Mutation Rates

Gene mutations are thought to occur more or less at random. In most cases, though certainly not in all, they apparently involve simply the change of a single base in the sequence of several hundred or more that are present in the DNA of the particular gene. A typical polypeptide chain may contain 100–500 amino acids, any one of which can be replaced by one of several others as a result of such a mutational event. So a vast number of distinct structural variants of any one protein may be generated by recurrent mutations at a single gene locus. Some of these variants, because of the alternation in their properties and functional activity induced by the specific amino-acid substitution, can be expected to give rise in either the homozygous or heterozygous state to some frank clinical abnormality. But many of the possible alterations in structure which could occur are likely to be of only moderate or minor functional significance and others may have no obvious consequences at all.

Considered in these terms, the significance of the mutation rates often quoted for human genes is difficult to assess. They have mainly been derived from studies on the population and familial incidence of rare inherited abnormalities such as haemophilia, retinoblastoma and neurofibromatosis (for review and discussion, see Penrose (1961)), and lead to estimates of mutation rates per gene locus per generation of around 10^{-5}. That is to say, they suggest that a fresh mutation for a gene determining one of these three conditions may be present in one in every 100,000 sperm or ova. If, however, one supposes that the mutations involved in causing these abnormalities usually represent single base changes resulting in specific defects in a particular enzyme or protein, one would expect that at any such locus there might be a number of different sites where the substitution of one base by another could cause a defect in the corresponding protein capable of inducing the particular clinical abnormality observed. There would probably also be a great many other sites in which a single base change would not have the same consequences. Furthermore, at any one site the effect that follows from a mutation will depend on the particular base that happens to be substituted. Thus only a fraction of all mutations involving single base changes at a particular gene locus might lead to the disease in question. And we have very little idea of what the magnitude of this fraction might be in specific cases. In general, however, it would appear likely that the mutation rates often cited for rare abnormalities in man are probably higher than the rate at which a single base at a particular site in a gene is altered by mutation. On the other hand, they probably underestimate the total mutation rate per locus per generation, since this would include base changes at all possible sites within the gene.

Watson (1965) suggests that the average probability of an error's giving rise to the insertion of an incorrect base during DNA replication may under optimal conditions be around 10^{-8} or 10^{-9}; and Kimura (1968) points out that this could imply a mutation rate for base substitutions of perhaps 5×10^{-7} or 5×10^{-8}/base pair/generation, assuming the number of cell divisions along the germ line from the fertilized ovum to a gamete in man to be roughly 50. Using such estimates, a number of interesting if speculative calculations are possible. For example, if we suppose that the mutation rate/base pair/generation is 5×10^{-8}; that 80% of base changes in a gene coding for a polypeptide chain result in single amino-acid substitutions; that the average polypeptide chain contains 300 amino acids and so is coded by a DNA sequence 900 bases long; and that there are perhaps 20,000 different polypeptide chains synthesized in the organism; then, on average, every newborn infant may be expected, as the result of a fresh mutation's having occurred in either of its parents, to synthesize at least one new structurally variant enzyme or protein (i.e., $2 \times (5 \times 10^{-8})(0.80)(9 \times 10^{2}) \cdot (2 \times 10^{4}) = 1.44$). This kind of calculation (see also Kimura, 1968) is of course largely guess-work, but the assumptions appear not unrealistic, and the result perhaps illustrates the extent to which fresh mutations may be continuously generating enzyme and protein diversity among individuals in a population.

b. Chance Effects

Quite apart from the question as to whether a particular mutant is relatively deleterious and so tends to be eliminated by natural selection, or confers some kind of selective advantage and so tends to spread, the odds against any new mutant allele's persisting in a population for many generations are very considerable. The new allele will on average be transmitted to only half the children of the individual who first receives it. There is, therefore, a distinct chance that it will not be transmitted to the next generation, and the odds in favour of its being lost by chance are compounded in successive generations. In a reasonably large stable population where each pair of parents is on average replaced by two children who become parents in the next generation, the probability that a new mutant will still be present after, say, 15 generations is only about 1 in 9 (Fisher, 1930). The odds in favor of the persistence of a mutant are somewhat greater if the population happens to be increasing in numbers when it appears, and are less if the population is declining. But, in general, the majority of new mutant alleles that appear are likely to be eliminated in the course of the next 10 or 20 generations in a more or less random manner. However, as we have seen, even if the mutation rates are low (e.g. 5×10^{-8}/base pair/ generation), there is always going to be an appreciable though changing reservoir or pool of rare variant forms of different enzymes or proteins in any sizeable human population at any given time. Their nature and incidence will be largely a matter of chance.

Furthermore the individual frequencies of those mutant alleles that do happen to persist in a population may, because of such random phenomena, vary considerably, quite apart from any selective effects that may be superimposed. So, very occasionally, one or other variant might become relatively common in a particular population purely by chance. This is particularly likely to occur if the population is small and relatively isolated. For instance, a variant form of serum albumin has recently been observed in as many as 25% of members of a group of North American Indians known as the Naskapi (Blumberg, Martin & Melartin, 1968) and in several closely related tribes. This variant has not been seen anywhere else in the world, although it would quite easily have been picked up by very widely used procedures. A number of similar examples of other peculiar enzyme and protein variants occurring with an unexpectedly high incidence in odd communities could be readily cited. They are probably most simply accounted for in terms of such chance fluctuations or what is known as "drift".

Chance effects may also be of great importance in situations where the numbers of an established population are severely reduced by some epidemic or other disaster and then subsequently increase again. The sample of alleles which happen to be carried by the survivors and so form the gene pool from which the population is reconstituted is unlikely to be in all respects exactly representative of the alleles in the original population. Similarly if a small group of individuals from one population migrate elsewhere and found a new community which expands, particular alleles may by chance be over- or under-represented. Under such circumstances marked changes in allele frequency may occur quite quickly. Relatively uncommon alleles in the original population may by chance become common in the population derived from it. Other alleles present in an appreciable frequency in the original population may be lost fortuitously and not appear in the derived one.

The role of selective forces in determining the incidence and distribution of the multiplicity of enzyme and protein variants that we observe in populations has therefore to be evaluated against this background of fortuitous and essentially haphazard effects inherent in the nature of the genetical structure of human populations and of the mutation process itself. Even if, as has often been extensively argued in the past, no allele is selectively neutral, the part that selective forces have played in determining its frequency may in many cases be effectively obscured by such chance effects.

c. Selection

Where two or more alleles determining structurally distinct forms of a particular enzyme or protein each occur with a relatively high incidence in some or all of the major ethnic population groups, it seems reasonable to suppose that selective forces at least to some degree have been important in determining their incidence and distribution. It seems biologically implausible to suppose that more than a small proportion of the many different enzyme and protein polymorphisms that evidently occur have come about purely fortuitously, although one may suppose that chance effects or drift could have been responsible for many of the detailed peculiarities in their distributions.

However, the elucidation of the selective forces which may have given rise to any particular polymorphism is turning out to be among the most difficult and intractable problems in human genetics. Various approaches seem possible but none has as yet been particularly rewarding, mainly perhaps because selective differences that may have been important in determining the polymorphism are possibly not of an order of magnitude capable of being demonstrated by currently available methods. The principal exception to this is the sickle-cell polymorphism and this is now beginning to emerge as a perhaps rather special and very unusual situation.

One way of approaching the matter is by the direct investigation of the functional properties of the structurally distinct forms of a polymorphic enzyme or protein, in the hope that this might define differences which are likely to be significant metabolically or in some other way which could be selectively important in certain sorts of environment. In a number of cases, it has been shown that common structural variants of a particular enzyme are indeed associated with marked differences in activity. But there is as yet little or no indication of the possible significance of such differences in relation to selection, except perhaps in the case of G6PD mentioned earlier. An obvious difficulty is that selection, as it affects most individuals, is presumably directed at complex physiological variables dependent on many different enzymes acting together. The important thing may therefore be the constellation of enzyme phenotypes of the individual rather than the characteristics of any single one of them.

Another approach is to try and find out whether particular alleles render individuals more or less susceptible to the development of particular disorders or disabilities, especially common ones. The general method is to compare the incidence of the allele in individuals affected by the particular condition with the total incidence in the population of which they are a part. A now well-established example of this kind of effect is the association of the ABO blood groups with certain gastrointestinal disorders (Fraser Roberts, 1959). Blood-group A individuals are somewhat more susceptible to gastric cancer than group O individuals. Group O individuals are more susceptible to peptic ulceration than group A. But the effects are relatively small and their significance in relation to selection for the different alleles in this polymorphism is difficult to assess. Nor have we any clear idea of exactly how these different blood-group antigens influence susceptibility to these diseases. A large number of different diseases, as well as differences in response to particular disease states such as acute and chronic infections, could well be studied in this way. And there are an increasing number of polymorphisms which might be tested. But generally there is no particular reason to expect one association rather than another, so that progress in this direction at the present time is likely to be somewhat fortuitous.

Where very wide variations in the frequencies of particular alleles are observed between different populations, it is of interest to ask whether their distribution is correlated in any obvious way with specific differences in the environments in which the populations live, or with any characteristic patterns of morbidity, in the hope that this might indicate the nature of the critical selective factors. This approach was valuable in directing attention to malaria in the case of the sickle-cell polymorphism, and provides suggestive evidence for the same kind of selection in the other haemoglobin polymorphisms as well as in the G6PD polymorphisms occurring in Africa and the Middle East. But it has not as yet appeared to be particularly helpful in other cases. And it might well in certain circumstances be rather misleading, if attention were arbitrarily focused on just one out of the multiplicity of environmental differences that frequently exist.

Another major line of attack is primarily demographic. The aim is to categorize individuals in one or more populations in terms of the various common allelic differences that are known, and then search for differences between them in the main parameters involved in selection, such as mortality and morbidity rates at various ages, and fertility; also by family analysis to investigate possible disturbances in segregation ratios and so on. Such data, although they may give only indirect information about specific selective factors in relation to particular polymorphisms, should in principle provide an assessment of the magnitude of any selective effects that are actually occurring in the given environmental situations. And this of course is fundamental to the whole problem. Such surveys are, however, extremely hard to mount on a scale which is both sufficiently large as to be likely to yield significant results, and yet sufficiently detailed and exact in the determination of the various demographic parameters as to yield precise answers. So far, although much suggestive information has been obtained (e.g., Morton, 1964; Reed, 1967, 1968a, 1968b), the results have perhaps mainly served to emphasize the difficulties involved in arriving at any certain conclusions about the biological implications of any particular polymorphism.

A general and inherent source of uncertainty in all these studies arises from the fact that the environment in which human populations live today or even during the last two or three generations is very different in many important aspects from that in which they lived in the past. In particular the incidence and age distribution of mortality and morbidity and its main causes have changed and are changing profoundly. Thus what may have been important selective agents in the past, and may well have shaped many of the polymorphisms that we see today, may now be of only minor or no significance. We are only looking at what is inevitably a changing situation over a very narrow period of time. Furthermore, as a general rule we have no means of knowing whether in any particular polymorphism we are, as is often assumed, dealing with a situation close to stable equilibrium due to heterozygous advantage, or with the steady increase of one particular allele at the expense of another, or with its progressive disappearance.

Thus, although there is little doubt about the occurrence of enzyme and protein polymorphisms as a wide-spread and general phenomenon in human populations, we have very little idea so far about the detailed nature of the selective effects that may have brought them about.

REFERENCES

ALLISON, A. C. (1964) *Cold Spring Harb. Symp. quant. Biol.* **29**, 137.

BLACK, J. A. & DIXON, G. H. (1968) *Nature, Lond.* **218**, 736.

BLUMBERG, B. S., MARTIN, J. R. & MELARTIN, L. (1968) *J. Am. med. Ass.* **203**, 180.

CRICK, F. H. C. (1967) *Proc. R. Soc.* B, **167**, 331.

DAVIDSON, R. G., FILDES, R. A., GLEN-BOTT, A. M., HARRIS, H., ROBSON, E. B. & CLEGHORN, T. E. (1965) *Ann. hum. Genet.* **29**, 5.

DETTER, J. C., WAYS, P. O., GIBLETT, E. R., BAUGHAN, M. A., HOPKINSON, D. A., POVEY, S. & HARRIS, H. (1968) *Ann. hum. Genet.* **31**, 329.

EVANS, D. A. PRICE & WHITE, T. A. (1964) *J. Lab. clin. Med.* **63**, 394.

FILDES, R. A. & HARRIS, H. (1966) *Nature, Lond.* **209**, 261.

FISHER, R. A. (1930) *The genetical theory of natural selection.* Clarendon Press, Oxford.

GIBLETT, E. R. (1968) *Ser. haemat., Kbh.* n.s. **1**, 3.

GILLES, H. M., FLETCHER, K. A., HENDRICKSE, R. G., LINDNER, R., REDDY, S. & ALLAN, N. (1967) *Lancet*, **1**, 138.

HARRIS, H. (1966) *Proc. R. Soc.* B, **164**, 298.

HARRIS, H. (1967) In: Crow, J. F. & Neel, J. V., ed. *Proc. III int. Congr. hum. Genet., Chicago, Illinois, September 5–10, 1966*, p. 207. Johns Hopkins Press, Baltimore, Md

HARRIS, H. (1968) *Br. med. J.* **2**, 135

HARRIS, H., HOPKINSON, D. A., ROBSON, E. B. & WHITTAKER, M. (1963) *Ann. hum. Genet.* **26**, 359

HOPKINSON, D. A. & HARRIS, H. (1966) *Ann. hum. Genet.* **30**, 167

HOPKINSON, D. A. & HARRIS, H. (1968) *Ann. hum. Genet.* **31**, 359

HOPKINSON, D. A., SPENCER, N. & HARRIS, H. (1964) *Am. J. hum. Genet.* **16**, 141

KALOW, W. & STARON, N. (1957) *Can. J. Biochem. Physiol.* **35**, 1305

KIMURA, M. (1968) *Nature, Lond.* **217**, 624

KRAUS, A. P. & NEELY, C. L., JR (1964) *Science, N.Y.* **145**, 595

LEHMANN, H. & HUNTSMANN, R. G. (1966) *Man's haemoglobins*. North-Holland, Amsterdam

LEWIS, W. H. P. & HARRIS, H. (1967) *Nature, Lond.* **215**, 351

LEWONTIN, R. C. & HUBBY, J. L. (1966) *Genetics, Princeton*, **54**, 595

LIVINGSTONE, F. B. (1967) *Abnormal hemoglobins in human populations*. Aldine, Chicago

MORTON, N. E. (1964) *Cold Spring Harb. Symp. quant. Biol.* **29**, 69

MOTULSKY, A. G. (1964) *Am. J. trop. Med. Hyg.* **13**, 147

PARKER, W. C. & BEARN, A. G. (1961) *Ann. hum. Genet.* **25**, 227

PARR, C. W. (1966) *Nature, Lond.* **210**, 487

PENROSE, L. S., ed. (1961) *Recent advances in human genetics*. Churchill, London

REED, T. E. (1967) *Am. J. hum. Genet.* **19**, 732

REED, T. E. (1968a) *Am. J. hum. Genet.* **20**, 119

REED, T. E. (1968b) *Am. J. hum. Genet.* **20**, 129

ROBERTS, J. A. FRASER (1959) *Br. med. Bull.* **15**, 129

ROBSON, E. B. & HARRIS, H. (1966) *Ann. hum. Genet.* **29**, 403

ROBSON, E. B. & HARRIS, H. (1967) *Ann. hum. Genet.* **30**, 219

SMITHIES, O., CONNELL, G. E. & DIXON, G. H. (1962) *Nature, Lond.* **196**, 232

SPENCER, N., HOPKINSON, D. A. & HARRIS, H. (1964) *Nature, Lond.* **204**, 742

SPENCER, N., HOPKINSON, D. A. & HARRIS, H. (1968) *Ann. hum. Genet.* **32**, 9

WATSON, J. D. (1965) *Molecular biology of the gene*. Benjamin, New York

WORLD HEALTH ORGANIZATION (1967) *Tech. Rep. Ser. Wld Hlth Org*. No. 366

YOSHIDA, A. (1967) *Proc. natn. Acad. Sci. U.S.A.* **57**, 835

Metabolic Polymorphisms and the Role of
Infectious Diseases in Human Evolution[1]

ARNO G. MOTULSKY[2]

The study of human evolution until recent years was limited to investigations of the gross structural characteristics of man and dealt primarily with
man's evolution from hominoid precursors. Such studies of long-term
evolutionary adaptations have resulted in our present concept of man's
ancestry. The human biologist in search of short-term evolutionary trends
in man will need to look elsewhere for a field of investigation. Genetically,
human evolution may be conceived as occurring through changes in the
frequency of genes in the human gene pool. As conceived today, the primary function of a gene is the determination of specific protein structure;
all other gene effects appear to be consequences of this primary effect. It
would therefore be profitable to study the gene frequency of genetically
controlled proteins and protein enzymes since such traits would represent
reasonably close primary gene products. Differential distribution of polymorphic traits[3] in different populations who have lived under varying environments should give some information about the possible adaptive values
of such traits. There is increasing evidence that in most cases the existence
of polymorphisms where the rarest trait exceeds a frequency of one or two
per cent or more in a population implies that the trait represents the end
result of some selective advantage for the trait carriers. Traits injurious to
their carriers lower fitness, lead to a smaller number of offspring, and therefore will be eliminated by natural selection. Fisher (1930) has pointed out
that mutations to neutral traits without advantage or disadvantage would
have a very small chance to become established in a population. Occasionally, however, chance or drift might lead to survival of neutral genotypes.

Natural selection, by causing differential survival of some genotypes at

[1]Investigations reported in this paper were supported by grants of the National Institute
of Health, Commonwealth Fund and the Rockefeller Foundation.
[2]John and Mary R. Markle Scholar in Medical Science.
[3]A polymorphism in man is defined as the simultaneous existence in the same population of
two or more distinct forms of a given characteristic in such proportions that the rarest cannot
be maintained by mutation alone. A metabolic polymorphism in man may affect the various
gene-controlled proteins and enzyme variants such as hemoglobins, transferrins, haptoglobins, enzyme deficiencies, etc.

Arno G. Motulsky, *a physician and medical geneticist, was born in Germany in 1923. He received his M.D. degree from the University of Illinois in 1947. At the present time he is head of the Division of Medical Genetics, University of Washington. His research concerns the various hereditary anemias of man, and for this purpose he has developed several methods for the study of biochemical traits in human blood. His interest in population genetics has focused on the relationship of genetics and disease in man, especially with reference to the role of malaria in the distribution of enzyme deficiencies in human populations. He has done genetic field work among populations in Greece and Africa. Some of his more recent publications in the area of population genetics are "Population Genetics of Glucose-6-Phosphate Dehydrogenase Deficiency of the Red Cell," with J. M. Campbell-Kraut, in B. S. Blumberg, Ed.,* Proceedings of the Conference on Genetic Polymorphisms and Geographic Variations in Disease *(New York, 1961); "Hereditary Red Cells and Malaria,"* American Journal of Tropical Medicine and Hygiene, *13:147–158 (1964); "Population Genetics in the Congo. I. Glucose-6-Phosphate Dehydrogenase Deficiency, Hemoglobin-S and Malaria," with J. Vandepitte and C. R. Fraser,* American Journal of Human Genetics, *18:514–537 (1966); and "Population Genetics of Mental Retardation," in G. A. Jervis, Ed.,* Expanding Concepts in Mental Retardation *(Proc. Joseph P. Kennedy, Jr. Foundation Symposium on Mental Retardation, Springfield, April, 1966).*

the expense of others, is recognized to be the most significant evolutionary agent. Natural selection in man has two broad aspects: differential fertility and differential mortality. Certain genes may lead to differential reproduction rates among different members of a population. Genes affecting fertility may interfere primarily with normal sperm, egg, or embryonic development, and thus lead to genetic death. Other genes, by affecting the emotional and social aspects of behavior, may interfere with reproduction in some individuals and not in others. It is not generally recognized that in western societies one-fifth to one-sixth of individuals of a given generation produce one-half of the next generation (Neel, 1958). In present western civilizations the genetic determinants of differential fertility are hard to untangle from cultural factors that lead to fertility reduction. Neel (1958) wisely pointed out the necessity of comparing fertility rates in civilized and primitive societies to elucidate the genetic component of differential fertility in man.

Emphasis will be placed in this discussion on differential mortality. An infant mortality of 50% or more is a common phenomenon in primitive societies. Obviously, if genes existed which protected against dying, they would have evolutionary survival value. Such genes would tend to increase in a population and, unless exerting some other unfavorable effect, would tend to replace their alleles. If, however, individuals die at random regardless of genotype, no significant change in gene frequency would be expected.

What are the environmental features leading to high childhood mortality? Infectious diseases ("pestilence") and starvation have long been known to be principal factors checking population growth (Malthus, 1798).

Starvation

Is there evidence that starvation-induced mortality is selective? Knowledge of the genetic control of human nutritional requirements is not yet far advanced. However, judging from mammalian experiments, it is fairly certain that individuals differ genetically in nutritional requirements for a variety of nutrients such as amino acids and vitamins (Williams, 1956; Schneider, 1958). There is also increasing evidence for the genetic control of obesity. Some individuals will be obese with identical caloric intake and energy expenditure while others will not (Teppermann, 1958). Since genes fundamentally control all enzyme-mediated reactions of intermediate metabolism, it is likely that death by starvation would not act at random, but certain genotypes would be more susceptible to famine-induced death than others. A human population that has lived through famines, as have most, is therefore likely to harbor genotypes which are relatively resistant to death by starvation.

A recently discovered polymorphism in man may conceivably be of pertinence. About three per cent of the Canadian and British population have a genetically determined relative deficiency of the serum enzyme, pseudocholinesterase (Kalow and Staron, 1957). This enzyme deficiency was discovered by noting that some patients developed prolonged apnea following administration of a drug, succinyldicholine, which requires the enzyme for breakdown. The fundamental lesion leading to pseudocholinesterase deficiency presumably is an altered enzyme protein molecule that no longer exerts its full enzymatic action. The primary function of pseudocholinesterase is not known. Its action on drugs such as succinyldicholine is most likely incidental to its unknown principal function. Assays for the enzyme utilize its property to overcome the effect of certain inhibitors. It has recently been observed that extract of potato peelings is a potent inhibitor of the abnormal enzyme (Harris and Whittaker, 1959). Since the potato is a relative newcomer to Western European foods (350 years), it is unlikely that the potato itself has played a role in producing this polymorphism. However, there may be other foods containing the unidentified ingredient of the potato peel which may have conditioned this polymorphism in the past.

Infectious diseases

Haldane (1949, 1957) pointed out that infectious diseases probably have been the main agent of natural selection of man during the past five thousand years. As soon as the invention of agriculture and urbanization made relatively dense populations possible, such selective agents as vertebrate predation ceased to be important and diseases, spread by overcrowding, took their place as agents of natural selection. When man was first exposed to the typical human infectious diseases, capacities acquired by earlier natural selection, such as the nimbleness of mind and agility of body necessary to outfight predators and escape from them, failed to protect him against infectious-disease mortality (Allison, 1960a). However, man's genotype usually was sufficiently heterogeneous that genes existed which protected their carriers against death from such diseases. With differential mortality of susceptible genotypes, consecutive epidemics of various infectious diseases acted as a sieve, concentrating resistant genes in those populations that were exposed. With a few exceptions (see below), the definite nature and action of such genes has remained unidentified.

Genetically determined resistance to mortality from infectious disease usually is not generalized but is highly specific for various diseases. Depending upon the preference for a given tissue and the mechanism of multiplication of the invaders, specific genes, acting on specific metabolic sequences and in different tissues, will protect the host against a given disease. Genetic resistance against disease produced by a given micro-organism usually does not protect against other micro-organisms. If micro-organisms are related, some protection usually carries over. An example of such a mechanism is the inherited virus-multiplication depressing factor in mice (Sabin, 1954). This single gene-controlled factor prevents infection by a group of related viruses (yellow fever, W. Nile fever, Japanese B encephalitis, St. Louis encephalitis, Russian spring tumor encephalitis). The identical gene, however, does not prevent infection by a different group of central nervous system viruses (Western and Eastern encephalomyelitis, poliomyelitis, rabies, lymphocytic meningitis, herpes virus, Rift Valley fever). All data (Gowen, 1948) indicate that the development of genetic resistance is a highly specific phenomenon acquired through evolutionary adaptation differing with the "medical history" of the population.

One would expect widespread diseases with high mortality to have had the most pronounced effect as selective agents. Age at the time of death from infectious disease also is important since only diseases which kill before reproduction will exert significant evolutionary effects. Degenerative diseases such as arteriosclerosis and cancer, which kill after reproduction has occurred, cannot be selective agents unless otherwise associated with fertility differentials.

Genetic factors causing resistance to infectious diseases should be carefully distinguished from acquired immunity. Immunity can be passive and is caused by direct transfer of circulating antibodies to the baby's circulation from the mother. Such passive immunity will disappear a few months after birth. Active immunity implies the development of antibodies after contact with the disease agent. In contrast, genetically determined natural resistance involves the inheritance of gene-controlled characters which protect the individual entirely or partially against the disease or its effects, regardless of whether or not prior contact with the disease agent has occurred. Although the capacity to develop immunity also has some genetic determinants, the present discussion excludes acquired immunity and is concerned only with inherited factors of natural resistance.

Evolution in host and parasite: myxomatosis as an evolutionary model

Carefully controlled studies in many mammalian species leave no doubt of the role of genetic factors in resistance to infectious disease. The literature in this field is considerable and has been summarized (Gowen, 1948 and 1951; Hutt, 1958).

The study of infectious diseases and their impact on the evolution of man is made more difficult by evolutionary adaptations in both host and parasite. At the same time as genetically determined host resistance develops, the parasite tends to become less virulent by mutation. The end result may be a highly attenuated micro-organism living in a genetically resistant host. Could it be that man's orphan viruses or viruses to which no specific disease can be assigned are examples of such evolutionary adaptations? Since the duration of a human generation is long as compared with that of man's parasites, evolutionary changes in micro-organisms producing human disease are probably frequently important in leading to less severe disease manifestation in a population. The main emphasis in this discussion, however, will deal with the changing genotype of the host.

Recent field studies on a viral disease of rabbits, myxomatosis, illustrate admirably the development of genetically determined host and parasite resistance and may be considered as a model of how genetic resistance to infectious disease develops (Fenner, 1959; Marshall and Fenner, 1958). The myxomatosis virus when first introduced into the Australian rabbit population was highly virulent and killed more than 95% of the infected animals. When the offspring of animals trapped in successive epidemics was infected with a uniform dose of a standard virus preparation, the percentage of animals killed decreased from year to year. There was an inverse relationship between mortality and the number of previous epidemics the animals' ancestors had suffered. These data indicated the development of a raised frequency of heritable resistance factors to myxomatosis. The heritability of these factors was demonstrated by breeding experiments. The exact nature of the genes producing genetic resistance to rabbit myxomatosis is not known. Active or passive immunity as the explanation of disease resistance to myxomatosis was carefully ruled out by the conditions of the experiments.

Evolutionary changes in the virus were demonstrated by infecting highly susceptible standard laboratory rabbits with different myxomatosis virus preparations isolated in successive epidemics. In these experiments using a genetically stable host, a decreasing mortality from year to year also

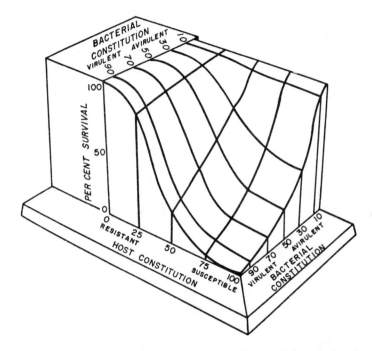

FIG. 1. Interaction of Host Constitution and Bacterial Constitution in Infectious Disease Mortality (redrawn after Gowen).

could be demonstrated, indicating the emergence of less virulent virus strains with successive epidemics.

The population dynamics in these epidemics involves differential mortality which kills susceptible genotypes and leaves the more resistant individuals to contribute more offspring. After several years, a large proportion of rabbits is therefore genetically resistant. Genetic changes of the virus probably result from the selective advantage enjoyed by those virus mutants which do not kill the host and therefore do not die with the host. Figure 1, redrawn from an article by Gowen (1952), illustrates the interaction of host susceptibility and parasite virulence and the effect on mortality, resulting from the interplay of both factors.

Infectious disease and balanced polymorphisms

In many instances genes which protect against an infectious disease may have lowered the fitness of the individual in other ways. This situation where a gene has both advantageous and injurious properties leads to "balanced polymorphism" where a fixed population frequency of the trait is reached (Ford, 1945; Sheppard, 1958; Allison, 1959a). This frequency represents a balance between the tendency for gene-frequency increase due to beneficial factors and for gene-frequency decrease from deleterious effects. It will be shown that three common red blood cell polymorphisms of man, sickling trait, thalassemia, and glucose-6-phosphate dehydrogenase deficiency, probably owe their present frequencies to selection by falciparum malaria.

It may be argued on general genetic principles and on the example of the polymorphisms conferring resistance to malaria, that most genes protecting against infectious diseases initially were rare and deleterious (Allison, 1960a; Sheppard, 1958). Furthermore, many presently known deleterious recessive and other genes, such as those for cystic fibrosis, phenylketonuria, spastic diplegia, schizophrenia, and hyperuricemia, exist in populations at such frequencies that chance or mutation pressure could not have accounted for their presently existing incidence (Penrose, 1957). They, therefore, must be considered true polymorphisms. Carriers of such traits presumably have enjoyed some advantage in the past. Since infectious diseases probably represent an important selective sieve in man's recent evolutionary history, it is not unlikely that many of these deleterious recessive genes may have been spread through populations by the selective action of epidemics. With disappearance of infectious disease in recent years, traits with deleterious effects (balanced polymorphisms), apart from their protection against death from infectious disease, will decrease in frequency once their selective advantage is removed. If most genes conferring resistance to infectious disease do indeed lower fitness, control of infectious diseases will lead to decreased frequency of many deleterious genes by removing their selective advantage. In medical terms, the result would be the decline of some genes that cause hereditary diseases. This is an important and somewhat paradoxical phenomenon demonstrating how environmental manipulation by improved hygienic conditions can lead to eugenic improvement.[4]

Those genes which confer resistance against infectious diseases but which otherwise have neutral effects, would be expected to have replaced their alleles through selective action. If the infectious disease disappears before such a trait has spread through the entire population, a given attained frequency would continue throughout future generations in contrast to the declining gene frequency observed with balanced polymorphisms. A gene that protects against infectious disease and that has additional beneficial effects would spread through a population very rapidly and would even continue to do so at lesser speed when the infectious disease is no longer present.

[4]It needs to be pointed out, however, that in rapidly expanding populations as exist today (doubling every 30 to 40 years) the actual number of patients with these diseases will actually increase, even though the percentage frequency of the genes determining these diseases will diminish.

Tuberculosis

When tuberculosis first strikes a population without previous contact with the disease, mortality is high. In 1890 tuberculosis was first introduced into the Qu'Appelle Valley Indian Reservation in Saskatchewan. The annual tuberculosis death rate reached the all-time high figure of almost 10% of the total population (Ferguson, 1955). More than one-half of the Indian families were eliminated in the first three generations of the epidemic. Twenty per cent of the mortality of the remaining population was also due to tuberculosis. After 40 years and three generations most susceptible individuals apparently died and the annual death rate had been reduced to 0.2%.

High mortality rates and acute disease had always been noticed when various populations were first exposed to tuberculosis (table 1). On genetic

TABLE 1. TUBERCULOSIS

Fulminant infection and high mortality on 1st exposure in:
American Indians
Negroes
Puerto Ricans
Eskimos
Polynesians
East Asians
Irish

principles, exposure over many generations should lead to death of susceptible individuals with survival of those individuals who are genetically resistant.

The Ashkenazi Jewish population of America represents a population that has survived the high tuberculosis attack rate of the crowded ghettos of Europe for many generations. Although the rate of tuberculosis infection, as determined by tuberculin testing, has been identical in Jews and Gentiles, tuberculosis mortality is significantly less in Jews than in Gentiles (Perla and Marmorston, 1941).[5] In contrast to the acute rapidly growing caseous type of tuberculosis with regional lymph node involvement and the high rate of tuberculosis meningitis during early exposure of a population to tuberculosis, populations with a history of contact over many generations develop a more chronic, fibrous tuberculosis, as seen now in Europe and in America. Yemenite Jews immigrated to Israel from an agricultural milieu with little tuberculosis. Tuberculosis among the Yemenites is of the rapidly progressing type with a high mortality (Dubos and Dubos, 1952). Statistics collected during World War II in the ghetto of Warsaw suggest that the genetically acquired relative resistance of Ash-

[5]Some of this difference, however, may be due to better medical care enjoyed by the Jewish group.

kenazi Jews to tuberculosis can be rapidly overcome in conditions of extreme crowding and starvation (Dubos and Dubos, 1952).

The importance of the host factor is underlined by the different reactions of human populations with different ancestral tuberculosis histories when exposed to the disease in the United States at the present time. Most of the populations listed in table 1, such as Indians, Eskimos, and Puerto Ricans, still demonstrate a much more acute type of disease when they contract tuberculosis.

Several family studies suggest hereditary susceptibility (Puffer, 1946). Since, in a contagious disease, environmental factors are difficult to separate from genetic determinants, these data are difficult to evaluate. All twin studies, however, have demonstrated increased concordance for the disease among monozygotic twins as compared with dizygotic twins (table 2).

In view of the suggestive nature of these human data, conclusive animal experiments on genetic resistance to tuberculosis are of special interest. By artificial selection several teams of investigators have succeeded in raising strains of rabbits which are either highly susceptible or resistant to tuberculosis (Laurie et al., 1951; Diehl, 1958). Resistance is unrelated to acquired immunity. It appears that resistance is caused by the inherited ability of the animal to phagocytose the tuberculosis organism, thus preventing spread (Lurie et al., 1951). The nature and action of the genes involved is unknown.

The combination of all evidence suggests strongly that the present relatively high resistance of Western populations to tuberculosis is genetically conditioned through natural selection during long contact with the disease. The decline in tuberculosis mortality had begun before discovery of the tuberculosis organism and before medical measures were taken and probably is partially due to selective mortality of susceptible population members.

Plague

Plague has been one of the great killers of populations in past times. During the European epidemic in the 14th century, a minimum of 25% of the entire European population died. In certain areas the mortality was much higher, as shown in table 3. Data from South Africa suggest that Europeans, as descendants of populations who survived the most severe plague epidemics, are more resistant to pneumonic plague than Negroes, while Negroes are more resistant than Asiatic Indians, Chinese and Malayans, who are most susceptible (Mitchell, 1927). Older accounts also mention the greater susceptibility of the Negro. The genetic significance of these data is doubtful in view of the differences in living conditions between the various populations.

TABLE 2. TUBERCULOSIS IN TWINS (VON VERSCHUER, 1959)

Year	Author	Monozygote Twins		Dizygote Twins		Concordance %	
		No.	Concordant	No.	Concordant	Mz	Dz
1936	Diehl and von Verschuer	80	52	125	31	65	24.8
1939	Uehlinger and Künsch	12	7	34	2	58.3	5.9
1943	Kallmann and Reisner	78	69	230	83	88.5	36.1
1944	Vaccarezza and Dutrey	20	13	43	5	65	11.6
1956	Mikami	26	14	9	3	53.8	33.3
1957	Simonds	93	27	218	39	29	17.9
1958	Koch	22	10	17	6	45.4	35.3
1958	Harvald and Hauge	50	10	167	18	20	10.8
	Total	381	202	843	187	53.0	22.2

TABLE 3. PLAGUE

	Location	Population Mortality	Source
14th Century epidemic	Cyprus	100%	Kollath (1951)
	Lübeck (Germany)	90%	"
	Silesia	80%	"
	Smolensk (Russia)	98%	"
	Avignon (France)	75%	"
	Marseille (France)	55%	"
	Venice (Italy)	75%	"
	Great Britain	66%	Pollitzer (1954)
Rats—no plague for 30 yrs.		91%	Sokhey and Chitre (1937)
Rats—severe plague in recent yrs.		10%	"

Although definite genetic resistance factors have not been proven in man, studies on rats suggest strongly that plague bacilli kill the more susceptible genotypes. Wild rats, the carriers of the human plague bacillus, were captured from many cities in India where the recent experience with plague differed greatly from city to city. These rats were inoculated with a standard dose of plague bacillus. Mortality in rats was inversely proportional to plague exposure, varying from 91% for rats from cities with no plague for the previous 30 years to 10% for rats from cities with severe plague up to two years before capture of the rats (Sokhey and Chitre, 1937). The design of these experiments appears to exclude active immunization and suggests the acquisition of genetic resistance.

Smallpox

There is good reason to believe that smallpox contributed to the rapid defeat of the Aztecs by Cortez and his conquistadores. The disease apparently was transmitted to the Indians in 1520 by a Negro in Cortez' army and spread very rapidly through the population. Early writers give a mortality figure of 3,500,000 for this epidemic (Kollath, 1951; Dubos, 1959). It appears that at least one-half of the Indian population died. A century or so later repeated outbreaks of smallpox decimated many American Indian populations in North America. Early in the 17th century the Massachusetts and Narragansett Indians were reduced from 30,000 and 9,000, respectively, to a few hundred. In the 19th century the Mandan population fell from 1,600 to 31 and very high mortality was noted among the Assiniboins, the Crows, the Plains tribes, and the Blackfeet (table 4). In fact, the spread of smallpox probably was one of the first examples of biological warfare. The European settlers realized the high susceptibility of the Indians to smallpox and purposely spread infected blankets (Dubos, 1959).

In Hawaii, in 1853, there were over 9,000 cases of smallpox with 6,000 deaths among a population of 70,000 (Hirsch, 1883). In 1707 after a long period of freedom from smallpox an epidemic in Ireland killed 18,000 out

TABLE 4. SMALLPOX

Year	Population	% Pop. Killed	Source
1520	Aztecs	appro. 50%	Dubos (1959)
1602–1620	American Indians (New England)	over 90%	Dubos (1959)
1707	Irish	36%	Perla and Marmorston (1941)
1837	American Indians (Mandan, Assiniboin, Crow)	20–95%	Dubos (1959)
1853	Hawaiians	8%	Hirsch (1883)
1863	Marquesas Islanders	25%	Hirsch (1883)

of 50,000 inhabitants (Perla and Marmorston, 1941). Protective immunization prevents smallpox in Western countries. The disease has become endemic rather than epidemic in areas of the world such as India where vaccination is not generally practiced. Case mortality—not population mortality—still averages 25%. The role of genetic resistance in this disease is difficult to evaluate but presumably has played a role.

Measles

Measles is a rather benign disease with negligible childhood mortality at this time in Western society. The measles virus produced severe disease which killed large numbers when first introduced into virgin populations without previous contact. Panum described the introduction of measles into the Faroe Islands in 1846 (Panum, 1940); three-quarters of the population became infected. Recent epidemics of this type have also occurred in isolated communities; for example, Eskimos of the Canadian Arctic suffered a mortality of as high as 7% (Dubos, 1959). The Tupari Indians of the Brazilian forest were first discovered in 1949 and numbered some 200 people. Six years later two-thirds of the group had died of measles introduced by rubber gatherers (Dubos, 1959). Measles had a very high mortality rate in various Pacific islands during the last century. A Hawaiian king and his queen died of measles in 1824 during a visit to England. Their attending physicians, including Sir Henry Halford, president of the Royal College of Physicians, found it hard to believe that a disease "which even a delicate London girl might bear could be so destructive to robust denizens of the Pacific" (Dubos, 1959). In 1848 every child born in Hawaii died of measles, pertussis or influenza. In 1874, 20,000 Fiji Islanders died of measles introduced from Sydney, Australia (Perla and Marmorston, 1941). Since adults suffer a markedly increased mortality whenever they contact measles, some of the above findings probably can be explained on that basis. It is possible however that modern populations with low measles mortality have acquired genetic resistance against death from measles. The possibility of diminished virulence of virus strains undoubtedly plays a role.

TABLE 5. PARALYTIC POLIOMYELITIS IN ISOLATED COMMUNITIES (SABIN, 1951)

Region	Year	Population Size	Attack Rate per 100,000
Chesterfield Inlet Arctic Eskimos	1949	275	21000
Sukkertoppen Greenland	1914	700	5300 dead many more paralyzed
Sukkertoppen Greenland	1932	700	2400
Kangamiut Greenland	1932	300	4300
Holsteinborg Greenland	1932	400	4000
St. Helena	1945	4000	1920
Nicobar Island India	1948	10000	8000
Guam	1899	8660	808 dead many more paralyzed
U.S.A.			20–100

Poliomyelitis

Genetic factors appear to play a role in susceptibility to paralytic polio-myelitis. Evidence along different lines is available: a. There is significantly increased occurrence of paralytic poliomyelitis in certain families in different years and in different generations (Aycock, 1942; Addair and Snyder, 1942). These data have been interpreted as compatible with single reces-sive gene inheritance (Addair and Snyder, 1942). Sabin's mouse data im-plicating a single pair of recessive genes in causing resistance to certain nervous system viruses are also of pertinence in this regard (Sabin, 1954). b. Twin studies show a high concordance rate among monovular twins (35% concordance in monozygous twins versus 6% in dizygous twins and 6% in other siblings) (Herndon and Jennings, 1951). c. Attack rates of par-alytic poliomyelitis are very high in isolated inbred communities (table 5). d. Attack rates of clinically recognizable poliomyelitis vary in different population groups living in the same neighborhoods and attending the same schools without segregation (table 6).

TABLE 6. POLIOMYELITIS IN HAWAII, 1938–1947 (SABIN, 1951)

Population	Mean Annual Attack Rate per 100,000
Total	5.5
Caucasian	10.2
Part Hawaiian	9.0
Japanese	3.9
Chinese	2.7
Filipino	1.6
Hawaiian	1.3

Although the case mortality of poliomyelitis usually is not high and does not compare in magnitude with some of the previously cited diseases, selective factors may play a role under primitive conditions where paralysis may be a serious handicap to life. As in other diseases, mutability of the virus, rather than the host, is again a most important factor in causing differences between epidemics.

Yellow Fever and Trypanosomiasis

It has been argued that genetic host factors have played a role in the natural history of yellow fever (Sabin, 1954). The highly endemic areas of Africa are for the most part inhabited by a population group that does not suffer from the severe clinical manifestations of the disease. Strangers coming to these areas usually die of a severe form of the disease. It is likely that prolonged exposure has killed off the susceptible genotypes. The occurrence among South American Indians of both mild and severe forms of yellow fever is in accord with the hypothesis that yellow fever was imported from Africa after the discovery of America and has not had sufficient time to kill off all susceptible individuals.

When Trypanosomiasis is first introduced into an area where it did not exist, it may kill one-third to two-thirds of the exposed population. After some years it becomes a much milder disease (Dubos, 1959). An interesting hemoglobin polymorphism exists in cattle which may have some bearing on trypanosomal resistance. Hemoglobin B in cattle is absent from the Mututu and N'Dama breeds of Nigeria which are more susceptible to trypanosomiasis than Zebu cattle (Bangham and Blumberg, 1958).

Malaria

This disease has had a profound influence on human events and mortality for at least 2,000 years (Boyd, 1949). Malaria is very widespread and even now is said to kill two million children every year. On evolutionary grounds, such a disease should be an important selective agent. Polymorphisms affecting body tissues necessary for malarial growth which exist in populations subjected to malaria for many generations (sickle-cell trait, thalassemia, glucose-6-phosphate dehydrogenase deficiency) would therefore be suspect of owing their distribution to the selective action of malaria.

Several varieties of malarial parasites produce human malaria. The most potent selective agent would be that variety of malaria associated with the highest mortality. Falciparum malaria is the most lethal type of malaria and is found most frequently in tropical and subtropical areas (figure 2). Vivax malaria is less likely to kill and is found more frequently in temperate zones. On evolutionary and parasitologic grounds, vivax malaria is the older species (Knowles et al., 1930; Boyd, 1949; Bray, 1957). The almost complete natural resistance of the West African Negro to vivax malaria is

FIG. 2. Distribution of Falciparum Malaria (see Boyd, 1949).

therefore of interest (Boyd, 1949). This resistance is unrelated to any of the known polymorphisms discussed below.

The host tissue in which malarial organisms primarily proliferate is red blood cells. The malarial parasite has many of the enzyme systems that exist in the red blood cell and depends on some of the enzymes and metabolites of the red cell for its normal metabolism (Trager, 1957; Geiman, 1951). The adaptation of the parasite to the red blood cell represents a finely balanced end result of evolutionary development. One would expect suboptimal growth of malarial parasites in red blood cells which deviate from the normal, since an abnormal red cell might be a less satisfactory host for malarial growth. Since the number of parasites in the red blood cells appears to be related to malarial mortality, patients with other than normal red cells would be expected to have a lower malaria mortality rate.

The development of acquired immunity in children living in holoendemic malarial areas must be understood to evaluate the effect of possible genetic resistance factors against malarial mortality. An infant in a holoendemic malarial area will be born with passive immunity acquired from the mother. Passive immunity disappears after the first one-half year of life. The child then becomes susceptible to malarial death between six months and the second to third year. At that time active immunity to the parasite begins to play a significant role in preventing mortality. Although older children may still be heavily infected with parasites, clinical illness becomes progressively milder and more rare so that such individuals appear to live more or less in harmony with the parasite (Macdonald, 1957; McGregor, 1959).

A study attempting to show the protective effect of a genetic trait on malarial mortality therefore should concentrate on young children in the age group of six months to three years.

Sickle-Cell Trait

This trait is due to the single dose of a mutant gene causing production of an abnormal hemoglobin molecule—the sickling hemoglobin. The sickling trait is widely spread throughout equatorial Africa and is also found in Greece, Turkey, and India (figure 3). Sickle-trait carriers have both normal and abnormal hemoglobin in all their red blood cells. For practical purposes the sickle-cell trait is not associated with disease. Mating of two sickle-trait carriers produces 25% offspring with the double dose of the sickling gene. Under primitive conditions of life, this condition—sickle-cell anemia—is usually lethal before reproduction. Frequencies of sickling trait as high as 40% occur in certain African populations. Since with every death of a child from sickle-cell anemia, two sickle genes are lost from the population, the explanation for high sickle-trait frequencies would demand an abnormally high mutation rate or reproductive overcompensation. Both possibilities have been ruled out by direct studies (Vandepitte *et al.*, 1955; Allison, 1956). An alternate explanation would postulate a selective advantage for sickle-cell trait carriers, balancing the loss of sickling genes from deaths due to sickle-cell anemia. It is now generally conceded that carriers of the sickle-cell trait are less likely to die from falciparum

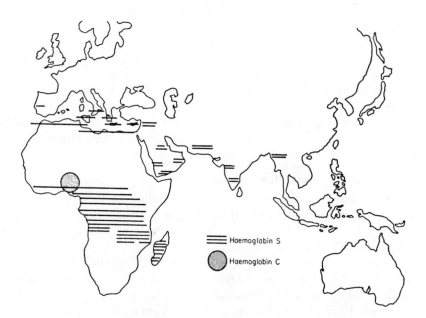

FIG. 3. Distribution of Hemoglobin S and Hemoglobin C.

TABLE 7. SICKLING AND MALARIA (VANDEPITTE AND DELAISSE, 1957)

| | | Sickling Trait | |
	Total	Number	%
Malarial children ($>$ 1000 parasites/mm^3)	386	53	13.7
Control	1180	286	24.4
X^2 = 18.9		P $<$ 0.01	

malaria. Properly controlled studies of young children have demonstrated protective action of the sickling trait (Vandepitte, 1959; Vandepitte and Delaisse, 1957; Allison, 1957; Raper, 1959). Table 7 shows representative data on the relationship of parasite counts to sickling trait in a highly malarial population, demonstrating the protective action of the sickling trait against malarial proliferation. Other less extensive data have shown definitely diminished malarial mortality of sickling-trait carriers (Lambotte-Legrand and Lambotte-Legrand, 1958). More such data are needed. It has been calculated that a mortality from malaria of 10%, if the deaths occur entirely among the nonsickling population, is sufficient to explain the persistence of the sickle-cell gene at the high observed frequency (Allison, 1956, 1957). If some trait carriers also die from malaria, the figure must be increased proportionately. The sickling trait does not necessarily protect against infection by malaria. In many studies the proportion of parasite carriers was found to be identical in sicklers and the normal population. From the evolutionary point of view, differential mortality is the important phenomenon and, indeed, has been found.

Control of malaria should lead to decline of the sickling gene by removing the selective advantage. Decrease of sickle-cell trait frequency in fact has been demonstrated in Negroes of the Dutch West Indies whose specific origin in Africa was known (Jonxis, 1959). The lowered frequency of the sickle-cell trait in American Negroes as compared with African Negroes is also suggestive. The literature on the relationship of sickling to malaria has been extensively reviewed (Vandepitte and Delaisse, 1957; Allison, 1957; Lehmann, 1959; Neel, 1956, 1957).

Other Abnormal Hemoglobins

Apart from Hemoglobin S, many hemoglobin variants have been discovered in the last few years. Most of these are rare mutants; those that can be definitely classified as polymorphisms are Hemoglobin C and Hemoglobin E.

Because of the limited geographic distribution of Hemoglobin C to a relatively small area of West Africa (figure 3) it has been postulated that the mutation leading to Hemoglobin C is more recent and may have developed from that of Hemoglobin S (Mourant, 1954). Hemoglobin C trait is not associated with illness. Since homozygous Hemoglobin C disease is far less lethal than sickle-cell anemia, the selective advantage enjoyed by Hemoglobin C carriers would need to be less for genetic balance to occur. Data failing to demonstrate a protective action of Hemoglobin C towards malarial parasite density (Edington, 1959), therefore, may not be crucial. A selective advantage towards falciparum malaria of lesser magnitude than that balancing the sickling gene would be extremely difficult to demonstrate by present methods. Such proof may be possible only by careful collection of extensive mortality statistics of the different genotypes. Such practices do not yet prevail in the malarial regions of Africa.

Hemoglobin E is found throughout Southeast Asia. Its incidence is about 10% in Siam, Burma, and North Malaya, and as high as 35% in Cambodia. All these are highly malarious countries. Hemoglobin E trait is innocuous and Hemoglobin E disease is relatively mild. No information on malarial protection is available.

Thalassemia

The thalassemia trait is a red cell abnormality common throughout the Mediterranean area, the Near East, Arabia, India, Southeast Asia, China, the Philippines, and Africa (figure 4). It is likely that the diagnostic term, thalassemia, includes several different genetic entities all producing a similar hematologic phenotype. A fairly well defined thalassemia-like condition is characterized by the presence of large amounts of fetal (F) hemoglobin in heterozygotes. Thalassemia in a yet unknown manner interferes

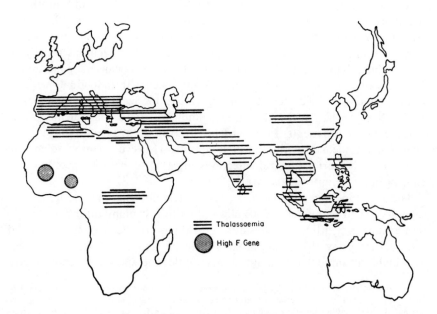

FIG. 4. Distribution of Thalassemia (see Chernoff, 1959).

with effective hemoglobin synthesis so that trait cells are deficient in hemoglobin. In some carriers the defect may be mild; in others significant anemia may occur. The homozygous condition—thalassemia major—is a highly lethal disease.

Considerations of the geographic distribution of thalassemia and the severe loss of thalassemia genes from thalassemia major suggested to Haldane the possibility of a selective advantage of the trait to malaria. Not many critical studies have been performed to test this hypothesis. Data collected in Sardinia suggest very strongly that thalassemia heterozygotes are indeed protected against malaria. Ceppellini and Carcassi studied two racially identical populations, one, living in a severely malarious area showed a high frequency (20%) of thalassemia, while the other population living in a non-malarial mountainous area, showed very little thalassemia (Ceppellini, 1959).

Glucose-6-Phosphate Dehydrogenase Deficiency (Primaquine Sensitivity)

This polymorphism (Beutler, 1959) was discovered several years ago when it was noted that 10% of American Negro soldiers developed severe blood destruction when given a new anti-malarial drug—primaquine. A series of brilliant investigations performed at the University of Chicago resulted in the demonstration that the defect causing susceptibility to drug-induced hemolysis was deficiency of a red cell enzyme—glucose-6-phosphate dehydrogenase (Carson et al., 1956). Abnormalities in glutathione metabolism of affected red cells were also described (Beutler, 1959). The trait is inherited as a sex-linked character with intermediate dominance (Childs et al., 1958).[6] Investigations in Italy and Israel demonstrated that favism (hemolytic anemia from ingestion of raw fava beans or inhalation of products of flowering fava bean fields) only occurs in individuals with an identical enzyme deficiency (Szeinberg et al., 1958a; Sansone and Segni, 1957). Many other drugs have been demonstrated to cause hemolysis in enzyme deficient patients. In the absence of drug or fava bean exposure, the trait is not associated with clinical symptoms. However, hemolytic anemia induced by bacterial or viral infections appears to be not uncommon in enzyme deficient individuals (Marks, 1959).

The development of a rapid screening test for the enzyme deficiency (Motulsky and Campbell, 1960) allowed us to test many different population groups in Seattle as well as during field trips to the Belgian Congo and Sardinia in spring, 1959, and to Alaska in the summer of 1959. Countries in which favism or drug-induced hemolytic anemia were reported are

[6]Population surveys on males (who have only one X chromosome) will give the gene frequency of the trait. Carrier females (heterozygotes) usually have intermediate but may have normal or low enzyme levels. Frequency data collected on females may therefore be misleading for genetic purposes. Incidence data on males are always bimodal and clearcut.

TABLE 8. COUNTRIES AND ISLANDS WITH FAVISM OR 8-AMINO-QUINOLINE
INDUCED HEMOLYTIC ANEMIA OR BOTH

Portugal	Rhodes	Turkey	India
Spain	Crete	Egypt	China
Southern Italy	Cyprus	Lebanon	Formosa
Minorca	Corfu	Iran	Burma
Sardinia	Bulgaria	Iraq	Java
Sicily	Greece	Israel	Mauritius
		(Non-Ashkenazi Jews only)	

listed in table 8 (Beutler, 1959; Motulsky and Campbell, 1960; Sansone *et al.*, 1959). Our studies and those of others on the distribution of the gene in different populations are summarized in table 9 (Besides the studies cited in the table see also Sansone *et al.*, 1959, and Vella, 1959a). The limitation of the trait to a wide belt of tropical Africa, the Mediterranean area, the Near East, Indian Southeast Asia and the Philippines—all ma-

TABLE 9. GLUCOSE-6-PHOSPHATE DEHYDROGENASE DEFICIENCY

Africans		
American Negroes	9-11%	Motulsky & Campbell (1960)
Leopoldville Negroes	18-23%	Vandepitte & Motulsky (unpublished)
Stanleyville Negroes	14-15%	Motulsky, Dherte & Ninane (unpublished)
Bayaka (S. Congo)	15-28% (average 20%)	Motulsky & Vandepitte (unpublished)
Bwaka (N. W. Congo)	6%	Motulsky (unpublished)
Watutsi	1-2%	Motulsky (unpublished)
Bahutu	7%	Motulsky (unpublished)
Bashi	14%	Motulsky (unpublished)
Pygmies	4%	Ninane & Motulsky (unpublished)
S. African Bantus	2-4%	Charlton & Bothwell (1959)
" Bushmen	1-2%	Bothwell (personal communication)
Nigerians	10%	Gilles, Watson-Williams & Taylor (1960)
Asiatics		
Asiatic Indians	3-8%	Vella (1959b), Motulsky & Campbell (1960)
Filipinos	12.7%	Motulsky & Campbell (1960)
Chinese	2%	Vella (1959), Beutler *et al.* (1959)
Japanese	0%	Motulsky & Campbell (1960)
Micronesians	0-1%	Blumberg, Campbell & Motulsky (unpublished)
Iranians	8.5%	Walker & Bowmann (1959)
Americans		
Eskimos (Alaskan)	0%	Motulsky & Campbell (1960)
American Indians	0%	Motulsky & Campbell (1960)
Peruvian Indians	0%	Best (1959)
Oyana Indians	16%	Keller *et al.* (in preparation)
Carib Indians	2%	Keller *et al.* (in preparation)

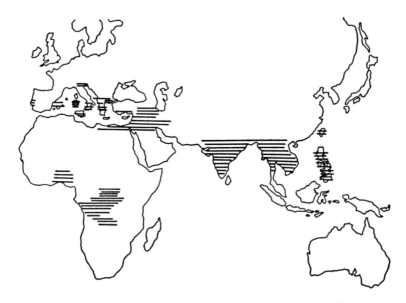

FIG. 5. Distribution of Glucose-6-Phosphate Dehydrogenase Deficiency.

larial areas (figure 5) is noteworthy. The trait has been found in South American Indians (Oyana), who have lived in malarial areas for many generations (Keller *et al.*, in preparation). It is absent in Eskimos and some American Indians (Motulsky and Campbell, 1960), and Peruvian Highland Indians (Best, 1959). It is also absent in northern Europeans, Japanese, and American Caucasoids (Motulsky and Campbell, 1960).

Israeli workers demonstrated (table 10), a trait frequency of as high as 60% among Kurdish Jews, 25% among the Baghdad and Persian Jews, and lower percentages among various groups of Sephardic Jews, such as those from North Africa. The trait was extremely rare in the Ashkenazi Jews, who have not lived in malarial environments for about two thousand years. A high frequency among the Kurdish Jews probably reflects a high degree of inbreeding in a relatively small isolate. Thalassemia also has a high frequency among the Kurdish Jews. No enzyme deficiency was found among the Falasha Jews of Ethiopia where there is no malaria (Sheba, C., personal communications).

TABLE 10. GLUCOSE-6-PHOSPHATE DEHYDROGENASE DEFICIENCY IN JEWS

Szeinberg, *et al.*, 1958b, and Sheba (personal communication)	
Kurdish Jews	60%
Persian and Iraqi Jews	25%
Turkish Jews	5%
Yemenite Jews	5%
N. African Jews	2%
Ashkenazi Jews	0.2%

Our field studies in Africa (with Dr. J. Vandepitte) demonstrated a high frequency of the trait in many tribes from malarial areas, and a low frequency in tribes with no malaria in the past (table 9). In view of the protective effect of sickling on malaria, the relationship of sickling and enzyme deficiency was of some interest. There was good correlation between frequency of sickling and enzyme deficiency. If the sickling rate of the tribe was low, then enzyme deficiency was also low and high enzyme deficiency frequency was associated with high sickling rates (figure 6). The only exception proved to be the Ituri Forest pygmies with a sickling rate of 31% while an enzyme deficiency rate of 4% was found. Sickling and enzyme deficiency affected different population members. The incidence of combined enzyme and sickling was not higher than expected by random coincidence of the two traits in an individual. The African data suggest that a common selective agent—malaria—has favored carriers of both these independent mutations.

Field data collected in Sardinia (with Dr. M. Siniscalco) demonstrated, as in thalassemia, extremely high rates of enzyme deficiency in some coastal areas which were highly malarious until recently and low rates in the nonmalarious central mountain areas (figure 7). Sardinia was until recently the most malarial area of Italy with the highest malaria mortality. Dr. Siniscalco and his group are carrying out further studies in Sardinia and other areas of Italy.

Data from Greece indicate that the enzyme deficiency is not infrequent (Zannos-Mariolea and Chiotakis, 1959). The trait is especially common in the Patras area of the Peloponnesus and in the Chalkidiki Peninsula.

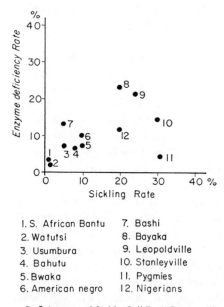

1. S. African Bantu
2. Watutsi
3. Usumbura
4. Bahutu
5. Bwaka
6. American negro
7. Bashi
8. Bayaka
9. Leopoldville
10. Stanleyville
11. Pygmies
12. Nigerians

FIG. 6. Enzyme Deficiency and Sickle-Cell Trait Frequencies in Africans.

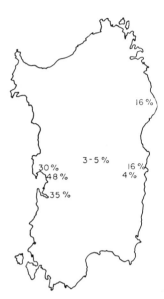

FIG. 7. Glucose-6-Phosphate Dehydrogenase Deficiency in Sardinia.
(Siniscalco and Motulsky, unpublished)

These areas are also principal foci for sickling and thalassemia and were known in the past for their high malarial endemicity. Greece is the only country in the world where sickling, thalassemia and glucose-6-phosphate dehydrogenase deficiency occur at significant frequencies in an identical population. Usually, either sickling, such as in Africa, or thalassemia, such as in the Mediterranean area, Near East, India and S. E. Asia, have been found in populations affected by glucose-6-phosphate dehydrogenase deficiency.

There is evidence that the enzyme deficiency in Africans may be produced by a different gene, since the mean level of enzyme among enzyme deficient Negroes is significantly higher than among enzyme deficient southern Europeans (Marks and Gross, 1959a; Motulsky unpublished). It is not unlikely, therefore, that there are at least two sex-linked mutations (alelles?) both producing glucose-6-phosphate dehydrogenase deficiency. It is also possible that the different mean enzyme levels present a phenotypic difference in red cells which in the two broad population groups constitute different genetic environments.

Glucose-6-phosphate dehydrogenase deficiency does not affect all red cells equally. The oldest red cells (red cell life span is 120 days) have a much more severe degree of enzyme deficiency than do the younger red cells (Marks and Gross, 1959a and b). Since vivax malaria more readily parasitizes young red cells, it is unlikely that enzyme deficiency protects against vivax malaria. Although the selective advantage of the enzyme deficiency probably relates to falciparum malaria, it is significant that a not uncommon malarial parasite—plasmodium malariae—occurring with the same approximate geographic distribution as plasmodium falciparum (Knowles *et al.*, 1930)—is said to parasitize preferentially the older red cells. In view of the marked degree of enzyme deficiency in old cells, more detailed studies in a population where plasmodium malariae is common might be of interest.

Malarial organisms require glutathione for *in vitro* growth (Trager, 1941; McGhee and Trager, 1950) and about 50% of the red cell's glutathione contributes to the cysteine requirement of malarial organisms (Fulton and Grant, 1956). Enzyme deficient red cells have a diminished amount of glutathione which is easily depletable (Beutler, 1959). There is some evidence that malarial organisms use the oxidative pathway of carbohydrate metabolism (Geiman, 1951) which is defective due to the enzyme deficiency. An enzyme deficient red cell would, therefore, be less likely to support optimal growth of malarial organisms than a normal red cell. To test this hypothesis directly, a field study was done with Dr. Vandepitte (unpublished) among 600 male Bayaka children less than 10 years of age in a holoendemic malarial area of the Southern Congo. Parasite counts of enzyme deficient and sickling children were compared with those of normal subjects. The results were inconclusive and failed to demonstrate a statistically significant effect of enzyme deficiency or sickling on parasite multiplication as measured by parasite density.

Allison and Clyde very recently have been able to demonstrate lower parasite levels in enzyme deficient children of East Africa in a field study limited to children between four months and four years (Allison and Clyde, 1960). For technical reasons Dr. Vandepitte and myself were unable to get many children in this age group, so that most of the individuals studied by us were five to ten years old. The difference in these results can be explained by immunity which in the older children blurs the critical differences. Using our test, Allison (1960b) also found a low frequency (1.7 to 2.9%) of enzyme deficiency in non-malarial areas and a high frequency (15 to 28%) in malarial areas of East Africa. All evidence therefore, strongly suggests that enzyme deficiency protects against malarial mortality.

If from the point of view of natural selection, enzyme deficiency were entirely neutral apart from its effect on malaria mortality, we would expect that the trait would have replaced its normal allele in some populations. Since the trait exists as a polymorphism in all populations studied so far, the protective effect against malarial mortality must be counterbalanced by an injurious effect. Although favism does not appear to exist in Africa, it is likely that the enzyme deficiency sometimes may be lethal by causing blood destruction during infection. Glucose-6-phosphate dehydrogenase deficiency, therefore, appears to be balanced in a population by resistance to malaria on the one hand, and by other infectious diseases (*e. g.*, common viruses) or foods (fava beans) which lower fitness on the other. Since the degree of lowered fitness in glucose-6-phosphate dehydrogenase deficiency undoubtedly is significantly less than that in sickle-cell anemia, the selective advantage necessary for balance needs to be only slight. The smaller the disadvantage of enzyme deficient carriers, the less of a selective advantage is required to explain the population frequencies of the trait. As in Hemoglobin C, critical studies may therefore be difficult to assemble. The detailed population-genetic implications of balance in a sex-linked trait such as the enzyme deficiency provide interesting problems and are under further study.

Evolutionary Adaptation in Malarial Parasites; Polygenes

Since malarial parasites are known to undergo evolutionary adaptations readily, the possibility of mutations leading to parasite strains adapted to optimal growth in genetically abnormal red cells needs to be considered. A strain of parasites better adapted to sickle cells or enzyme deficient cells might theoretically emerge, although it has not been identified. Allison (1957) cites the following reasons why the normal strain might survive at the expense of the mutant. Subjects with the trait are always in the minority and usually represent a small fraction of the population. Furthermore, there is evidence in the sickle-cell trait that the normal parasite strain forms gametocytes[7] readily so that the normal strain would have as good a chance of infecting mosquitos as the postulated mutant strain.

The demonstration that three specific traits, each under control of a single gene, appear to protect against malaria suggests how different genes might interact to produce a polygenic system of disease resistance. In fact, most data on genes conferring resistance to infectious disease in animals suggest the operation of polygenes. Judging from the relative resistance of the West African to vivax malaria, it is likely that many other yet unidentified genes exist which protect against malaria. Present-day success in isolating specific genes of a possible polygenic complex is a hopeful development indicating that analysis of polygenic traits in man may sometimes be approached with direct rather than statistical methods.

Comments

In contrast to the rather extensive data on the polymorphisms which presumably owe their distribution to malarial mortality, no data could be given on the nature of genes conferring resistance to other diseases. The reason obviously is the limited present-day possibility to study infectious diseases, such as smallpox or plague. Another reason relates to the pathophysiology of infection. With present methods we detect metabolic polymorphisms of body tissues that are easily obtainable, such as blood. When the various components of blood play a minor role or none at all in the production or dissemination of a disease, blood polymorphisms are unlikely to

[7]That form of the parasite's life cycle which is ingested by the mosquito vector.

play a role in disease resistance. Present techniques do not allow direct examinations of polymorphisms which affect the internal tissues of many individuals in a population. Sometimes, however, blood cells may carry vestigial enzyme systems whose principal function is exerted in other organs. Study of such systems may allow a more ready approach to some of these problems.

Serum Protein Polymorphisms and Infectious Disease

Recently discovered polymorphisms of serum proteins are the haptoglobins and transferrin system. Haptoglobins are hemoglobin binding alpha 2 globulins, controlled by a single pair of autosomal genes (Smithies and Walker, 1955). Three electrophoretically distinguishable main varieties of haptoglobins exist (2 homozygotes and 1 heterozygote). Genetic suppression of haptoglobins occurs as a polymorphic trait among African and American Negroes (Giblett, 1959). Hemoglobin binding differs in the three varieties (Nyman, 1959) and is absent in ahaptoglobinemia. Since the amount of haptoglobin increases in a variety of acute and chronic diseases, the fundamental physiologic function of haptoglobins may not be related to hemoglobin binding. In view of this adaptive response to disease in the individual, one might speculate that the basis of haptoglobin polymorphism may represent differential adaptation to infectious disease in the past.

Transferrins are beta globulins which bind plasma iron. Ten percent of American Negroes and Australian Aborigines carry an electrophoretically detectable transferrin variant (Type CD1) (Giblett *et al.*, 1959). Other, more rare transferrin variants have also been described in Caucasoid and Negro populations (Giblett *et al.*, 1959). The basis for transferrin polymorphism is not apparent from considerations of iron metabolism since the different variants bind iron to the same extent (Turnbull and Giblett, 1960). The demonstration that transferrin is a potent inhibitor of bacterial and viral multiplication (Martin and Jandl, 1959) suggests that this polymorphism may owe its origin to the selective action of infectious diseases in the past.

Summary

Infectious diseases in man often produce a huge mortality when first striking virgin populations. Extensive evidence suggests that, apart from, and in addition to, immunity, genes exist which confer inherited resistance to a given infectious disease. The history and epidemiology of some of the great pandemics of man are reviewed with special reference to genetic resistance factors. It is considered likely that infectious diseases were one of the most potent agents of human natural selection in the past. Starvation is briefly discussed as another powerful agent of natural selection.

A variety of chemical and enzymatic variants (metabolic polymorphisms) exists in human populations at frequencies which only could have been reached with a selective advantage in the past. It is suggested that the present incidence of some metabolic polymorphisms has been caused by infectious diseases as selective agents.

Host tissues essential for parasite growth or defense against the invading micro-organism would be the most probable site of such polymorphisms. For example, genetically defective red cells, by limiting proliferation, might protect the host against an invader requiring red cells for multiplication. The only widespread and lethal human infectious disease of the red cell is malaria. There is good evidence that red cell defects, such as the sickle-cell trait and probably the thalassemia trait, protect against death from falciparum malaria. Recent personal investigations are reviewed in detail, suggesting that another common red cell variant in tropical and subtropical populations—glucose-6-phosphate dehydrogenase deficiency of the red cell —also protects against falciparum malaria mortality.

The genetic implications of disease resistance are discussed in reference to the disappearance of infectious disease now and in the future. It is shown that the frequency of genes for disease resistance that have otherwise injurious effects, will decline in the future. Improved hygienic conditions, therefore, will lead to reduction of some human genes which produce illness. This effect has been observed with the sickling gene. Genes for disease resistance which otherwise are neutral will remain in the population at the frequency reached when the selecting infectious disease disappeared.

LITERATURE CITED

ADDAIR, J., AND L. H. SNYDER 1942 Evidence for an autosomal recessive gene for susceptibility to paralytic poliomyelitis. J. Hered., *33*: 307–309.

ALLISON, A. C. 1956 The sickle cell and hemoglobin C genes in some African populations. Ann. Hum. Genetics, *21*: 67–89.

———— 1957 Malaria in carriers of the sickle cell trait and in newborn children. Exp. Parasit., *68*:418–447.

———— 1959 Metabolic polymorphism in mammals and their bearing on problems of biochemical genetics. Am. Naturalist, *93*:5–16.

———— 1960a Natural selection in human populations. U. of Kansas Science Bull. (in press)

———— 1960b Glucose-6-phosphate dehydrogenase deficiency in red blood cells of East Africans. Nature *186*:431–432.

ALLISON, A. C., AND D. F. CLYDE 1960 Malaria in African children with deficient glucose-6-phosphate dehydrogenase in erythrocytes. Brit. Med. J. *i:1346–1348*.

AYCOCK, W. L. 1942 Familial aggregation in poliomyelitis. Am. J. Med. Sci., *203*:452–465.

BANGHAM, A. D., AND B. S. BLUMBERG 1958 Distribution of electrophoretically different haemoglobins among some cattle breeds of Europe and Africa. Nature, *181*:1551–1552.

BEST, W. R. 1959 Absence of erythrocyte glucose-6-phosphate dehydrogenase deficiency in certain Peruvian Indians. J. Lab. Clin. Med., *54*:791.

BEUTLER, E. 1959 The hemolytic effect of primaquine and related compounds: a review. Blood, *14*:103–139.

BEUTLER, E., M. K. Y. YEH, AND T. NECHELES 1959 Incidence of the erythrocytic defect associated with drug sensitivity among Oriental subjects. Nature, *183*:684–685.

BOYD, M. F. (editor) 1949 Malariology. W. B. Saunders, Philadelphia, 2 vols.

BRAY, R. S. 1957 Studies on the Exo-erthrocytic Cycle in the Genus Plasmodium. London School of Hygiene and Tropical Medicine. Memoir No. 12. H. K. Lewis & Co., London.

CARSON, P. E., C. L. FLANAGAN, C. E. ICKES, AND A. S. ALVING 1956 Enzymatic deficiency in primaquine-sensitive erythrocytes. Science, *124*:484–485.

CEPPELLINI, R. 1959 Blood groups and haematological data as a source of ethnic information. In Medical Biology and Etruscan Origins. Edited by G. E. W. Wolstenholme and C. M. O'Connor. Little, Brown and Company, Boston.

CHARLTON, R. W., AND T. H. BOTHWELL 1959 The incidence of glutathione instability of the red blood cells in the South African Bantu. S. Afr. J. of Med. Sc., *24*:88–89.

CHERNOFF, A. 1959 The distribution of the thalassemia gene. Blood, *14*:899–912.

CHILDS, B., W. ZINKHAM, E. A. BROWNE, R. L. KIMBRO, AND J. V. TORBERT 1958 A genetic study of a defect in glutathione metabolism of the erythrocyte. Bull. Johns Hopkins Hosp., *102*:21–37.

DIEHL, K. 1958 Gestaltungsfaktoren bie der Tuberkulose in besonderer Berücksichtigung der Individualität des befallenen Organismus. Handbuch der Tuberkulose, I:519. Thieme, Stuttgart.

DUBOS, R. 1959 Mirage of Health. Harper & Brothers, New York.

DUBOS, RENE, AND JEAN DUBOS 1952 The White Plague: Tuberculosis, Man and Society. Little, Brown and Company, Boston.

EDINGTON, G. M. 1959 Some observations on the abnormal haemoglobin diseases in Ghana. In Abnormal Haemoglobins. Edited by J. H. P. Jonxis and J. F. Delafresnaye. Blackwell Scientific Publications, Oxford.

FENNER, F. 1959 Myxomatosis in Australian wild rabbits—evolutionary changes in an infectious disease. The Harvey Lectures 1957–1958. Academic Press Inc., New York.

FERGUSON, R. G. 1955 Studies in Tuberculosis. University of Toronto Press, Toronto.

FISHER, R. A. 1930 The Genetical Theory of Natural Selection. Oxford University Press, Oxford.

FORD, E. B. 1945 Polymorphism. Biol. Rev., *20*:73–88.

FULTON, J. D., AND J. P. GRANT 1956 The sulphur requirements of the erythrocytic form of *Plasmodium knowlesi*. Biochem. J., *63*:274–282.

GEIMAN, Q. M. 1951 The cultivation of malarial parasites. In Parasitic Infections in Man. Edited by H. Most. Columbia University Press, New York.

GIBLETT, E. R. 1959 Haptoglobin types in American Negroes. Nature, *183*:192.

GIBLETT, E. R., C. G. HICKMAN, AND O. SMITHIES 1959 Serum transferrins. Nature, *183*:1589–1590.

GILLES, H. M., J. WATSON-WILLIAMS, AND B. G. TAYLOR   1960   Glucose-6-phosphate dehydrogenase deficiency trait in Nigerians. Nature, *185*:287.

GOWEN, J. W.   1948   Inheritance of immunity in animals. Ann. Rev. Microbiology, *2*:215–254.

———   1951   Genetics and disease resistance. In Genetics in the 20th Century. Edited by L. C. Dunn. Macmillan, New York.

———   1952   Humoral and cellular elements in natural and acquired resistance to typhoid. Am. J. Hum. Genet., *4*:285–302.

HALDANE, J. B. S.   1949   Disease and Evolution. Supplem. La Ricerca Scientifica, *19*:68–76.

———   1957   Natural selection in man. Acta Genetica et Statistica Medica, *6*: 321–332.

HARRIS, H., AND M. WHITTAKER   1959   Differential response of human serum cholinesterase types to an inhibitor in potato. Nature *183*:1808.

HERNDON, C. N., AND R. G. JENNINGS   1951   A twin-family study of susceptibility to poliomyelitis. Am. J. Hum. Genet., *3*:17–46.

HIRSCH, A.   1883   Handbook of Geographical and Historical Pathology. The New Sydenham Society, London.

HUTT, F. B.   1958   Genetic Resistance to Disease in Domestic Animals. Comstock Publishing Associates, Ithaca, New York.

JONXIS, J. H. P.   1959   The Frequency of Haemoglobin S and Haemoglobin C Carriers in Curaçao and Surinam. In Abnormal Haemoglobins, edited by J. H. P. Jonxis and J. F. Delafresnaye. Blackwell Scientific Publications, Oxford.

KALOW, W., AND N. STARON   1957   On distribution and inheritance of atypical forms of human serum cholinesterase, as indicated by dibucaine numbers. Canad. J. Biochem. Physiol., *35*:1305–1320.

KELLER, J., X. PI-SUNYER, T. PLAUT, AND P. A. MARKS. Incidence of erythrocyte glucose-6-phosphate dehydrogenase deficiency among different population groups in Surinam (in preparation).

KNOWLES, R., R. S. WHITE, AND B. M. DAS GUPTA   1930   Indian Medical Research Memoirs. Memoir No. 18. Thacker, Spink & Co., Calcutta.

KOLLATH, W.   1951   Die Epidemien in der Geschichte der Menschheit. Verlag der Grief, Wiesbaden.

LAMBOTTE-LEGRAND, J., AND C. LAMBOTTE-LEGRAND   1958   Notes complémentaires sur la drepanocytose. I. Sicklemie et malaria. Ann. Soc. Belge Med. Trop., *38*:45–53.

LEHMANN, H.   1959   Variations in human haemoglobin synthesis and factors governing their inheritance. Brit. Med. Bull., *15*:40–46.

LUIRE, MAX B., PETER ZAPPASODI, A. M. DANNENBERG, JR., AND G. H. WEISS   1951   On the mechanism of genetic resistance to tuberculosis and its mode of inheritance. Am. J. Human Genet., *4*:302–314.

MACDONALD, G.   1957   The Epidemiology and Control of Malaria. Oxford University Press, New York.

MALTHUS, T. R.   1798   Population: The First Essay. University of Michigan Press, Ann Arbor, Michigan. 1959 reprint.

MARKS, P.   1959   Proceedings of first Macy conference on genetics. Princeton (in press).

MARKS, P., AND R. GROSS 1959a Erythrocyte glucose-6-phosphate dehydrogenase deficiency: evidence of differences between Negroes and Caucasians with respect to this genetically determined trait. J. Clin. Inv., *38*:2253–2262.

―――― 1959b Drug induced hemolytic anemias and congenital galactosemia: examples of genetically determined defects in erythrocyte metabolism. Bull. N.Y. Acad. Med., *35*:433–449.

MARSHALL, I. D., AND F. FENNER 1958 Studies in the epidemiology of infectious myxomatosis in rabbits. J. Hygiene, *56*:288–302.

MARTIN, C. M., AND J. H. JANDL 1959 Inhibition of virus multiplication by transferrin. J. Clin. Inv., *38*:1024.

MCGHEE, R., AND W. TRAGER 1950 The cultivation of *Plasmodium lophurae in vitro* in chicken erythrocyte suspensions and the effects of some constituents of the culture medium upon its growth and multiplication. J. Parasitol., *36*:123–127.

MCGREGOR, I. A. 1959 The hyperendemic malaria of the Gambia. In Current Medical Research. Med. Res. Council. Her Majesty's Stationery Office, London.

MITCHELL, J. A. 1927 Plague in South Africa; historical summary (up to June 1926). Publications of South African Inst. Med. Res., *3*:89.

MOTULSKY, A. G., AND J. M. CAMPBELL 1960 Rapid detection of glucose-6-phosphate dehydrogenase deficiency in red cells. Distribution and frequency of the trait and its possible relationship to malaria. Blood (in press).

MOURANT, A. E. 1954 The Distribution of the Human Blood Groups. Blackwell Scientific Publications, Oxford.

NEEL, J. V. 1956 Genetics of human hemoglobin differences: problems and perspectives. Ann. Human Genet., *21*:1–30.

NEEL, J. V. 1957 Human hemoglobin types: their epidemiologic implications. New England J. Med., *256*:161–171.

NEEL, J. V. 1958 The study of natural selection in primitive and civilized human populations. In Natural Selection in Man. Wayne State University Press. Detroit, Michigan. Reprinted from Human Biol., *30*:43–72 (1958).

NYMAN, M. 1959 Serum haptoglobin, methodological and clinical studies. Suppl. 39. The Scand. J. of Clin. and Lab. Inv.

PANUM, P. L. 1940 Observations made during the Epidemic of Measles on the Faroe Islands in the Year 1846. F. H. Newton, New York.

PENROSE, L. S. 1957 Mutation in Man. In Effect of Radiation on Human Heredity. World Health Organization, Geneva.

PERLA, DAVID, AND JESSIE MARMORSTON 1941 Natural Resistance and Clinical Medicine. Little, Brown and Company, Boston.

POLLITZER, R. 1954 Plague. World Health Organization, Geneva.

PUFFER, R. R. 1946 Familial Susceptibility to Tuberculosis. Harvard University Press, Cambridge, Massachusetts.

RAPER, A. B. 1959 Further observations on sickling and malaria. Trans. Roy. Soc. Trop. Med. Hyg. *53*:110–117.

SABIN, A. B. 1951 Paralytic consequences of poliomyelitis infection in different parts of the world and in different population groups. Am. J. Pub. Health, *41*:1215–1230.

―――― 1954 Genetic factors affecting susceptibility and resistance to virus diseases of the nervous system. In Genetics and the Inheritance of Integrated Neurological and Psychiatric Patterns. Edited by D. Hooker and C. C. Hare, Williams & Wilkins Co., Baltimore.

SANSONE, G., AND G. SEGNI 1957 Sensitivity to fava beans. Letter to the Editor, Lancet, *273*:295.

SANSONE, G., A. M. PIGA, AND G. SEGNI 1959 Il Favismo. Edizione Minerva Medica, Genova.

SCHNEIDER, H. A. 1958 What has happened to nutrition? Perspectives in Biol. and Med., *1*:278–292.

SHEPPARD, P. M. 1958 Natural Selection and Heredity. Hutchinson & Co., London, England.

SMITHIES, O., AND N. F. WALKER 1955 Genetical control of some serum proteins in normal human serum. Nature, *176*:1265–1266.

SOKHEY, S. S., AND R. B. G. D. CHITRE 1937 L'immunité des rats sauvages de l'Inde vis-à-vis de la peste. Bull. Off. Int. Hyg. Publ., *29*:2093–2096.

SZEINBERG, A. C. SHEBA, AND A. ADAM 1958a Enzymatic abnormality in erythrocytes of a population sensitive to vicia faba or haemolytic anaemia induced by drugs. Nature, *181*:1256.

——— 1958b Selective occurrence of glutathione instability in red blood corpuscles of the various Jewish tribes. Blood, *13*:1043–1053.

TEPPERMAN, J. 1958 Etiologic factors in obesity and leanness. Perspectives in Biol. and Med., *1*:293–306.

TRAGER, W. 1941 Studies on conditions affecting the survival *in vitro* of a malarial parasite (*Plasmodium lophurae*). J. Exper. Med., *74*:441–462.

——— 1957 The nutrition of an intracellular parasite; avian malaria. Acta Tropica, *14*:289–301.

TURNBULL, A., AND E. GIBLETT 1960 The binding and transport of iron by unusual transferrins. Clin. Res. Proc., *8*:133.

VANDEPITTE, J. 1959 The incidence of haemoglobinoses in the Belgian Congo. In Abnormal Haemoglobins. Edited by J. H. P. Jonxis and J. F. Delafresnaye. Blackwell Scientific Publications, Oxford.

VANDEPITTE, J. M., W. W. ZUELZER, J. V. NEEL, AND J. COLAERT 1955 Evidence concerning the inadequacy of mutation as an explanation of the frequency of the sickle cell gene in the Belgian Congo. Blood, *10*:341–350.

VANDEPITTE, J., AND J. DELAISSE 1957 Sicklemie et paludisme. Ann. Soc. Belge Med. Trop., *37*:703–735.

VELLA, F. 1959a Favism in Asia. Med. J. Australia, *46*:196–197.

——— 1959b Susceptibility to drug induced haemolysis in Singapore. Med. J. Malaya, *13*:1–11.

VON VERSCHUER, O. F. 1959 Genetik des Menschen. Urban & Schwarzenberg, Berlin, Germany.

WALKER, D. G., AND J. E. BOWMAN 1959 Glutathione stability of the erythrocytes in Iranians. Nature, *184*:1325.

WILLIAMS, R. J. 1956 Biochemical Individuality, John Wiley, New York.

ZANNOS-MARIOLEA, L., AND P. CHIOTAKIS 1959 Favism in Greece. Paper read before the European Congress on Hematology, London (typescript).

Population Genetics of Haemoglobin Variants, Thalassaemia and Glucose-6-Phosphate Dehydrogenase Deficiency, with Particular Reference to the Malaria Hypothesis

M. SINISCALCO,[1] L. BERNINI,[1] G. FILIPPI,[2] B. LATTE,[3]
P. MEERA KHAN,[1] S. PIOMELLI,[4] and M. RATTAZZI[1]

Introduction

The present status of knowledge of haemoglobin variants, thalassaemia and glucose-6-phosphate dehydrogenase (G-6-PD) deficiency has been summarized on several occasions in recent years. Excellent reviews of all aspects of these questions are now available (Allison, 1965; Baglioni, 1962; Fessas, 1965; Ingram, 1963; Itano, 1965; Motulsky, 1965; Rucknagel, 1964; Silvestroni & Bianco, 1963).

Consideration of the recent exhaustive accounts of Silvestroni & Bianco (1963) on the world distribution of haemoglobin variants and of thalassaemia, that of Motulsky (1965) on G-6-PD deficiency, together with the elegant monograph of Rucknagel & Neel (1961) on the dynamics, at a population level, of the genes controlling these conditions, and the discussions of the same subject by Livingstone (1964) and by Allison (1965) may serve as ideal introductions to the subject of the present report—that is, the hypothesis that malaria may have been the common ecological factor that was responsible for the selection of these three groups of inherited abnormalities of the red cells.

The idea that all individuals might not be equally liable to malarial infection was first proposed by Haldane (1949) to explain the preponderance of thalassaemia in the Mediterranean basin. A few years later, Allison (1954) reported the interesting finding that persons who carry the sickle-cell trait are indeed more resistant to subtertian malaria than those who do not. Although this claim has been disputed in some instances, it is now supported by a most impressive body of evidence, including data of the

[1]Department of Human Genetics, University of Leiden, Leiden, Netherlands.
[2]Department of Genetics, University of Rome, Rome, Italy.
[3]Town Hospital, Nuoro, Italy.
[4]Department of Genetics, University of Rome. Present address: Department of Pediatrics, New York University–Bellevue Medical Center, New York, N.Y., USA.

Reproduced by permission of the publisher and M. Siniscalco from Bulletin of the World Health Organization, *34:379-393 (1966).*

MARCELLO SINISCALCO, a human geneticist, was born in Naples, Italy in 1924. He received his M.D. from the University of Naples in 1948. Since 1962 he has been Professor of Human Genetics at the University of Leiden, Holland. His research interests cover a broad spectrum of human genetics including population genetics, biochemical genetics, linkage studies, and somatic-cell genetics. His population-genetic work has been done primarily on the island of Sardinia, but currently he is engaged in an analysis of the genetic structure of two Indian tribes in Andhra Pradesh as part of the International Biological Program.

Some of his publications in the field of population genetics include "A focus of congenital hemeralopia in Sardinia. Familial and population data," with others, Boll. Oculistica, *39*: 891–905 (1960); "Favism and thalassaemia in Sardinia and their relationship to malaria," with others, Nature, *190*: 1179–1180 (1961); "Failure to detect linkage between Xg and other X-borne loci in Sardinians," Ann. Hum. Genet. *29*: 231–252, (1966); "Studies on African Pygmies. I. A pilot investigation of Bahinga Pygmies in the Central African Republic (with an analysis of genetic distances)," Am. J. Hum. Gen., *2*: 252–274, (1969).

three following kinds (Allison, 1965): (1) direct demonstration, in areas where malignant malaria is still endemic, that young children who are heterozygous for the sickle-cell gene have lower *Plasmodium falciparum* parasite counts than do children without this trait; (2) the finding of a low incidence of sickling carriers among the cases of fatal malarial infections observed in the same areas; and (3) the overlap between the world distribution of the sickle-cell trait and that of malignant malaria.

Motulsky (1960) was the first to report population data that showed that the distribution of G-6-PD deficiency in the Eastern Hemisphere also overlaps that of malignant malaria, and suggested that even this well-known example of sex-linked polymorphism owes its establishment to the higher fitness of enzyme-deficient genotypes in a malarial environment.

On the other hand, studies of the malaria parasite counts in groups of normal and enzyme-deficient children yielded contradictory results. Thus, while Allison & Clyde (1961), in East Africa, and Harris & Gilles (1961) in West Africa found significantly lower *P. falciparum* counts in young enzyme-deficient children, Kruatzachue and his co-workers (1962), Motulsky (1965), and Edington & Watson-Williams (1965) failed to do so.

Since, however, it is known that, for instance, a protein-deficient diet (Pérez et al., 1964) and hypothyroidism (personal unpublished data) can lower G-6-PD activity, it is uncertain how much of this disagreement may be due to misclassification of the G-6-PD phenotypes, which reasonably may be expected, especially when dealing with indigenous African populations, in which enzyme deficiency is not as complete as in Caucasians, and when only screening tests have been used for the diagnosis.

Equally contradictory have been the conclusions of studies on malarial parasite counts in the heterozygous carriers of haemoglobin C and haemoglobin E genes (Edington & Laing, 1957; Thompson, 1962; Edington & Watson-Williams, 1965; Brumpt & Brumpt, 1958; Kruatzachue et al., 1961). Investigations of these types have been impossible to perform on other haemoglobin variants because of their rarity, or on the different forms of thalassaemia because their correct diagnosis, which requires elaborate laboratory studies, is unreliable under field-work conditions. Moreover, the common occurrence, in the primitive areas where malaria is still prevalent, of environmental and biological stress factors may distort the haematological picture of thalassaemia.

Thus, it is not surprising that the only evidence for the relationship between malaria, thalassaemia and G-6-PD deficiency should come exclusively from population studies showing a positive correlation between the incidence of these genes and malaria morbidity or, rather, past malaria morbidity; since, for the reasons outlined above, the correct classification of these abnormalities of the red cells is difficult in primitive areas where malaria is still endemic. Nevertheless, even these types of studies present difficulties, since they can have very little value unless the following three essential requirements are met: (1) a reasonable degree of ethnic homogeneity among the "genetic isolates" chosen for the study; (2) the availability of accurate historical data as well as general vital statistics, data on the malaria morbidity, mating patterns, and rates of consanguinity in order to be able to estimate the relative importance of drift, migration and natural selection as potential causes of any genetic heterogeneity that might be found among neighbouring isolates; and (3) the availability of control populations within the malarial areas under consideration; that is, the existence of "malaria-free islands" or human settlements of appreciable size, inhabited by persons of the same ethnic group, who have been living in genetic isolation for a very long time. Populations of nomadic habits are clearly useless for investigations of this type.

Since these conditions are met only very rarely, it is impossible to attempt an over-all evaluation of the "malaria hypothesis" from the data available on the world distribution of the genes under discussion here. Instead, we prefer to report here, for the first time in full, the data that we collected in Sardinia and have so far published only in part (Siniscalco et al., 1961; Siniscalco, 1964; Adinolfi et al., 1960), since we feel that they represent one of the few sets of data among the population studies so far published on the subject that meet the requirements outlined above.

Historical Background

The island of Sardinia is the ideal place for such population studies. The degree of isolation to be found there is still very high for most of its villages and, as shown in Fig. 1, its geography is such that, within a few hundred square miles, one can easily find isolated settlements with very high malaria morbidity in the past and others that have been practically free from this disease.

Furthermore, it can reasonably be assumed that the populations of many

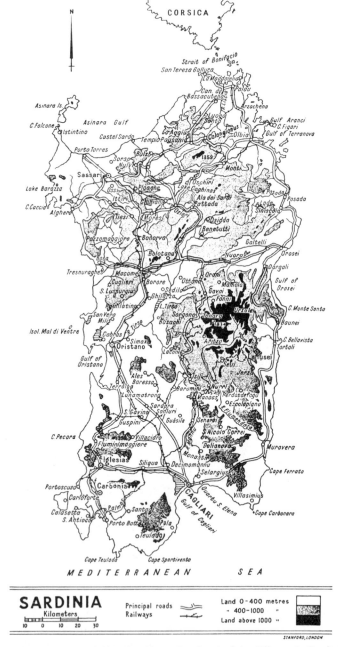

FIG. 1. Locations and Altitudes Above Sea-Level of the Villages in Sardinia
Cited in the text. Reproduced from Logan, J. A. (1953) *The Sardinian
project*, with the permission of the Johns Hopkins Press, Baltimore, Md.,
USA.

Sardinian villages must have remained free from external admixtures over a very long period. Ancient Greek colonization was, in fact, limited to Olbia (the northern portion of the island), while the Romans and Carthaginians only exploited the coastal regions for grain, as is proved by the way localized areas along the south-west coast in which archaeological remains of their towns can now be found, as at the excavations at Nora. Later, the Vandals and Goths simply overran the island, which subsequently became the scene of struggles among the Pisans, Genovese and Saracens, none of whom, however, were there in sufficient numbers or for a long enough time to alter, significantly, the genetic structure of the autochthonous population. Even the Spanish, who ruled the island from 1297, when Pope Boniface VIII awarded it to James II of Aragon, never really cared to penetrate the rocky paths leading to the interior of the island, which remained half-forgotten, with its primitive villages consisting of huts clustered around the ancient nuraghi, and where little social or economic change took place until the accession of the House of Savoy in 1720.

The nuraghi, those beautiful and impressive Bronze Age stone constructions, are scattered in thousands all over the island, showing how well organized the Sardinians were, even in prehistoric times, and how concerned with defence against invaders. It can hardly be denied, however, that as a result of these invasions, a few sets of "external" genes must have entered the Sardinian genetic pool from time to time.

Malaria has probably been endemic in Sardinia since prehistoric times, according to the Roman historian Livy, and has remained one of the most important causes of infant mortality until the beginning of the present century. This disease was not eradicated completely until after the Second World War, when the Rockefeller Foundation, by a massive and well-planned anti-malaria campaign (Logan, 1953), successfully completed the pioneering work of Fermi and Missiroli, the distinguished Italian malariolgists who had struggled against the disease in Sardinia for decades and made available to posterity the most accurate and complete information that could be desired on the malaria morbidity of every Sardinian village. This information, together with the very ancient church records and the first-class vital statistics on the island that have been available since the Savoy accession in 1720, enabled us to draw valid conclusions about the population structure and dynamics of the Sardinian isolates chosen for our studies.

When we began our investigations, the existence of thalassaemia and G-6-PD deficiency in Sardinia already had been well established (Carcassi, Ceppellini & Pitzus, 1957; Larizza et al., 1958), and the preliminary studies of Ceppellini (1955) had shown that the incidence of thalassaemia in two non-malarial villages in the Gennargentu Mountains was strikingly low as compared with that found in two lowland villages in formerly very malarial parts of the eastern coast of the island.

We repeated these studies in a total of 19 villages and extended them to G-6-PD deficiency (Siniscalco et al., 1961) and were able to demonstrate that there was, indeed, a very close positive correlation between the present-day frequencies of thalassaemia and G-6-PD deficiency and former malaria morbidity as reported by Fermi (1938).

Additional studies of the same kind have been made in the past few years; new data are now available and are discussed in detail below.

Summary of the Sardinian Population Data

Table 1 shows the frequency of the gene for G-6-PD deficiency, Gd(−), and that for β-thalassaemia, Th(+), in 52 Sardinian villages, based on more than 6000 observations of unrelated individuals. The estimated frequencies of the two traits refer to the youngest generation, since the data for each village were obtained from a random sample of its school-boys.

These estimates are better summarized in Fig. 2, in which a positive correlation between the frequency of thalassaemia and that of G-6-PD deficiency is evident. This correlation seems to be described reasonably well by a straight regression line up to a maximum frequency level of about 24%. The correlation fades off above this level, since the gene for thalassaemia, being lethal in the homozygous condition, can never reach equilibrium levels as high as those that are possible for the much less unfavourable gene for the enzyme deficiency.

Basic considerations of population genetics (Livingstone, 1964) make it evident that:

(1) Levels of gene frequency such as those reported for the villages in the Sardinian plains can be explained only by assuming a higher fitness of the heterozygous genotype for thalassaemia and of the heterozygous and, perhaps, the hemizygous and homozygous genotypes for enzyme deficiency.

(2) A "migration" hypothesis is grossly inadequate to account for the very high frequencies of these traits found in the plains, since they could not be expected, even if one were to assume, against all historical evidence, that the autochthonous populations had been totally replaced by equally numerous groups of immigrants, all of whom were carriers of thalassaemia and/or G-6-PD deficiency.

(3) If the genetic heterogeneity between the lowland villages and those in the high mountains is attributable to different adaptive values of the carrier genotypes, malignant malaria is the obvious ultimate factor, since the two environments are known to have differed from each other for centuries and, until only about twenty years ago, almost exclusively in respect to mortality and morbidity from malaria.

The negative correlation with altitude above the level of 400 m is clearly demonstrated in Fig. 3, where the average frequencies (± 3 times their sampling errors) of the two traits are reported for each group of villages of similar altitude. This correlation is therefore positive when gene frequen-

TABLE 1. INCIDENCE OF GLUCOSE-6-PHOSPHATE DEHYDROGENASE DEFICIENCY AND THE β-THALASSAEMIA TRAIT IN 52 SARDINIAN VILLAGES

Village[a]	Altitude (metres)	Gd(−)[b]		TH(+)[c]	
		Number tested	Percentage positive	Number tested	Percentage positive
1. Assemini	6	108	20.4	108	11.0
2. Marrubiu	7	98	32.6	98	28.6
3. Cabras	9	200	35.0	100	28.0
4. Terralba	9	100	30.0	—	—
5. Carloforte	10	99	5.0	99	5.0
6. Decimomannu	10	100	26.0	100	25.0
7. S. Giusta	10	42	30.9	—	—
8. Pula	15	100	15.0	100	16.0
9. Tortoli	15	50	16.0	—	—
10. Orosei	19	180	13.0	308	18.8
11. Torpe	24	100	22.0	100	38.0
12. Irgoli	26	100	15.0	100	32.0
13. Galtelli	40	175	12.0	235	21.2
14. Siniscola	42	195	11.3	97	24.4
15. Barisardo	50	98	15.3	98	18.4
16. Teulada	50	101	16.9	100	25.0
17. S. Gavino	53	100	26.0	—	—
18. Capoterra	54	92	16.3	92	20.6
19. Siliqua	66	100	26.0	100	22.0
20. Vallermosa	70	86	20.9	86	21.0
21. Monastir	83	94	23.4	94	21.2
22. Nuraminis	86	100	25.0	100	17.0
23. Villamar	108	100	23.0	100	18.0
24. Guspini	137	99	28.2	99	34.4
25. Domusnovas	152	100	22.0	100	19.0
26. Gonnosfanadiga	156	49	24.5	—	—
27. Usini	190	99	6.1	99	14.2
28. Ottana	195	72	8.0	72	28.0
29. Senorbi	204	101	24.7	101	13.8
30. Tresnuraghes	257	86	12.7	86	31.4
31. Sedilo	288	100	22.0	96	18.8
32. Serrenti	307	100	21.0	100	25.0
33. Arbus	311	95	35.7	95	32.8
34. Abbasanta	315	97	18.5	92	22.8
35. Dualchi	321	75	21.3	100	18.0
36. Suni	333	98	14.3	100	25.0
37. Lode	335	820	28.2	820	27.6
38. Gergei	374	92	18.5	92	13.2
39. Borore	399	100	9.0	99	33.2
40. Benetutti	406	100	9.0	100	12.0
41. Bolotana	472	93	11.8	93	21.4
42. Luras	508	100	7.0	98	23.0
43. Lula	521	100	7.0	100	19.0
44. Isili	523	100	9.0	100	17.0
45. Bitti	549	193	5.1	193	12.2
46. Lanusei	595	100	4.0	—	—
47. Ala dei Sardi	663	80	22.5	80	20.0
48. Orune	745	97	6.1	97	14.4
49. Gavoi	777	98	3.0	98	10.2
50. Desulo	891	313	3.0	320	3.8
51. Tonara	935	148	4.0	102	4.8
52. Fonni	1000	100	3.0	—	—

[a] These villages are arranged and numbered in order of their increasing altitude. The data were obtained from random samples of the schoolboys of each village.

[b] Glucose-6-phosphate dehydrogenase deficiency.   [c] Thalassaemia.

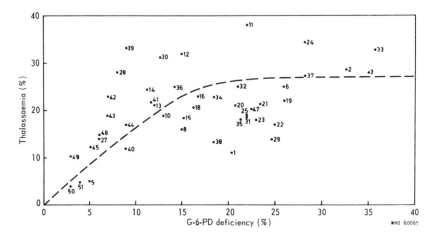

FIG. 2. Distribution of G-6-PD Deficiency and of the Thalassaemia Trait in Sardinia. Estimates of the gene frequencies obtained from random samples of schoolboys. The numbers identify the villages listed in Table 1.

cies are compared with the relative incidence of malaria, as we demonstrated in a series of villages for which direct estimates of past malaria morbidity were available (Siniscalco et al., 1961), and as is made clear by the data presented in Table 2.

A few villages included in Fig. 2 and 3 (Carloforte, Usini, Lode and Ala dei Sardi) require special mention.

Carloforte is the only village on the beautiful little island of San Pietro, which is adjacent to the very malarial plains of the south-western Sardinian coast. This island was first settled, about AD 1700, by a small group of

TABLE 2. SPLEEN AND PARASITE SURVEYS OF SCHOOLCHILDREN IN 66 SARDINIAN VILLAGES BETWEEN NOVEMBER 1947 AND MARCH 1948, BY VILLAGE ALTITUDE ABOVE SEA-LEVEL[a]

Altitude above sea-level (metres)	Spleen examinations			Parasite examinations					
	Number examined	Number with palpable spleen	Spleen rate (%)	Number examined	Number positive, by *Plasmodium* species			Total number positive	Parasite rate (%)
					vivax	falciparum	malariae		
0–50	2858	996	34.8	2779	59	56	1	116	4.2
51–100	2015	595	29.5	1861	39	29	1	69	3.7
101–200	1282	288	22.5	1289	58	29	1	88	6.8
201–300	1215	226	18.6	1201	26	11	0	37	3.1
301–400	1094	212	19.4	1144	12	2	0	14	1.2
401–600	1540	215	14.0	1539	12	10	0	22	1.4
601–800	1481	277	18.7	1481	18	9	0	27	1.8
801–1000	1430	106	7.4	1371	8	1	0	9	0.7
All altitudes	12915	2915	22.6	12665	232	147	3	382	3.0

[a] Reproduced from Logan, J. A. (1953) *The Sardinian project*, with the permission of the Johns Hopkins Press, Baltimore, Md., USA.

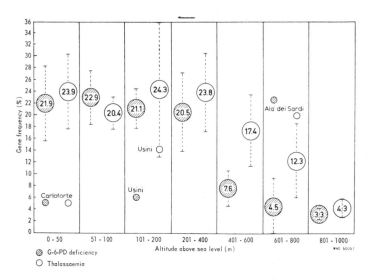

FIG. 3. Incidence of G-6-PD Deficiency and of the Thalassaemia Trait in relation to altitude above sea-level. The figures in each of the large circles are the averages of the gene frequencies found in the villages that fall within the corresponding altitude groupings (0–50 metres, 51–100 metres, etc.). These villages may be found easily in Table 1, in which they are arranged in sequence according to their altitude above sea-level. The smaller circles refer to the three villages of Carloforte, Usini and Ala dei Sardi, which are considered separately, as explained in the text.

Genovese fishermen who had come from the island of Tabarca, near the North African coast and, when expelled by the Bey of Tunis, had requested and received the hospitality of the King of Sardinia. This group, now numbering about 7000, kept itself in close isolation from the rest of Sardinia until very recently. It is thus not surprising that, despite the heavy malaria morbidity that was reported in the area until a decade ago, only a few genes for G-6-PD deficiency and thalassaemia can be found among them, and they had clearly been derived from Sardinian ancestors, as could be proved by genealogical studies.

Usini is a small village not far from the north-west coast of Sardinia, where settlements of Genovese and Spanish origin are found. It is in this part of the island that the influence of the Catalan language on the local dialect can be observed most easily. Here, intermixture with the inland population has been more massive and frequent, and the intermediate values of gene frequencies today are an obvious consequence. Carcassi (1962) has shown that the same situation prevails for Alghero and its neighbouring villages.

Lode and Ala dei Sardi are two villages that are especially useful to demonstrate the relationship between high gene frequencies and the former prevalence of malaria. Beyond any doubt, these villages originated in the remote past; both of them are noted as sizable settlements in a 16th-century map of Sardinia drawn by Igrazio Donati and now kept in the Vatican Museum. Until a few years ago, their isolation was very strict, since a high mountain, difficult to cross even today, separates them from the coastal villages. Fermi (1938) reported a very high malaria morbidity for both of these villages, despite the relatively high altitude of one of them (Ala dei Sardi); the gene frequencies for both thalassaemia and G-6-PD deficiency are particularly high, unlike those observed in a neighbouring village (Bitti) located on the very summit of the mountain and reported by Fermi (1938) as being relatively free of malaria. All of these villages, together with others in the interior plains, such as Abbasanta and Guspini (see Table 1), again indicate that the suggestion that ethnic heterogeneity is a possible main cause of differences in gene frequencies is certainly not a likely one. Moreover, the blood-group distributions in the interior, coastal and mountain regions are remarkably similar, all showing an unusually high incidence of the M gene and a very low frequency of Rh-negative individuals, which seems to differentiate the Sardinian population from the general European population (Ceppellini, 1955).

Interaction between G-6-PD Deficiency and Thalassaemia at the Individual and Population Levels

As expected, some individuals were found in Sardinia who carry both thalassaemia and G-6-PD deficiency. The association of both of these conditions in the same individual does not appear to involve a more serious red-cell defect, as has been demonstrated directly by chromium-51 studies that have showed that the reduction of red-cell survival time in these individuals is of the same order as that reported for carriers of G-6-PD deficiency alone (Bernini et al., 1964).

Indeed, there are reasons to believe that the association of these two defects in the same person may, rather, produce higher biological fitness in him. For example, we have reported (Siniscalco et al., 1961) that the frequency of severe haemolytic crises from exposure to fava beans (clinical favism) is less among carriers of both the enzyme deficiency and thalassaemia than among carriers of the enzyme deficiency alone. Since G-6-PD activity is always increased in carriers of thalassaemia,[1] it was thought that some kind of compensation for the enzyme deficiency exists in the presence of thalassaemia.

The probability that there is a higher fitness in the carriers of both of these genes is suggested by the finding that the number of such persons in the general population appears to exceed that which would be expected by calculation from the estimated gene frequencies for these traits in each

---

[1]Thalassaemic red cells, although smaller than normal ones, have the same enzymatic activity; thus, carriers of the Th (+) gene, who are polycythaemic, have relatively higher G-6-PD activity per blood-volume unit (Piomelli & Siniscalco, 1966, to be published).

village. While this excess is not significant within any village, it clearly becomes so when the data of the 21 villages studied for that purpose are pooled (Table 3).

TABLE 3. CALCULATION OF THE EXCESS OF DOUBLE CARRIERS
IN 21 SARDINIAN VILLAGES

Village No.[b]	Double carriers: Gd(−) & Th(+)[a]				
	Persons tested	Double carriers found	D_i expectation	X_i deviation from expectation	$\dfrac{X_i}{\sqrt{D_i}}$
2	98	8	7.80	+0.20	+0.072
3	100	11	8.42	+2.58	+0.889
11	100	5	6.76	−1.76	−0.677
14	97	2	2.44	−0.44	−0.281
15	98	3	2.49	+0.51	+0.323
16	100	3	3.68	−0.68	−0.354
24	99	12	7.90	+4.10	+1.459
28	72	4	1.38	+2.62	+2.230
31	96	3	2.63	+0.37	+0.228
33	95	9	9.31	−0.35	−0.100
34	92	3	3.44	−0.44	−0.237
36	98	4	3.13	+0.87	−0.492
37	820	65	55.00	+10.00	+1.350
38	92	2	2.10	−0.10	−0.071
39	99	2	0.24	+1.76	+3.592
40	100	2	1.00	+1.00	+1.000
42	98	2	1.42	+0.58	+0.487
43	100	2	1.54	+0.46	+0.371
44	100	1	1.38	−0.38	−0.323
45	193	3	0.96	+2.0	+2.082
47	80	3	3.24	−0.24	−0.133

$$\sum \left(\frac{X_i}{D_i}\right) = \begin{array}{r} +14.575 \\ -\ 2.176 \\ \hline +12.399 \end{array}$$

$$\sum \left(\frac{X_i}{D_i}\right)\bigg/ \sqrt{N}\ ;\ \frac{+12.399}{\sqrt{21}} = \frac{12.399}{4.58} = 2.70;\ P<0.01$$

[a] Gd(−) = glucose-6-phosphate dehydrogenase deficiency; TH(+) = thalassaemia.
[b] See Table 1 for numbering of villages.

Other Haemoglobin Variants and Different Forms of Thalassaemia in Sardinia

To date, there has been no systematic search for haemoglobin variants or for other forms of thalassaemia in Sardinia, although the rarity of haemoglobin variants can be inferred from the scarcity of case reports during the last ten years. On the other hand, the presence of the so-called α-thalassaemia at an appreciable frequency appears probable, from our findings, of a certain number of thalassaemic families without elevated haemoglobin A_2 (Carcassi, Ceppellini & Siniscalco, 1957), from the report of

Silvestroni & Bianco (1963) of two cases of Bart's haemoglobin among Sardinians living in Rome and from the report of Fiaschi, Campanacci & Naccarato (1964) of a case of thalassaemia-haemoglobin H disease in a haematological patient in the medical clinic of the University of Cagliari.

We have performed an extensive study of nearly 1200 random blood samples collected in villages of the southern plains of Sardinia already known for their high frequency of thalassaemia and G-6-PD deficiency, and not a single instance of variant haemoglobin or of high foetal haemoglobin was found among them. Recently, however, the occurrence has been reported of a fast-moving haemoglobin variant that appears to be due to a mutation on the α-haemoglobin chain, probably similar to one that has been described for haemoglobin "Mexico" (Baglioni & Sulis, personal communication). Moreover, an investigation of the distribution of haemoglobin A_2 levels among a random sample of apparent carriers of thalassaemia in an area where this condition is present in about 30% of the population revealed the occurrence of a normal level of haemoglobin A_2 in about 4% of the cases. Further examination of these individuals and of their families led us to the conclusion that they had to be considered instances of α-thalassaemia, although the presence of minor quantities of haemoglobin H could be established in only two of these individuals. If this conclusion is correct, it follows that the incidence of α-thalassaemia in the given area is of the order of 1% ($0.30 \times 0.04 = 0.012$); thus only a minor classification error is involved when population-screening for β-thalassaemia is performed by red-cell fragility and blood-film studies alone (Table 4).

The absence of the sickling trait in Sardinia is particularly noteworthy in view of its appreciable frequency in the neighbouring Mediterranean areas (Greece, North Africa, the Middle East, Sicily and, in general, southern Italy) and of its well-established adaptive value in a malarial environment (Allison, 1965).

It has been reported, however, that the incidence of the haemoglobin S gene tends to be correlated inversely with that of β-thalassaemia in those populations in which both of these genes occur with appreciable frequency (Barnicot et al., 1963). This phenomenon has been interpreted as a consequence of the frequently poor adaptability of the genotype that combines haemoglobin S and thalassaemia, thus leading to the elimination of the gene that happened to be the rarest when selective mechanisms of the present type became operative.

Consequently, it may be postulated that, when thalassaemia and G-6-PD deficiency were introduced into Sardinia, the haemoglobin S gene may still have been uncommon in the Mediterranean basin and that it was therefore entirely eliminated from the Sardinian gene pool in the long run, while the other two genes successfully established themselves among the populations of the malarial plains. An alternative explanation is the assumption that

TABLE 4. DISTRIBUTION OF HAEMATOLOGICAL PARAMETERS IN A RANDOM GROUP OF 235 ADULT SARDINIAN MALES (157 NORMALS, 78 WITH THALASSAEMIA)[a]

		Total No.	Means	Standard Deviation	Standard errors	Discrimination threshold	Classification error
Haemoglobin	N	157	13.33	1.513	0.121	12.43	27%
	T	78	11.56	1.460	0.165		
Red cell number	N	157	4.543	0.545	0.043	4.77	34%
	T	78	5.019	0.601	0.068		
Hematocrit	N	156	42.56	3.571	0.286	40.96	32%
	T	78	39.17	3.972	0.450		
Mean corpuscular volume	N	156	93.33	9.828	0.787	85.19	21%
	T	78	77.05	9.823	1.112		
Mean corpuscular haemoglobin	N	157	29.53	3.854	0.308	25.70	16%
	T	78	22.87	2.854	0.323		
Mean corpuscular haemoglobin concentration	N	156	31.46	2.859	0.229	30.37	35%
	T	78	29.36	2.629	0.298		
Red cell fragility[b]	N	127	0.413	0.024	0.002	0.378	7%
	T	61	0.348	0.021	0.003		
Haemoglobin A_2	N	157	2.329	0.392	0.031	3.37	0.7%
	T	78	5.186	0.680	0.077		

[a] Key: N = normal individuals; T = persons with thalassaemia (i.e., parents of persons with Cooley's disease). Data from Carcassi, Ceppellini & Siniscalco (1957).
[b] Red cell fragility is expressed here as the concentration of NaCl required to produce 50% of haemolysis.

265

both G-6-PD deficiency and thalassaemia are much older mutations than the haemoglobin S gene, and that the enzyme deficiency appeared in the Mediterranean basin when Sardinia had already split off from the mainland and isolated its population.

At any rate, the absence of haemoglobin S from Sardinia is a good piece of evidence for the hypothesis that the Sardinians, unlike their neighbours, must have been genetically isolated for a very long time after the original arrival among them of the genetic raw material upon which natural selection must have been acting for at least 2000 years.

An inverse correlation similar to that described for thalassaemia and sickling appears to exist between the genes for haemoglobins S and C in populations in which both of these variants are present (Allison, 1965). On the other hand, in Greece, a positive correlation has been reported between G-6-PD deficiency and β-thalassaemia (Allison et al., 1963) and, in Greece as well as in Africa, between G-6-PD deficiency and the sickling trait (Motulsky, 1960; Allison, 1965).

These interactions between different genes at a population level are an obvious illustration of the important role that natural selection must evidently play in the maintenance of the genic load of human populations.

Population Dynamics of G-6-PD Deficiency and Thalassaemia

In an attempt to provide a unitary explanation of the world distribution of genes known to involve genetic adaptability in a malarial environment, Zaino (1964) considered the possibility that they may have begun to have an adaptive value more than 50,000 years ago, when Europe was still bridged by land to Africa and malaria was probably already a strong factor in natural selection. When the glaciers receded, the Mediterranean basin was flooded, and its inhabitants either remained isolated along the newly formed seacoasts and islands or migrated towards the Middle and Far East into India, China, and, eventually, the Americas. In their new ecological niches, these genes were exposed to different selective pressures and, under the combined influences of selective pressures such as consanguinity, mutation, interaction between genes, migration, drift, and differential survival, independently reached the diverse equilibrium frequencies that we observe today.

There is naturally no hope of collecting evidence for or against these fascinating arguments. However, it seems to us to be irrelevant to establish whether a set of genes in a given area arose by mutation or migration, as long as it is made clear that their maintenance over an appreciably long time would not be possible without the intervention of selective mechanisms such as those proposed in the classic works of Fisher, Haldane and Wright to explain the occurrence of stable genetic polymorphism.

Livingstone (1964) has recently discussed, in detail, the population dynamics of the genes for thalassaemia, haemoglobin variants and G-6-PD deficiency and has presented the general equations for calculating the time required to attain genetic equilibrium in different selective models. In order to do so, he had to hazard some estimates of the fitness of the various genotypes for each of the genetic systems under consideration. These estimates were probably not far from reality for the haemoglobin S and β-thalassaemia genes, but we cannot agree with those proposed for the G-6-PD deficiency gene, which Livingstone supposed to have a positive adaptive value during malarial periods only in the female heterozygotes. This is, indeed, the obvious conclusion if the gene frequencies observed today are considered "equilibrium frequencies," but there are no strong reasons to assume that this is necessarily the case. Unlike the situation with the thalassaemia gene, whose lethality in homozygous conditions acts as a strong buffering factor to maintain the system under stable equilibrium at maximum levels of gene frequencies between 0.10 and 0.15, the G-6-PD deficiency gene, which certainly does not produce serious handicaps in its carriers, might well have been in a condition of transient equilibrium in malarial areas. Indeed, we feel that, at least in Sardinia, a situation similar to that presented in Table 5 is closer to reality.

TABLE 5. THEORETICAL FITNESS VALUES OF THE DIFFERENT GENOTYPES
FOR THE THREE SELECTIVE MODELS HYPOTHESIZED IN FIG. 4

	Males		Females		
	G-6-PD deficient	Normals	G-6-PD deficient	G-6-PD inter-mediate	Normals
Model I	1.04	0.96	1.04	1.00	0.96
Model II	1.01	0.95	1.01	1.00	0.95
Model III	0.98	0.94	0.98	1.00	0.94

In other words, we assume that, when malaria was killing about one-half of the Sardinian population before reproductive age, the fittest genotypes may well have been the male hemizygote and the female homozygote, despite the slight risk of disease caused by the enzyme deficiency itself, since it is reasonable to suppose that it might, for instance, have been the susceptibility of the enzyme-deficient red cells of these abnormal individuals to haemolysis that made the growth and multiplication of malarial parasites in their blood more difficult than in the blood of normal individuals. In such a situation, the greater the enzyme deficiency, the greater the protection against malaria.

Under such a selective model, the ultimate fate of the enzyme-deficiency gene would, in the long run, have been its complete fixation, had the ecological factor that was responsible for the differential fitness not been removed (Fig. 4). This assumption does not appear to be unduly absurd when one considers the extremely high frequency of this gene in some other

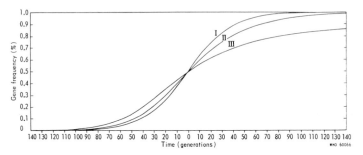

FIG. 4. Theoretical curves describing the population dynamics of the G-6-PD deficiency gene in Sardinia under three selective models. The Roman numerals refer to the three selective models presented in Table 5. The curves were calculated by W. S. Volkers, Department of Human Genetics, University of Leiden, according to the equations proposed by Livingstone (1964).

populations such as the Iraqi Jews, among whom frequencies of this enzyme deficiency in males range from 25% to 52%, and among the Kurdish Jews, among whom frequencies as high as 70% have been recorded (Szeinberg, 1963).

Nevertheless, it must be borne in mind that such attempts to express, in quantitative terms, the effects of natural selection in human populations are usually over-simplifications, since they can be done only by considering each genetic system separately. There are obvious difficulties in treating, in mathematical terms, more complicated models that would take into account the interactions between various gene systems. This is particularly true when different genes have been selected for by a common ecological factor. In the present instance, it is obviously possible that the greater fitness of the combined genotype Gd(−)Th(+), to which reference was made above, and the total absence of the haemoglobin S gene from Sardinia undoubtedly must have had their weight in influencing the gene-frequency distributions shown in Fig. 2 and 3 and in Table 1.

Moreover, although there are adequate grounds for believing that malaria was the principal ecological factor responsible for the selection of these genes, the possible existence of other genetic and environmental selection factors should not be disregarded. Consanguinity, for example, must have been an important counteracting selective agent for the accumulation of thalassaemia genes in Sardinian isolates, since the increased homozygosity that follows close inbreeding would undoubtedly help in the elimination of the lethal genes. The effects of consanguinity on G-6-PD deficiency must, instead, have been quite the opposite in the presence of malaria, if the fitness estimates shown in Table 5 are correct. To avoid the disturbing effects of consanguinity, we deliberately avoided the inclusion of villages that were of very different sizes and therefore likely to involve significant differences in inbreeding coefficients that were otherwise known to be quite constant in all Sardinian villages of ancient formation and longstanding genetic isolation (A. Moroni, personal communication).

Possible Genetic Heterogeneity Between G-6-PD Deficiency and Thalassaemia Pertinent to Different Areas

Unlike the case with haemoglobin variants, for which it can always be established with certainty whether or not one is dealing with a specific mutation (that is, with a specific amino-acid substitution in the haemoglobin molecule), one can never be sure of the genetic homogeneity of the several forms of thalassaemia and G-6-PD deficiency that are reported from different parts of the world. The numerous and not always concordant attempts to classify different forms of thalassaemia (Fessas, 1965) and G-6-PD deficiency (Motulsky, 1965) demonstrate the limitations of our knowledge in this matter, evidently because of the very fact that the diagnoses of thalassaemia and G-6-PD deficiency are made at a level far removed from the primary product of the genes.

For the foregoing reasons, it appeared unwise, at present, to attempt an over-all evaluation of all of the population data so far published. It might well be, for example, that the β-thalassaemia observed in Sardinia is quite different from those reported in Greece or in the Far East, or even from that observed in the Ferrara district. The same is true for G-6-PD deficiency even within the so-called subclasses of the G-6-PD deficiency of the Caucasian type, which involves a complete red-cell enzyme deficiency and no obvious electrophoretic differences, and the G-6-PD deficiency of the African type, which involves a partial red-cell enzyme deficiency associated with not yet well understood changes in the electrophoretic enzyme pattern.

In view of the contrasting conclusions that have been drawn from linkage studies performed in Sardinia and elsewhere to establish the linear sequence on the X chromosome of the G-6-PD gene and other X-borne loci (Siniscalco, 1964; Siniscalco, Filippi & Latte, 1964), we have stressed the possibility that the G-6-PD deficiency may be genetically heterogeneous in different populations. We have also reported some population data that suggest a non-random distribution of colour blindness of the deutan type among enzyme-deficient and normal individuals in Sardinia, suggesting that the neutral or slightly detrimental gene that is responsible for this defect in colour vision has also enjoyed protection in a malarial environment because of its very close linkage with the highly adaptive gene for G-6-PD deficiency (Siniscalco, 1963). Similar observations on the non-random distribution of colour blindness and G-6-PD deficiency have been reported in Israel (Adam, 1963).

If these findings are confirmed, it could mean that, in fact, it is not the G-6-PD deficiency gene alone that confers a higher fitness upon its carriers, but a "successful gene complex" (such as, for example, the

G-6-PD deficiency mutant plus a sex-linked modifier that is capable of reducing the red-cell enzyme deficiency to a minimum), which natural selection would have maintained because of its strong adaptive value in a malarial environment. If this were the case, the occurrence in some populations of a genetic mechanism, such as chromosomal inversion, that is capable of reducing crossing-over, could be suspected. This would help to explain how the selective advantage of the G-6-PD deficiency gene may differ from one population to another, if not from one family to another.

Conclusions

The data that have been reported to date offer striking examples of the possibilities and limitations of studies on human population genetics. On the one hand, we find the availability of true natural populations living in all sorts of ecological situations and therefore exposed to all degrees of natural selection, a detailed knowledge of historical and prehistorical migrations, mating systems, inbreeding, public health and vital statistics, and the facilities for obtaining direct estimates of different genetic fitnesses and exhaustive information concerning their interaction at the individual and population levels. On the other hand, we must consider the difficulty in establishing the relative importance of the numerous identifiable factors of evolution, the time and expense required to collect an amount of data sufficient to justify significant conclusions, the technical difficulties in performing critical studies in areas in which certain forces of natural selection, such as malaria, are still active, and the rapidly increasing impact of modern civilization on human ecology and therefore upon the relative fitness of a given genotype within the time interval of even a single human generation.

However, from all of the foregoing, a conclusion clearly emerges; namely, that it is necessary to learn a great deal more about the genetic structure of our own species and on the biological fitness of the principal genotype combinations in the existing "natural environments" as well as in the new environments that our civilization is likely to produce.

ACKNOWLEDGMENTS

The authors wish to express their gratitude to Professor G. Montalenti, whose encouragement, help and advice were the main factors that made possible the fulfilment of the present research project.

The collection of the blood samples would not have been possible without the generous and unlimited help of the health and administrative authorities of the districts of Cagliari and Nuoro and of the medical officers and school headmasters of the numerous Sardinian villages visited from time to time.

Our gratitude goes also to the members of all of the families investigated, to the Directors of the Ophthalmological Clinic, the Paediatric Clinic, the Institute of Medical Pathology and the Institute of Genetics of the University of Cagliari for their hospitality to the members of the team during the collection of data.

Special thanks go to Mr. R. Palmarino, whose thorough and excellent technical assistance was of substantial significance throughout the entire investigation.

The investigations reported in the present paper were supported by grants allocated to the Department of Genetics, University of Rome, by the Rockefeller Foundation (RF 62009, 1962–64), the National Science Foundation (G 19710, 1961–63) and the Italian National Research Council (Commissione di Genetica, 1960–64).

REFERENCES

ADAM, A. (1963) In: Goldschmidt, E., ed., *The genetics of migrant and isolate populations*, Baltimore, Williams & Wilkins, p. 111

ADINOLFI, M., BERNINI, L., CARCASSI, U., LATTE, B., MOTULSKY, A. G. & SINISCALCO, M. (1960) *Atti Accad. naz. Lincei R. C.*, **28**, Cl. Sc., 716

ALLISON, A. C. (1954) *Brit. med. J.*, **1**, 290

ALLISON, A. C. (1965) *Population genetics of abnormal haemoglobins and glucose-6-phosphate dehydrogenase deficiency.* In: Jonxis, J. H. P., ed., *Abnormal haemoglobins in Africa*, Oxford, Blackwell, p. 365

ALLISON, A. C. & CLYDE, D. F. (1961) *Brit. med. J.*, **1**, 1346

ALLISON, A., ASKONAS, B. A., BARNICOT, N. A., BLUMBERG, B. S. & KRIMBAS, C. (1963) *Ann. hum. Genet.*, **26**, 237

BAGLIONI, C. (1962) *Correlation between genetics and chemistry of human haemoglobins.* In: Taylor, J. H., ed., *Molecular genetics*, Academic Press, New York. p. 405

BARNICOT, N. A., ALLISON, A. C., BLUMBERG, B. S., DELIYANNIS, G., KRIMBAS, C. & BALLAS, A. (1963) *Ann. hum. Genet.*, **26**, 229

BERNINI, L., LATTE, B., SINISCALCO, M., PIOMELLI, S., SPADA, U., ADINOLFI, M. & MOLLISON, P. L. (1964) *Brit. J. Haemat.*, **171**, 10

BRUMPT, L. C. & BRUMPT, V. (1958) *Bull. Soc. Path. exot.*, **51**, 217

CARCASSI, U. (1962) *Giornate di studio sulla microcitemia*, Rome, Istituto Italiano di Medicina Sociale, pp. 78, 182

CARCASSI, U., CEPPELLINI, R. & PITZUS, F. (1957) *Boll. Ist. sieroter. milan.*, **36**, 206

CARCASSI, U., CEPPELLINI, R. & SINISCALCO, M. (1957) *Haematologica*, **42**, 1635

CEPPELLINI, R. (1955) *Cold Spr. Harb. Symp. quant. Biol.*, **20**, 252

EDINGTON, G. M. & LAING, W. N. (1957) *Brit. med. J.*, **2**, 143

EDINGTON, G. M. & WATSON-WILLIAMS, E. J. (1965) *Sickling, haemoglobin C, glucose-6-phosphate dehydrogenase deficiency and malaria in western Nigeria.* In: Jonxis, J. H. P., ed., *Abnormal haemoglobins in Africa*, Oxford, Blackwell, p. 393

FERMI, E. (1938) *La malaria in Sardegna*, Sassari, Stamperia Libreria italiana e straniera, vol. 2

FESSAS, P. (1965) *Forms of thalassaemia*. In: Jonxis, J. H. P., ed., *Abnormal haemoglobins in Africa*, Oxford, Blackwell, p. 71

FIASCHI, E., CAMPANACCI, L. & NACCARATO, R. (1964) *Minerva med.*, **55**, 1717

HALDANE, J. B. S. (1949) *Ricerca scient.*, suppl. 2, p. 12

HARRIS, R. & GILLES, H. M. (1961) *Ann. hum. Genet.*, **25**, 199

INGRAM, V. I (1963) *The haemoglobins in genetics and evolution*, New York, Columbia University Press

ITANO, H. A. (1965) *The synthesis and structure of normal and abnormal haemoglobins*. In: Jonxis, J. H. P., ed., *Abnormal haemoglobins in Africa*, Oxford, Blackwell, p. 3

KRUATZACHUE, M., CHAROEULARP, P., CHOUGHSUPHAJAISIDDHIM, T. & HARINASUTA, C. (1962) *Lancet*, **2**, 1183

KRUATZACHUE, M., NA-NAKORU, S., CHAROEULARP, P. & SUWOMAKUL, L. (1961) *Ann. trop. Med. Parasit.*, **5**, 468

LARIZZA, P., VENTURA, S., MATTIOLI, G., SULIS, E. & ARESU, G. (1958) *Haematologica*, **43**, 517

LIVINGSTONE, F. B. (1964) *Am. J. hum. Genet.*, **16**, 435

LOGAN, J. A. (1953) *The Sardinian project*, Baltimore, Johns Hopkins Press

MOTULSKY, A. G. (1960) *Hum. Biol.*, **32**, 28

MOTULSKY, A. G. (1965) *Theoretical and clinical problems of glucose-6-phosphate dehydrogenase deficiency: its occurrence in Africans and its combination with haemoglobinopathy*. In: Jonxis, J. H. P., ed., *Abnormal Haemoglobins in Africa*, Oxford, Blackwell, p. 143

PÉREZ, N., CLARK-TURRI, L., RABJILLE, E. & NIEMEYER, H. (1964) *J. biol. Chem.*, **239**, 2420

RUCKNAGEL, D. L. (1964) *Ann. N.Y. Acad. Sci.*, **119**, 436

RUCKNAGEL, D. L. & NEEL, J. V. (1961) *The hemoglobinopathies*. In: Steinberg, A. G., ed., *Progress in medical genetics*, New York, Grune & Stratton, p. 158

SILVESTRONI, E. & BIANCO, I. (1963) *Le emoglobine umane*, In: Gedda, L., ed. *De genetica medica*, Rome, Istituto Mendel, vol. 4, pp. 197–336

SINISCALCO, M. (1963) *Linkage data for G6PD deficiency in Sardinian villages*. In: Goldschmidt, E., ed., *The genetics of migrant and isolate populations*, Baltimore, Williams & Wilkins, p. 106

SINISCALCO, M. (1964) In: Geerts, S., ed., *Genetics today. Proceedings of the XIth International Congress of Genetics, The Hague, 1963*, London, Pergamon Press, vol. 2, p. 851

SINISCALCO. M., BERNINI, L., LATTE, B. & MOTULSKY, A. G. (1961) *Nature (Lond.)*, **190**, 1179

SINISCALCO, M., FILIPPI, G. & LATTE, B. (1964) *Nature (Lond.)*, **204**, 1062

SZEINBERG, A. (1963) *G6PD deficiency among Jews—genetic and anthropological considerations*. In: Goldschmidt, E., ed., *The genetics of migrant and isolate populations*, Baltimore, Williams & Wilkins, p. 69

THOMPSON, G. R. (1962) *Brit. med. J.*, **1**, 682

ZAINO, E. C. (1964) *Ann. N.Y. Acad. Sci.*, **119**, 404

Sickle-Cell Trait in Human Biological and Cultural Evolution

STEPHEN L. WIESENFELD

Medical anthropology has been moving increasingly into central areas of anthropological theory. Alland (*1*) has demonstrated some of the ways in which medical anthropology may serve "as a major link between physical and cultural anthropology, particularly in the areas of biological and cultural evolution." The purpose of this article is to examine the relationship between the sickle-cell trait, malaria, and agriculture in east and west Africa so as to derive hypotheses regarding concomitant human biological and cultural evolution.

Malarial infection, both natural and experimental, and mortality from such infection are consistently lower in individuals having the sickle-cell trait (*2, 3*). The normal population has reduced fertility rates as compared to the "sickler" population in endemic areas (*4*). Also, the distribution of the sickle-cell trait in tropical Africa parallels that of subtertian malaria (*3, 5*), so it is reasonable to believe that malaria is the selective agent producing high frequencies of the sickle-cell trait in the area of sub-Saharan Africa stretching from the east coast to Gambia on the west coast. Livingstone (*5*) proposed that malaria in west Africa became hyperendemic when large tracts of tropical rain forest were reclaimed for agriculture, by multiplying the number of breeding places for the *Anopheles gambiae* species complex, which contains major vectors of hyperendemic malaria (*6*).

Two important areas in the interaction of the sickle-cell trait, malaria, and agriculture have not been examined previously. First, not all agricultural systems have the same effect on the development of malaria and of high frequencies of the sickle-cell trait. The data presented here show that agricultural systems do differ in this respect, with one, the Malaysian agricultural system (*7, 8*), having a greater effect than any other. Second, the effects of changes in the frequency of an adaptive gene on the incidence of the disease selecting for it have not been fully examined. Computer models were developed to determine the nature of the interaction of the sickle-cell trait and malaria, and it was found that increasing frequencies of the sickle-cell trait cause reductions in malaria parasitism by reducing the

Stephen L. Wiesenfeld was born in Oakland, California in 1945. He received his B.A. in 1966 and his M.D. in 1970 from the University of California. He researched and wrote the selected article while still a medical student at the University of California, San Francisco Medical Center. The paper was awarded first prize in 1967 by the California Academy of Preventive Medicine. The author has since continued his research in the field of population genetics, investigating the relationship between the abnormal hemoglobins and G-6-PD deficiency and disease in the San Francisco area.

number of people capable of being infected in a population. Both of these arguments are critical to the hypothesis that the development and differentiation of the Malaysian agricultural system is intimately bound to changes in the gene pools of populations using this agricultural system. The action of high frequencies of the sickle-cell trait is to reduce the environmental limitation of malarial parasitism on these populations, thus allowing more human energy to flow into the development and maintenance of the Malaysian agricultural system. A number of lines of evidence are presented here to support the hypothesis.

Methods and Materials

Data, for the communities of this survey, on the degree of dependence on agriculture, the type of crop regime, and the manner of crop production were obtained from the "World Ethnographic Survey," a continuing series in *Ethnology*. Each community was scored from 0 to 9 according to the reported degree of its dependence on agriculture (0, dependence 0 to 5 percent; 1, 6 to 15 percent; 2, 16 to 25 percent; 3, 26 to 35 percent; 4, 36 to 45 percent; 5, 46 to 55 percent; 6, 56 to 65 percent; 7, 66 to 75 percent; 8, 76 to 85 percent; and 9, 86 to 100 percent).

Data on the frequency of the sickle-cell trait for various tribes in west and east Africa were used only if the sample size was considered adequate. Otherwise, data were used only if confirming studies had been made, in which case the results were pooled.

The mathematical models relating hyperendemic malaria and the sickle-cell gene frequency were set up on the basis of a Fortran IV program and run on the IBM 1401 computer.

Agriculture and the Sickle-Cell Trait

Data from 60 communities in east and west Africa met our criteria; they are plotted in Fig. 1, according to the percentage of individuals with the sickle-cell trait in the community. Data for tribes with agricultural dependence scores of 3 and below were combined, as were those for tribes with scores of 8 and 9.

The 50th percentile (median), the 25th percentile, and the 75th percen-

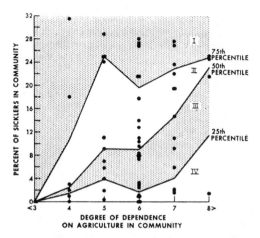

FIG. 1. Relationship between percentage of individuals with sickle-cell trait and degree of dependence on agriculture, in 60 communities.

tile found for each degree of dependence on agriculture show that a greater dependence on agriculture is associated with a higher frequency of the sickle-cell trait. The 25th and 75th percentiles represent the amount of variation around the median. Tribes are ranked either I, II, III, or IV according to which region of Fig. 1 they occupy. The sample of communities having less than 35-percent dependence on agriculture is too small to be considered significant, but communities falling in this region probably have very low frequencies of the sickle-cell trait. There is a trend toward stabilization of the median from 46- to 65-percent dependence on agriculture, but the widest variations in the trait frequencies are found here. Cultures with greater than 66-percent dependence on agriculture show a marked increase in the frequencies of the sickle-cell trait, with a reduction in the amount of variation around the median.

For any degree of dependence on agriculture there is a wide variation in the frequencies of the sickle-cell trait in various communities, indicating that different relevant variables may differentiate cultures in ranks III and IV from those in ranks I and II. Table 1 shows the distribution of the crop

TABLE 1. THE NUMBER OF COMMUNITIES IN
EACH RANK, ACCORDING TO CROP (SEE FIG. 1).

Crop	Rank			
	IV	III	II	I
Cereals	15	13	8	7
Root and tree crops	0	1	3	6

TABLE 2. RELATIONSHIP BETWEEN CROP REGIME AND FREQUENCY
OF THE SICKLE-CELL TRAIT IN 53 COMMUNITIES

Crop	Percentage of individuals with sickle-cell trait						
	0– 4	5– 9	10– 14	15– 19	20– 24	25– 29	30 and up
Cereals	19	9	5	2	0	8*	0
Root and tree crops	0	0	0	1	6	2	1

*These communities practice extensive irrigation and have high malarial parasitism (see text).

complexes among the four ranks of Fig. 1. Fifty-three of the cultures for which data were available were classified according to their main crop, but this classification does not mean that members of one crop complex may not have crops from another. Cereals, representing crops found in the Sudanic agricultural complex, have a wide distribution but are associated mainly with low frequencies of the sickle-cell trait. Root and tree crops from the Malaysian agricultural complex are associated almost exclusively with higher frequencies of the sickle-cell trait. A more revealing analysis of the relationship between crop regime and frequency of the sickle-cell trait is shown in Table 2. Cereals show wide variation but, for the most part, are associated with low frequencies of the trait. The eight communities in the 25- to 29-percent sickler group do not conform to the normal distribution that would be expected on the basis of data in Table 1. Most of these communities are in east Africa and are societies practicing intensive irrigation of crops. Root and tree crops are clearly associated with very high frequencies of the trait.

Also, note that communities with root and tree crops tend to have a greater economic dependence on agriculture than communities using cereals (Fig. 2).

Malaysian Agricultural Complex

According to Murdock (7), the predominant root crops in east and west Africa are yams and taro and not the Guinea yam as has been suggested by Livingstone and Chevalier (9). The Guinea yam does not have the status of a staple in either of the areas considered. The main tree crops are bananas and coconuts. These root and tree crops all belong to what Murdock calls the Malaysian agricultural complex that was introduced into Madagascar and the coast of east Africa from certain parts of southeast Asia a few centuries before Christ.

The ancient Malayo-Polynesian-speaking Mongoloid peoples of the coastal areas of southeast Asia were excellent sailors who carried their culture eastward into the islands of the Pacific and westward to Madagascar

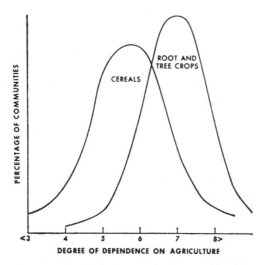

FIG. 2. Normal distribution of the degrees of dependence on agriculture in communities using cereals and in those using root and tree crops.

(*10*). The Malagasy are Mongoloid peoples found on Madagascar who speak a language of Malayo-Polynesian stock which is very closely related to the language of the Ma'anyan of Southeast Borneo (*11*). These people once occupied a coastal position and were great navigators. Southeast Borneo lies along the Sabaean Lane, a water route of great antiquity connecting Indonesia, Malaya, the Philippines, southeast China, and India. Dyen (*12*) has applied lexicostatistical methods to Dahl's work and estimates that the separation of the Malagasy from the Ma'anyan occurred at least 1900 years ago.

It is believed that the Ma'anyan brought with them from Borneo swidden agricultural techniques, dry rice, and root and tree crops. The sweet potato also appears to have been introduced at this time. Evidently the wet-rice or paddy-rice cultivation prominent on the mainland of southeast Asia and established on Java and Sumatra around the beginning of the Christian era had not yet been introduced into Borneo. Wet rice appears never to have been adopted by the Ma'anyans, who were, in succeeding centuries of the Christian era, displaced from their original coastal position into the adjacent interior of the island by tribes practicing paddy-rice cultivation (*7*).

The Azanians on the coast of east Africa adopted parts of the Malaysian agricultural complex during the first century of the Christian era, according to early written evidence (*13*). Taro (*Colocasia antiquorum*), yams (*Dioscorea alata, D. bulbifera, D. esculenta*), bananas, and coconuts are still prominent in this area (*14*). These Cushites of Azania carried the agricultural complex west into the interior of Africa and southwest into Uganda (7, p. 210).

A solid band of tribes located in what is called the "yam belt" runs across sub-Saharan Africa from the east to the west coast. In all these tribes the Malaysian agricultural complex figures importantly (7). Prior to the spread of this complex the Sudanic agricultural complex, with many cereal crops, predominated over native agriculture in sub-Saharan Africa (7). Communities practicing agriculture with the Sudanic set of crops were limited to the fringes of the tropical rain forest, because these crops were poorly suited to the rain-forest environment. Also, these communities never attained a high degree of dependence on agriculture in their economies (see Fig. 2). Throughout the "yam belt" there is evidence of the recent development of the Malaysian agricultural complex, which is well suited to the tropical forest environment, and of the recent penetration of the tropical rain forest by Negroid peoples using this complex; this penetration caused the displacement of hunting and gathering pygmy peoples (15). It is estimated that the occupation and displacement occurred within the last 2000 years.

Although the crops of the Sudanic complex play only minor roles in the societies under consideration, their planting and harvesting are attended by elaborate ritual, indicating great antiquity. In west Africa the Sudanic complex is more important in the north, where is appears to have arrived at an early date. Also, a number of crops were probably domesticated originally in west Africa, among them the Guinea yams and the oil palm (16). However, the Malaysian complex is of great importance in the south, where the tropical forest predominates. Wet rice is cultivated in many parts of west Africa; it appears to have been introduced by Arabs around A.D. 1500 (17).

Murdock also believes that the Malaysian complex was an important factor in the large expansion of the Bantu peoples through much of tropical Africa, and in the development of forest states in west Africa, due to the suitability of the crops for the tropical forest environment. As noted above, slash-and-burn cultivation is conducive to the development of breeding places for the *Anopheles gambiae* species complex, and thus to the development of intense malaria parasitism. With the shift from use of the tropical forest for hunting and gathering to swidden cultivation, there was a change in the nature of breeding places available to various species of mosquitoes and therefore a shift from *A. funestus* to *A. gambiae* as the major vector of malaria. This shift is significant because the behavior of the species comprising the *A. gambiae* species complex is much more conducive to the development of hyperendemic malaria.

Livingstone (5) noted a north-south gradient in the frequencies of the sickle-cell trait, with greater frequencies in the south. All along the "yam belt," especially in the central Sudan and in Nigeria, there is a north-south gradient in the distribution of Malaysian crops (7), paralleling the gradient of the sickle-cell trait. Figure 3 shows the distribution of the Malaysian agricultural complex and the distribution of sickle-cell-trait frequencies higher than 5 percent. There is a striking overlap of these areas.

The relationship between agriculture and malaria was noted over a century ago when outbreaks of malaria followed irrigation work (18). Angel (19) has demonstrated that in the eastern Mediterranean during prehistoric times porotic hyperostosis, or a thickening of the spongy marrow space of the skull due to sicklemia or one of the thalassemias, is associated frequently with early farmers who lived in marshy areas in Greece, but rarely with paleolithic hunters. The frequency of the disease also appears to have decreased in areas where farming methods improved.

Genes and Socioeconomic Adaptations

The introduction of the Malaysian food complexes into sub-Saharan Africa brought about major changes in human ecological niches. Introduction of the crops allowed Negroes to penetrate the tropical rain forest and allowed their populations to increase to new equilibrium levels, making this band across the continent, especially in west Africa, the most densely populated part of Africa. As Geertz (20) has demonstrated in Indonesia, a population adopting a new agricultural system which provides a greater and more certain food supply not only undergoes expansion but also tends to involute. More people are centered in one area, and people tend to move less, due to vastly increased needs of husbandry to keep the agricultural system yielding at maximum capacity. Malaria increases in

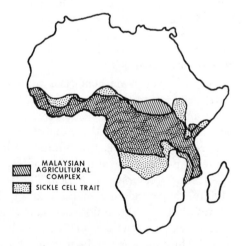

FIG. 3. Distribution of the Malaysian agricultural complex and distribution of sickle-cell-trait frequencies higher than 5 percent.

such a situation, for infected individuals remain constantly close to unin-
fected individuals and the probability of transmission by mosquito increases.

The new agricultural system allows expansion and involution of the
population and, at the same time, is the ultimate cause of an increase in
malaria parasitism. The population growth puts increased pressure on the
agricultural system for greater food production, but the parasitism takes
its toll through mortality and morbidity, thus reducing the total energy
available for agricultural production. The sickle cell presents a biological
solution to a cultural problem by providing many members of the population with genetic immunity, thereby allowing more human energy to flow
into agricultural production to meet the demands of an increasing population.

These concepts may be formulated into a principle regarding the interaction of biology and culture in man. Where a socioeconomic adaptation
causes a change in the environment, the frequency of a gene will change
in proportion to the survival value the gene confers on the carriers in the
new ecosystem. Increasing frequencies of an adaptive gene remove environmental limitations and allow further development of the socioeconomic adaptation. The environmental conditions crucial to the transmission
of malaria are also crucial to the economy, but the sickle-cell trait removes
a limitation for agricultural development and maintenance by reducing the
number of people capable of being infected by malaria in a community.
The gene frequency and the socioeconomic adaptation continue to develop
in a stepwise fashion until either the limit of the gene frequency or the
limit of the socioeconomic adaptation is reached.

Evidence to support the hypothesis is found in Table 2 and Fig. 2, which
show that societies with the greatest dependence on agriculture have the
highest frequencies of the sickle-cell trait.

Mathematical Models: Concepts

It has been possible to develop mathematical simulations relating the
dynamics of the sickle-cell trait to the dynamics of hyperendemic malaria.
The models show that an increase in the sickle-cell trait reduces the intensity of malaria parasitism by reducing the proportion of the population
capable of being infected. The models are offered as evidence that the
sickle-cell trait removes an environmental limitation on the development
of the Malaysian agricultural complex in Africa. The model presented is
believed to be valid for hyperendemic malaria in sub-Saharan Africa.
The variables used reflect basic epidemiologic determinants of malarial
transmission and are taken from actual field studies. Before considering
the models we should consider certain factors of the epidemiology of
malaria.

Malaria is a three-factor disease, with mosquitoes and mammals serving
as hosts, and is a disease caused by infection with one of the four malaria

parasites; however, only two of them, *Plasmodium falciparum* and *P. ovale*, are important in areas where the sickle-cell trait is found. *Plasmodium falciparum* is usually associated with stable, hyperendemic malaria, while *P. vivax* and *P. ovale* are dominant in areas with unstable, epidemic malaria with marked seasonal variations. Stable malaria, as found in sub-Saharan Africa, is associated with the following characteristics. The mosquito, usually from the *Anopheles gambiae* species complex, bites man frequently, and its probability of survival through the period of development of the parasite in its salivary gland is good. An infective mosquito with a long life span is a greater epidemiologic hazard than a short-lived counterpart. The high longevity of the mosquito in sub-Saharan Africa is due mainly to the humid and stable climatic conditions.

The main vectors for malaria in sub-Saharan Africa are the species of the *Anopheles gambiae* complex and *A. funestus*. The *A. gambiae* species complex is mainly associated with agricultural societies and hyperendemic malaria, while *A. funestus* is associated with areas of unchanged tropical rain forest, hunting and gathering cultures, and lower levels of endemic malaria. This difference in ecological niche is due solely to a difference in the breeding places of the various species, as noted above. The freshwater breeders of the *A. gambiae* species complex will multiply in many types of water but prefer open, sunlit pools of the type created in the tropical rain forest when slash-and-burn agriculture is practiced. *Anopheles funestus* is found in swamps with heavy vegetation, in vegetated river edges, and in other bodies of water that are not in direct sunlight (*21*). Such breeding places are found most readily in the unchanged tropical forest. It is possible to use these mosquitoes as "ecological labels" for different human socio-economic adaptations, a concept introduced by Audy (*22*).

Figure 4 shows the relationship between the sporozoite rate (or the proportion of mosquitoes, in a population of mosquitoes, with malaria parasites in their salivary glands) and the selective advantage of the sickle-cell trait in seven African communities. The sporozoite rate is a good index of the amount of malaria in an area where it is endemic because the sporozoite rate depends (i) on the number of infected people in a population capable of producing infection in a mosquito, (ii) on the frequency with which a mosquito will bite men, (iii) on the time it takes the malaria parasite to develop within the mosquito, and (iv) on the mortality rate of the mosquitoes (see Appendix, Eq. 3). The selective advantage of the population with the sickle-cell trait is determined according to Livingstone (*23*), and in Africa is represented as some value greater than 1, because 1 is taken as the selective advantage of the population lacking the sickle-cell trait. The greater the selective advantage of the sickler, the more the sicklers will contribute to the composition of future generations. The basic prerequisite for determining the selective advantage of the heterozygote is that the frequency of the sickle-cell gene be in equilibrium with the amount of malaria in the area being considered.

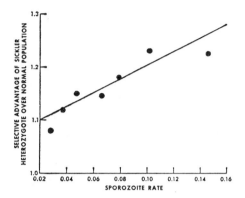

F IG. 4. Relationship between the intensity
of endemic malaria (the sporozoite rate) and
the selective advantage of the sickler hetero-
zogote over the normal population
(see Appendix).

The best-fitting straight line was determined (see Appendix, Eq. 1). The selective advantage of the heterozygote clearly tends to increase as the amount of malaria in an area increases.

Mathematical Models: Results

Figures 5 and 6 are mathematical simulations of various epidemiological conditions and show the dynamics of endemic malaria in response to the presence of the sickle-cell trait in the population.

One more epidemiological concept must be introduced before the models are discussed. This is the basic reproduction rate, or the number of human infections produced by an earlier human infection (see Appendix, Eq. 5). If one case of malaria gives rise to two other cases at a later date, then the basic reproduction rate is 2. The basic reproduction rate is directly proportional to the sporozoite rate and serves as a very good index of the intensity of malaria. If the basic reproduction rate ever falls below 1.0, malaria will not continue to be transmitted in the community, so 1.0 represents the critical value for the continuance of malaria. If the basic reproduction rate "continuously exceeds this critical value, malaria will persist but its epidemiological characteristics will be largely determined by the biting habit and longevity of the mosquito" (24).

The logic of these models may be presented with a minimum of mathematics. The frequency of the sickle-cell gene in a population is assumed to be initially very low, and the longevity of the mosquito and its man-biting habit are taken to be high. For a given sporozoite rate in a particular area it is possible to determine what the selective advantage of the sickler population is, according to the relationship shown in Fig. 4 and in Eq. 1 of the Appendix. For each generation the frequency of the sickle-cell trait

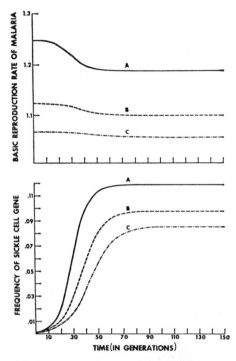

FIG. 5. Mathematical models of the interac-action of the sickle-cell trait and malaria in populations where the percentage of human individuals producing the form of parasite capable of infecting mosquitoes is (A) 3.6 percent; (B) 1.8 percent, or (C) 1.0 percent, or where mosquito bites man once every (A) 2 days, (B) 4 days, or (C) 10 days.

is determined for the next generation on the basis of the selective advantage of the sickler (Appendix, Eq. 2). Given this new frequency of the gene, it is possible to see by what degree the population capable of being infected is reduced, hence the value for the basic reproduction rate of malaria for that generation may be determined (Appendix, Eq. 6). Since the basic reproduction rate is directly proportional to the sporozoite rate and the sporozoite rate is directly proportional to the selective advantage of the sickler heterozygote, it is possible to derive an expression relating the selective advantage to the basic reproduction rate (Appendix, Eq. 7), and a new value for the selective advantage is found, which is used in determining the frequency of the gene in the next generation. These three steps—determining (i) the new gene frequency, (ii) the new basic reproduction rate, and (iii) the new selective advantage of the heterozygote—are followed for each generation.

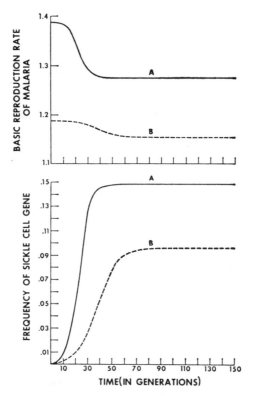

FIG. 6. Mathematical models of the interaction
of the sickle-cell trait and malaria where the
daily mosquito mortality is
(A) 5 percent and (B) 10 percent.

Comparison of the corresponding curves of the top and bottom graphs of Fig. 5 shows that an increase in the frequency of the sickle-cell gene causes a reduction in the basic reproduction rate of hyperendemic malaria by reducing the proportion of people capable of being infected. High sporozoite rates in the mosquitoes, caused by the fact that a relatively high proportion of the human population is capable of infecting mosquitoes, allow development of proportionately high equilibrium frequencies of the gene, which in turn cause marked reductions in the basic reproduction rate (Fig. 5, top, curve *A*). When the proportions of human populations capable of infecting mosquitoes are lower, this allows the development of lower equilibrium frequencies, and this in turn causes lesser reductions in the basic reproduction rate of malaria (Fig. 5, top, curves *B* and *C*). Notice also that it takes more time for lower frequencies to reach equilibrium because of the lower pressures of malaria on the population as a whole.

Figure 5 also shows the effect of varying the mosquito man-biting habit —its effect on the frequencies of the gene developed, and its subsequent effect on the reduction of malaria in a population. The frequencies with which mosquitoes bite man vary a great deal with various species and have profound effects on the nature of malaria in a region. Stable, hyperendemic malaria is associated with high frequencies of biting. Where mosquitoes bite man regularly, the number of infected mosquitoes (the sporozoite rate) is high, the selective advantage of the heterozygote is high, and the frequency of the gene rises rapidly to a high equilibrium value (Fig. 5, bottom, curve *A*). This increase in the gene frequency causes a marked reduction in the basic reproduction rate of malaria. Lower values for the man-biting habit of mosquitoes allow lower frequencies of the gene to develop, with proportionately milder reductions in the basic reproduction rate (Fig. 5, bottom, curves *B* and *C*).

Increases in the man-biting habit of the mosquito have significant effects on the intensity of malaria and hence of the frequency of the sickle-cell trait. In many parts of the world an increase in the mosquitoes' man-biting habit is caused by the displacement or reduction in numbers of cattle or other animals that previously served as hosts for the mosquito. This displacement is caused by growth in human population, so the mosquito must turn increasingly to man as the major host. Epidemics caused by such a situation have been reported in India (*25*), and it has been speculated that man has caused the displacement of original hosts for many parasites in densely populated parts of the world, especially in the tropics.

Figure 6 shows the effect on the interaction of the sickle-cell trait and malaria of varying the longevity of the mosquito. The daily mortality of mosquitoes depends mainly on humidity and on seasonal climatic variations. Throughout tropical Africa there is relatively little seasonal climatic variation and the relative humidity is constantly high, so the daily mortality of mosquitoes is low. Figure 6, top and bottom, curve *A*, shows a simulation in which a low value was used for mosquito daily mortality, one typical for hyperendemic malaria in Africa, which was determined from actual field studies (*26*). The model stabilized after about 40 generations or 1000 years, and the equilibrium values approximate very closely the values determined for malaria and the sickle-cell trait in Uganda, where values for all the parameters in this model were obtained. MacDonald determined the basic reproduction rate to be around 1.25, and various field studies on frequencies of the sickle-cell trait show values ranging from 27 to 39.1 percent (*27*). For the sickle cell, the gene frequency is half the trait frequency.

On the fringe of the tropical rain forest large areas of savanna are found, where the humidity is lower and where there is more seasonal climatic variation. These conditions cause a reduction in the longevity of the mosquito, the results of which are represented in Fig. 6, top and bottom, curve *B*. The model reaches equilibrium in about 60 generations or 1500 years, and the equilibrium values reached are similar to the values

reported for northwestern Nigeria, where the values for the parameters of
the model were obtained (5, 28). The time necessary for the development
of equilibrium in all models is well within 2000 years—the interval since
the Malaysian agricultural system was introduced into Africa.

Conclusion

The particular agricultural adaptation we have been considering is the
ultimate determinant of the presence of malaria parasites in the intracellular environment of the human red blood cell. This change in the cellular environment is deleterious for normal individuals, but individuals
with the sickle-cell gene are capable of changing their red-cell environment so that intense parasitism never develops. Normal individuals suffer
higher mortality rates and lower fertility rates in a malarious environment
than individuals with the sickle-cell trait do, so the latter contribute proportionately more people to succeeding generations.

In the case of an intensely malarious environment created by a new
agricultural situation, the viability of the normal individual is reduced and
there is selection for the individual with the sickle-cell trait; this means
that the nature of the gene-pool of the population will change through
time. This biological change helps to maintain the cultural change causing
the new cellular environmental change, and the biological change may
allow further development of the cultural adaptation, which in turn increases the selective pressure to maintain the biological change. In this
way it is possible to see human biology and culture interacting and differentiating together in a stepwise fashion. If valid, the hypothesis developed
here serves to demonstrate the important role that disease may have in
human evolution.

Appendix

The mathematical analysis. The purpose of this mathematical analysis
is to allow the development of models in which various factors may be
varied to determine their effects on the process being considered. The
mathematics of the epidemiology of malaria has been developed by Mac-
Donald (29), and the basic statements about malaria in this simulation are
his. Equations relating to changes in the gene frequency are derived from
the Hardy-Weinberg law and the notation of Wright used by Livingstone
(23).

1) The equation for the relationship between the sporozoite rate and
the selective advantage of the sickler over the normal individual shown
in Fig. 4 is

$$w_{12} = 1.075 + 1.289s \tag{1}$$

where w_{12} is the selective advantage of the heterozygote sickler and s is the
sporozoite rate, or the proportion of mosquitoes capable of transmitting
the parasite to man.

2) The expression for the change in gene frequency for one generation is derived from the Hardy-Weinberg law and is given as follows:

$$q_{i+1} = \frac{1}{\left(\dfrac{w_{11}(1 - q_i)}{w_{12i}q_i}\right) + 2} \tag{2}$$

where q is the gene frequency, i is the number of the generation being considered, and w_{11} is the selective advantage of the normal population.

3) The expression for the sporozoite rate is given by MacDonald as

$$s = \frac{p^n ax}{ax - \log_e p} \tag{3}$$

where p is the probability of a mosquito's surviving through the first day of life; p^n is the probability of its surviving through n days, or the time that it takes for the parasite to develop within the mosquito; a is the number of times a mosquito bites a man in one day; and x is the proportion of the human population capable of infecting mosquitoes.

4) To Eq. 3 I have added the following factors: $2q$, the proportion of the human population incapable of being infected due to genetic immunity, and $\alpha = (1 - 2q)$, or the proportion of the population capable of being infected due to lack of genetic immunity. Rewritten, the sporozoite rate is

$$s = \frac{p^n ax\alpha}{ax\alpha - \log_e p} \tag{4}$$

5) MacDonald defines z_o, or the basic reproduction rate, as

$$z_o = \frac{p^n}{p^n - s} \tag{5}$$

Substituting the value of s from Eq. 4 into Eq. 5, we get

$$z_o = 1 - \frac{ax(- 2q)}{\log_e p} \tag{6}$$

6) The selective advantage of the heterozygote may be expressed in terms of the basic reproduction rate through substitution of the expression for s in Eq. 1 into Eq. 5; the substitution yields

$$w_{12i+1} = 1.289 p^n [1 - (1/z_o)] + 1.075 \tag{7}$$

REFERENCES AND NOTES

1. A. ALLAND, *Amer. Anthropol.* **68,** 40 (1966).
2. A. ALLISON, *Ann. N. Y. Acad. Sci.* **91,** 710 (1961): A. Motulsky, *Amer. J. Trop. Med. Hyg.* **13,** 147 (1964).

3. A. ALLISON, *Brit. Med. J.* **1**, 290 (1954).

4. D. RUCKNAGEL AND J. NEEL, in *Progress in Medical Genetics*, A. G. Steinberg, Ed. (Grune and Stratton, New York, 1956), p. 158; H. Archibald, *Bull. World Health Organ.* **15**, 842 (1956).

5. F. LIVINGSTONE, *Amer. Anthropol.* **60, 13**, (1958).

6. What has been called *Anopheles gambiae* is now thought to be a complex of five or more sibling species [G. Davidson, *Riv. Malariol.* **43**, 167 (1964)]. Two of these species (*A. melas* and *A. merus*) are saltwater breeders and three unnamed species are freshwater breeders. The relative role of each species in producing endemic malaria has not been evaluated [P. Russell, L. West, R. Manwell, G. MacDonald, *Practical Malariology* (Univ. of Oxford Press, London, 1963)]. This group of species is referred to in this article as the *A. gambiae* species complex.

7. G. MURDOCK, *Africa: Its Peoples and Their Culture History* (McGraw-Hill, New York, 1959).

8. The term *Malaysian agricultural complex* is Murdock's. I have adopted Murdock's scheme for the origin and development of this complex.

9. F. LIVINGSTONE, *Amer. Anthropol.* **60**, 551 (1958); A. Chevalier, *Rev. Botan. Appl. Agr. Trop.* **26**, 26 (1946).

10. R. LINTON, *The Tree of Culture* (Knopf, New York, 1958), p. 45.

11. O. DAHL, *Arhandl. Egede-Inst.* **3**, 1 (1951).

12. I. DYEN, *Language* **29**, 577 (1953).

13. E. SCHOFF, in *The Periplus of the Erythraean Sea,* E. H. Schoff, Ed. (London, 1912).

14. I. BURKILL, *Proc. Linnean Soc. London* **164**, 12 (1953).

15. N. VAVILOV, *Chron. Biotan.* 533 1 (1949); C. Seligman and B. Z. Seligman, *Pagan Tribes of the Nilotic Sudan* (London, 1932).

16. A. CHEVALIER, *Rev. Botan. Appl. Agr. Trop.* **32**, 14 (1952).

17. D. PAULME, *Les Gens du Riz* (Paris, 1954).

18. W. BAKER, T. DEMSTER, H. YULE, *Records Malariol. Surv. India* **1**, 1 (1929) (reprint of 1847 article).

19. J. L. ANGEL, *Science* **153**, 760 (1966).

20. C. GEERTZ, *Agricultural Involution* (Univ. of California Press, Berkeley, 1965).

21. D. WILSON, in *Malariology,* M. F. Boyd, Ed. (Saunders, Philadelphia, 1949) pp. 800–809.

22. J. R. AUDY, *Brit. Med. J.* **1**, 960 (1954).

23. F. LIVINGSTONE, *Amer. J. Human Genet.* **16**, 435 (1964).

24. G. MACDONALD, *The Epidemiology and Control of Malaria* (Oxford Univ. Press, London, 1957), pp. 39–40.

25. S. CHRISTOPHERS, *Sci. Mem. Officers Med. Sanit. Dep. Govt. India* 46.

26. G. DAVIDSON AND C. DRAPER, *Trans. Roy. Soc. Trop. Med. Hyg.* **47**, 522 (1953); G. Davidson, *ibid.* **49**, 339 (1955).

27. G. MACDONALD, *Proc. Roy. Soc. Med.* **48**, 295 (1955); H. Lehmann and A. Raper, *Nature* **164**, 494 (1949); A. Allison, *Trans. Roy. Soc. Trop. Med. Hyg.* **48**, 312 (1954). Gene frequencies represent, almost exclusively, heterozygotes.

28. H. ARCHIBALD, *Bull. World Health Organ.* **15,** 695 (1956); A. Allison, *Ann. Human Genet.* **21,** 67 (1956).
29. G. MACDONALD, *The Epidemiology and Control of Malaria* (Univ. of Oxford Press, London, 1957).
30. I thank N. L. Petrakis, P. Mustacchi, S. L. Washburn, J. N. Anderson, and F. L. Dunn for aid and helpful criticism. The research reported in this article was supported in part by U.S. Public Health Service fellowship grant 5T5 GM 43-05 and in part by U.S. Public Health Service grant FR-00122 for computing services at San Francisco Medical Center, University of California. The research was also supported in part by PHS research grant CA 05485-07.

Patterns of Survival and Reproduction in the United States: Implications for Selection

DUDLEY KIRK

The outstanding human biological fact of our time is the rapid multiplication of our species. In this country, and most of the Western world, man has freed himself from many of the selection pressures that have kept his numbers down since his beginnings. Now this achievement is spreading to the rest of the world. The result is what is dramatized as the "population explosion."

At the same time growth has not been unchecked. Western populations have undergone a revolution in their vital processes of birth and death, as fundamental as the industrial revolution and modernization in its impact on the individual. The vital revolution, or the demographic transition, as it is more academically described, is the transition from wastefully high birth and death rates to the much more efficient lower birth and death rates that now prevail in the more advanced countries. The following discussion relates to the United States, but similar developments are occurring in all advanced countries.

In this country, as in all developed countries, *both* deaths and births are now largely *controlled,* on the one hand by mastery of the physical environment and on the other by the voluntary choices of couples on the number of their offspring.

In mortality, of course, controlled means *postponed,* not eliminated altogether. But for present purposes, postponement beyond the ages of reproduction is equivalent to immortality as a factor in natural selection.

Similarly, the reduction in average number of offspring is chiefly the result of voluntary choice by millions of couples to restrict family size. Surveys show that over 90 per cent of white couples of proved fertility practice birth control at some time during their married life. This practice is not always efficient, but collectively in terms of statistical averages it is very effective in restricting family size. In 1800, the average American woman passing through her reproductive life had seven children; today she has *three* or less.

It seems reasonable to suppose that such major changes should have very important genetic effects. In what follows, demographic changes will be discussed in relation to natural selection and selection intensity.

Reproduced by permission of the publisher and Dudley Kirk from Proceedings of the National Academy of Sciences, *59:662–670 (1968).*

Dudley Kirk, *a demographer, was born in Rochester, New York in 1913. He received his B.A. from Pomona College in 1934 and his M.A. and Ph.D. from Harvard University in 1938 and 1946 respectively. Presently he is Professor of Demography at the Food Research Institute and Department of Sociology, Stanford University. As a demographer in the United States Department of State, he served as a member of the United States Delegation to the United Nations Population Commission, 1948–1953. He is a past president of and currently on the Board of Directors for both the American Eugenics Society and the Population Association of America. His research has focused on the demography and population dynamics of Western industrialized nations. He has also written extensively on the world population problem, especially with reference to the rapidly growing populations of underdeveloped countries. Author of several books on demography, some of his more recent articles include "Factors Affecting Moslem Natality," in B. Berelson, et al., Eds., Family Planning and Population Programs (Proceedings of the International Conference in Family Planning Programs, Geneva, August 1965); "Natality in the Developing Countries: Recent Trends and Prospects 1969," in S. J. Behrman, et al., Eds., Fertility and Family Planning:—A World View (University of Michigan Sesquiantennial Symposium, Ann Arbor, 1968).*

In any society natural selection occurs because different genotypes produce different numbers of offspring and because different proportions of offspring survive to the age of reproduction. It also follows that differentials in reproduction and survival are important only to the extent that they involve important differences in genotype. Consequently I am here discussing not the actual genotypic selection, which is, of course, extraordinarily difficult to measure, but *possible* selection intensity under different demographic conditions, that is, the *opportunity* for natural selection.

I shall discuss first the relation of demographic changes to survival, and second their relation to reproduction.

Survival

There has been a dramatic reduction in mortality and hence presumably in selection intensity. While the fact is generally understood, the magnitude of the gains is probably not often realized. Historical experience and projection for the near future are shown in Table. 1

The reduction of mortality is most spectacular for females. In 1840 only two thirds of white females born in 1840 reached age 15 and only about one half reached age 45. According to conservative projections, 97.5 per cent of females born in 1960 will live to age 15 and 96 per cent to age 45. Fewer and fewer now die before the end of the reproductive years —only about 4 per cent instead of 50 per cent as in 1840.

Somewhat less progress has been made in reducing deaths among males. The estimates for survivorship of male cohorts born in 1960 are less than 97 per cent to age 15 and about 93 per cent to age 45. Furthermore, male

TABLE 1. PER CENT SURVIVING TO AGE 15, 30, AND 45, UNITED STATES,
WHITE MALES AND FEMALES, BY YEAR OF BIRTH, 1840–1960.

Year of birth	Age 15		Age 30		Age 45	
	Males	Females	Males	Females	Males	Females
1840	62.8	66.4	56.2	58.1	48.2	49.4
1880	71.5	73.1	65.7	67.4	58.3	61.1
1920	87.6	89.8	83.4	88.0	79.8	85.8
1960	96.6	97.5	95.1	96.9	92.9	95.9

Data from Jacobson, Paul H., *Milbank Memorial Fund Quarterly*, **42**, 36 (1964).

These are cohort data by year of birth and therefore involve projections, especially for those born in 1960. Already by 1965, current life table values had exceeded or were approximating the projections given in this table.

reproductivity generally starts later and continues beyond age 45 with less abrupt termination than for women. The effect of mortality is greater at these higher ages. Consequently, selection intensity is somewhat greater for males, but still minimal in comparison with the universal situation in the past. Barring major catastrophes, further reduction of selection intensity for both sexes is almost certain, but of course within a narrow range—it is already so close to zero.

Most of the force of natural selection is directed at maintaining stability, i.e., cancelling the deterioration of the organism's adaptations to its environment that otherwise would occur. Has relaxation of this "protective" selection occurred? Surely it must have. The force of this is somewhat ameliorated by several factors.

(1) The major saving of life has been in the reduction of infectious and epidemic diseases through environmental means. It does not seem that genetically determined resistance to most infectious diseases (to the extent that it exists) is an adaptation of great consequence in modern society. Several of the great epidemic and endemic diseases—smallpox, cholera, plague—were mastered and largely eliminated in the 19th century. Since then, great progress has been made against those remaining: pneumonia and influenza, typhoid fever, tuberculosis, syphilis, diarrhea and enteritis, and the communicable diseases of childhood. Organic diseases are now the great killers—cardiovascular-renal diseases and cancer account for about three fourths of all deaths. These organic diseases may be more dependent on genetic inheritance, but they are not so important in earlier life and therefore do not much affect survival to and through the reproductive period.[1]

Largely because other causes of death have been so reduced, deaths by accident and violence are major causes of death prior to age 45. It is certainly nothing new to have negative selection of persons prone to accident and violence. This selective factor remains, though unhappily it seems unlikely that fatal automobile accidents will early lead to a genetically superior race of better drivers!

(2) Deaths related to specific inherited defects have not been reduced nearly so much as the total. Thus two thirds of female deaths and over 60 per cent of male deaths up to age 30 are now due to congenital malformations and diseases of early infancy, which often involve immaturity.[2] These include an important component of genetic defects. The most serious of such defects are still being removed from the population through deaths.

(3) Were it possible to do so, the survival factor should be computed from *conception* rather than from live births. Fetal wastage and stillbirth, especially of the malformed, is still high. The wastage in early pregnancy is believed to be still very high, presumably much more important than subsequent fetal wastage and mortality in eliminating gross malformations and genetic anomalies.

It would be unwise to take much comfort from these ameliorative factors. Undoubtedly many persons of genetically weak constitution are now surviving who earlier would have succumbed to disease or to a harsher environment. Furthermore, determined efforts are being made to reduce infant deaths, including those involving genetic defects. It is reasonable to expect that an increasing proportion of individuals having a weak physical constitution and/or carrying deleterious mutations will survive and reproduce. The direction is not at issue—only the specific nature and rapidity of the changes. Measurement of the latter is extremely difficult and beyond the scope of this paper.

On the other hand, environmental gains are more than compensating for any genetic deterioration, as evidenced by the gains in physical stature, by the reductions in morbidity, and by the falling death rates themselves.

Even more important, what is deteriorating is our fitness for the physical environment of the *past*, not that of the present or the future. The genetic qualities required for survival and effective performance in a peasant or pastoral society are presumably very different from those required for best performance in our primarily urban and sedentary life. We are losing our adaptability for the former but certainly much less for the life of the present, and perhaps even less for that of the future.

Reproduction

It is a commonplace that in modern societies natural selection by deaths has been replaced by the social selection of births. This frequent observation is faulty in two ways. First, selection by number of progeny is just as "natural" as selection by deaths; and second, selection by number of progeny has usually been the more important element in the past, as it is in the present. Thus Sphuhler's comparisons of selective intensity from mortality and from fertility components show the latter factor to be more important in a majority of the numerous modern and premodern populations covered.[3]

Demographic changes in four areas affect the opportunity for natural selection through differential fertility. These are (1) mating and marriage, (2) childlessness, (3) number of offspring, and (4) age at childbearing and mean length of generation.

(1) *Mating and marriage:* The married state has become more prevalent in the United States over the last two generations, but particularly since 1940. The per cent of adults in the principal ages of reproduction who are married has risen substantially. Thus, of all men at ages 14–44, 50.5 per cent were currently married in 1940 and 61.2 per cent in 1960. The corresponding data for women are 58.7 per cent in 1940 and 68.9 per cent in 1960, in each case a gain of over 10 per cent.[4] This increase is due to three factors: a larger percentage of men and women marry; they marry earlier; and they spend more of their potentially reproductive years in marriage.

To illustrate, in 1966, 95.0 per cent of all women at ages 35–44 (i.e., at the later years of childbearing) had ever married, as compared with 88.6 per cent in 1910 and 91.4 per cent in 1940. The median age at first marriage of men fell from 26.1 in 1890 to 22.8 in 1960; that of women from 22.0 in 1890 to 20.3 in 1960. This earlier first marriage plus less widowhood and higher rates of remarriage have led to more years in the reproductive ages spent in marriage despite a rise in divorce.

Childbearing is not entirely restricted to married couples. Illegitimacy in the United States has risen rapidly. Since 1940, the estimate of per cent of illegitimate births has doubled for whites and increased by over 60 per cent for non-whites. National estimates, based on the official statistics of states that report illegitimacy, show 3.4 per cent of white births and 24.5 per cent of nonwhite births to be illegitimate in 1964.[5] The true figures are doubtless higher. A reasonable inference is that mating outside of marriage, as well as marriage itself, has increased.

Thus the potential reservoir of parents has been increasing, and this was in part responsible for the "baby boom." What it means genetically is that a larger percentage than previously of each cohort is exposed to the "risk" of childbearing, especially through earlier and almost universal marriage. We are in this way "reverting" to the situation prevailing in premodern societies, where marriage and mating are almost universal. In this country, age at marriage is lower than in most other Western countries, but of course higher than in most premodern societies.

(2) *Childlessness:* In our society there are three reasons why adults fail to have children: biological infecundity, failure to mate or marry, and choice, i.e., the practice of birth control. The second and third are unimportant in most premodern societies, including our own at an earlier stage. In the United States 100 years ago only a tenth of American women living through the childbearing period were childless. With less marriage, later marriage, and more practice of birth control, 23 per cent

of women born in 1909 were childless. Now, again with more and earlier marriage, and probably greater success in the medical treatment of sterility, only 7 per cent of married women and 12 per cent of all women at age 35 are currently childless.

Let us consider the joint effect of mortality and childlessness, which are genetically the same: the persons concerned have no descendants. Combining these effects, about half of women born in 1840 did not participate in the reproduction of the next generation. Now the combined effect is only 15 per cent. Thus at current experience some 85 per cent of each female cohort born now not only survives but also produces offspring.

(3) *Number of offspring:* As noted earlier, the average number of children born per woman is much lower than in the earlier years of our history—three or less versus seven. Surviving children per woman are much more comparable in number, thanks to the great saving of life at earlier ages.

While mortality has declined continuously, natality has not followed a continuous decline since the 1930's. As result of the "baby boom" after World War II, the average number of children per woman rose from a low of 2.3 (for the female cohort born in 1909) to estimates of about 3 for those born in 1940, who have not yet completed their childbearing.

From the point of view of natural selection the variance is more important than the average (see Table 2). Contrary to "common sense," the possible selective intensity of the natality component did not decline with the long secular decline in family size. This anomaly is explained by the fact that reduction of the mean number of children does not necessarily reduce the variance. On the contrary, in the United States the opposite has been the case. Thus the index of selection intensity for the fertility component[6] was 0.710 for women born 1871–1875, who averaged 3.5 chil-

TABLE 2. NUMBER OF OFFSPRING OF SELECTED COHORTS OF
WOMEN IN THE UNITED STATES.

	Per Cent Distribution by Parity			
	Women of Completed Fertility			Expectations of married white women under age 40 in 1962
Parity	Born in 1871–75	Born in 1909	Born in 1928 (at age 36)	
0	17	23	11	4
1	16	22	12	7
2	14	22	24	26
3	12	13	22	28
4	10	8	14	21
5	7	4	8	7
6 and over	24	8	9	7
Total	100	100	100	100

dren. For women born in 1909 the index was 0.881, with an average of only 2.3 children, the lowest number for any U.S. cohort.

With the recovery of the birth rate since 1940 the index will certainly decline sharply, although the women concerned have not completed their childbearing. There will almost certainly not be much increase at higher parities that would contribute heavily to variance since birth rates at the higher reproductive ages are now very low. Finally, if the expectations of married white women under age 40 in 1962 are realized (with 75% in the range 2 to 4),[7] the index for this group will decline to 0.26.

Since 1957 the "baby boom" has receded, and the birth rate in 1966 was the lowest for any year since 1936. This is related to changes in the age structure of the population and may or may not portend a reduction in family size. The recent drop certainly is not just the result of the introduction of the *pill;* during the 1930's couples were just as successful in birth control with other methods. But the fact that birth control methods are becoming easier and more reliable makes family planning more effective. This factor will probably further restrict the number of high parities.

(4) *Parental age:* The intrinsic rate of natural increase is determined not only by the number of children, but also by the length of generation. A couple marrying early and having their children early have a higher rate of reproduction than one marrying and having their children late, *even though* they have the same number of children. Variance in age at mariage and length of generation is a source of differential fertility and of selection.

Recent demographic trends have also reduced this source of differential fertility. As noted earlier, marriage is earlier; also intervals between marriage and first and subsequent births have been reduced so that births are increasingly concentrated in the early childbearing years. The average length of female generation has declined from 28 in the 1930's to 26 at the present time, with less variance around the mean. The median American woman is married before she can vote, has her first child at age 21, and her third and last at age 27. The average childbearing period for women has been reduced to about seven years.

This concentration of childbearing in young ages and at low parities has direct genetic influence since a number of genetic disorders are correlated with age of mother, birth order, and number of children per family. These include new mutations, such as Down's syndrome, and genetic factors in combination with environmental factors, such as Rh-erythroblastosis, congenital malformations of the circulatory and nervous systems, cerebral palsy, etc. Newcombe's Canadian data suggest twice as high an incidence for the latter type of disorder for older mothers of high parity as compared with young mothers of low parity.[8] Reviewing Japanese experience between 1947 and 1960 (i.e., earlier childbearing and smaller families), Matsunaga estimates a reduction value of one third for mongolism, more than one half for Rh-erythroblastosis, and on the order of one tenth for the remaining defects.[9] While these lack specificity for the experience in this country, they suggest some order of magnitude for gains deriving from our growing concentration of childbearing at lower ages and lower parities.

Futhermore, reduction of the number of siblings limits the opportunity for consanguineous marriage, and more broadly inbreeding, and hence cuts the number of births with defects arising from such marriages.

Differential Fertility

In the preceding I have been discussing the situation in the U.S. population as a whole, often more specifically the white population.

Within the total population there have been very major differences in reproduction or differential fertility between social, economic, religious, and ethnic groups. Historically there were, for example, enormous differences between the urban dweller and rural farm resident, the uneducated and the college graduate, the professional and the laborer, etc. These differentials evoked fear that we have been "breeding from the bottom," the least capable producing the most children.

Whatever their earlier significance, socioeconomic differentials in fertility have contracted since World War II, and especially between 1950 and 1960. There is abundant evidence in the censuses of this narrowing of socioeconomic differentials between 1950 and 1960 for such characteristics as rural or urban residence, occupation, income, and education.[10] Most elements in the population have shared in the increase in fertility, but especially the better educated and upper income and occupational groups. This has narrowed the differentials.

To take a single, rather dramatic example: in 1950, white women at age 40-44 with less than eight years of schooling who had never been married had had over twice as many children as women with four or more years of college. In 1960, this margin had been cut to 50 percent more, and with current trends it will decline by 1970 to below 40 percent more. Women of intermediate educational levels fall between these extremes, and because of general upgrading of education there is now less variance about the mean. Thus the completed fertility of married women with four or more years of college was only 59 per cent of the average for all married women born in 1901-5; it rose to 73 per cent for women born 1916-20; and will rise further to 89-93 per cent for women born 1926-30, the latter figure depending on assumptions about future fertility of this younger cohort. The gap between the most educated and the average is much less than before.

This reduction is not so true of two other fertility differentials—those by religion and by race.

In this country Catholic fertility has been consistently higher than non-Catholic, and this difference has probably not been narrowed in recent years. The genotypic significance of this difference is obscure. In any case, there is now rapid change in Catholic attitudes about family planning and there may well be a convergence in birth rates by religion.

The position of the nonwhite population (chiefly Negro) is quite different from the white. The opportunity for natural selection remains much higher both as regards mortality and natality. Twice as high a proportion of Negroes die before reaching age 15 and age 45. Negro women have more children, more widely scattered through the childbearing period, but more Negro women are childless in marriage. The result is greater variance in number of offspring and more opportunity for natural selection.

Differential fertility is also much greater among Negroes than among whites. Reproduction of urban, educated, and middle-class Negroes is similar to that of whites, whereas fertility of low-income and rural Negroes is much higher than that of whites of comparable socioeconomic status. At present the Negro birth rate is declining as the low-fertility elements become a larger part of the Negro socioeconomic structure. The same forces are at work among Negroes as among non-Negroes—reduction of mortality, reduction of family size, reduction of fertility differentials—but in each case the process has not gone so far as in the remainder of the population. In the population as a whole there is and will be for some time a growing proportion of Negroes—the per cent of nonwhite births increased from 14.4 in 1950 to 15.8 in 1960 and to about 16.9 in 1965.[11]

Summary and Conclusions

Demographic trends in the United States are strongly in the direction of reducing the opportunity for natural selection. This is the result of several convergent elements.

(1) The mortality component in selection intensity has been dramatically reduced. The genetic load is doubtless being increased by the survival of deleterious mutations but environmental changes have made some hereditary defects irrelevant, such as susceptibility to now-curable diseases.

(2) A larger proportion of persons in the reproductive ages are married, and hence exposed to the "risk" of childbearing, than a generation ago. There is also a retreat from childlessness and perhaps some decrease in physiological sterility. About 85 per cent of each white female cohort born now lives to adulthood and participates in the reproduction of the next generation.

(3) Since 1940 the average number of children per couple has risen somewhat after the long secular decline in fertility ending in the 1930's. There is a growing concentration of couples with two to four, and especially two and three, children. The variance and hence the opportunity for natural selection through variability in numbers of offspring is much reduced.

(4) The average length of generation has fallen and the variance has di-

minished, owing to the concentration of births to parents in their twenties and early thirties.

(5) Differential fertility between socioeconomic groups is generally declining. Potential selectivity in terms of residence, occupation, income, and education is thereby diminished. An important exception is the fertility differential by race. Negro fertility and mortality are declining but in both cases are still substantially higher than those for whites. The opportunity for natural selection is higher in the Negro population.

There is a growing homogeneity in reproduction. The white population of the United States appears to be moving toward a situation in which the great majority live to adulthood, marry, and have two or three children. To the extent that this occurs, each generation will be close to a genetic carbon copy of its predecessor, aside from mutations that are absorbed by the genotype.

At the same time it must be emphasized that present fertility trends could change rapidly, as they have in the past. Furthermore, their genetic implications could be modified, for example, by an increase in assortative mating, bringing a rearrangement of genes without necessarily changing their frequency in the total population.

The levels and trends of mortality have been more stable than those of fertility. Barring catastrophe, very low mortality at the younger ages will be a continuing feature of our civilization. A relaxation of selection intensity of the degree and durability now existing among American and Western peoples has surely never before been experienced by man.

The potential results are not clear and will require far more sophisticated analysis. In the short run present demographic trends are reducing the incidence of serious congenital anomalies because of the younger average age at childbearing, the lower average parities, and the reduction of consanguineous marriages. And surely in the foreseeable future the possibility of medical and environmental correction of genetic effects will far outrun the effects of the growing genetic load. The longer-run effects depend upon the increase in mutational loads and how serious these may prove to be in a rapidly changing environment more and more created by man himself.

REFERENCES

1. A good summary of changes in causes of death is given by Spiegelman, Mortimer, *Significant Mortality and Morbidity Trends in the United States since 1900* (Bryn Mawr, Pa.: American College of Life Underwriters, 1964).

2. Computed from U.S. Public Health Service, *Vital Statistics of the United States, 1965* (Washington, D.C.: Government Printing Office, 1967), vol. 2, *Mortality,* pt. B, Tables 1–4 and 1–9.

3. SPUHLER, J. N., "Empirical studies on human genetics," *Proceedings of the UN/WHO Seminar on the Use of Vital and Health Statistics for Genetic and*

Radiation Studies, September 5–9, Geneva, 1960 (New York: United Nations, 1962).

4. These and following illustrative data are from the appropriate U.S. censuses and from U.S. Department of Commerce, Bureau of the Census, *Current Population Reports,* Series P-20, 159 (Washington, D.C.: Government Printing Office, Jan. 25, 1967).

5. U.S. PUBLIC HEALTH SERVICE, *Natality Statistics Analysis,* 1964, Series 21, 11 (Washington, D.C.: Government Printing Office, February 1967).

6. After CROW, J. F., *Human Biology,* **30,** 1–13 (1958). The index of selection intensity of the fertility component (I_f) is the variance in the number of children born divided by the square of the average number born per parent.

7. FREEDMAN, R., D. GOLDBERG, and L. BUMPASS, *Population Index,* **31,** 3–20 (1965).

8. NEWCOMBE, H. B., *Ann. Human Genetics,* **27,** 367–382 (1964).

9. MATSUNAGA, E., *J. Am. Med. Assoc.,* **198**(5), 120 (1966).

10. U.S. DEPARTMENT OF COMMERCE, BUREAU OF THE CENSUS, *U.S. Census of Population,* 1950 (Washington, D.C.: Government Printing Office, 1955), Special Report P-E No. 5C, *Fertility;* and *U.S. Census of Population, 1960* (Washington, D.C.: Government Printing Office, 1964), subject reports, final report PC(2), pt. 3A, *Women by Number of Children Ever Born.*

11. Figures for 1950 and 1960 from U.S. Department of Health, Education and Welfare, "White-nonwhite differential," in *Indicators,* September 1965; data for 1965 from U.S. Public Health Service, *Vital Statistics of the United States,* 1965 (Washington, D.C.: Government Printing Office, 1967), vol. 1, *Natality,* Tables 1–3. Data for the two earlier years had been corrected for under-registration of births.

The Small Population

Overview

Most of the mathematical models of population genetics assume an infinite, panmictic population the gene frequencies of which are subject to the measurable systematic changes induced by mutation, natural selection, and migration. However, as was discussed in Chapter II, human populations do not conform to these theoretical conditions; they do not mate randomly, nor are they infinitely large. Rather, there are hundreds of human populations in the world today which fulfill the definition of a *deme;* that is, a small, relatively isolated group.

Some of the very early workers dealing with populations recognized the fact that humans live in groups whose extension is restricted, and understood some of the consequences of this universal living pattern. In 1928 the Swedish scientist S. Wahlund considered the case of two closely related small panmictic populations which exchanged individuals across their boundaries. He termed these populations *isolates.* In the next year another Swedish worker, G. Dahlberg (1929), pointed out that the number of consanguineous matings in a population was closely related to the size of a population. He devised a method (the Dahlberg formula) whereby the size of an isolate could be estimated from the number of first-cousin marriages occurring within the group.[1]

---

[1] The formula is $N - 1 = 2b(b - 1)/c$, where c is the frequency of cousin marriages, b is the average number of children reaching adulthood and marrying, and N is the size of the isolate. Because the formula assumes that cousin marriage is neither favored nor disfavored by social custom, but is entirely random among all matings, it is inapplicable to many societies.

301

"The Sewall Wright Effect"

The formal geneticist Sewall Wright (1931) developed models which mathematically dealt with the population as a *finite* unit of evolution. Wright proposed that a finite population is subject to a special force of evolution which he called *genetic drift*. The term genetic drift, or "the Sewall Wright effect," refers to the random fluctuation or "drift" of gene frequencies from generation to generation in a small, relatively isolated population.

In practice, as Derek Roberts points out in the first of the two articles by him included in this chapter, the term *random drift* is now used to cover various interrelated genetic phenomena, all of which are random processes affecting gene frequencies. However, the term genetic drift as it is used here, unless otherwise indicated, refers only to the deviations in gene frequencies that occur as a result of the sampling process involved in passing gametes from one generation to the next in a small population. In this manner, a number of local populations, although originally derived from the same parental population, may come to differ from one another to varying degrees. The random nature of genetic drift puts it in a different class from the three determinate evolutionary forces—mutation, natural selection, and gene flow—all of which exert systematic pressures on a population's gene pool. In contrast, genetic drift exerts nonsystematic pressure, and the changes in gene frequencies are therefore indeterminate in direction (see Chapter I).

Sewall Wright (below) has proposed that chance, in the form of genetic drift, could be of evolutionary significance for a species. He is of the opinion that the random genetic changes in a number of small semi-isolated populations provide raw material to the total species in the form of "genetic complexes" upon which natural selection acts. Yet, in order for this process of "intergroup selection" to operate sucessfully, there must be not only "sufficient *isolation* to permit local differentiation," but also "sufficient possibility of excess *migration* from centers in which superior genetic complexes have appeared (by mutation) and have been improved by local selection." Only when these conditions are fulfilled are there genetic consequences of importance to the species as a whole. Wright's theory is rejected by the English geneticists Ronald Fisher and E. B. Ford. They have argued that control of evolution by selection is so direct and complete as to preclude any significant role of random processes above the level of mutation. The controversy has been set forth in a number of papers, and in 1947 Fisher and Ford published data which they maintained was "fatal" to the Wright effect. The article included in this chapter is Wright's rebuttal to the two geneticists; in it he discusses his theory of genetic drift as it had been mathematically defined in his many previous papers.

The theory of genetic drift as developed by Sewall Wright is a *prob-*

TABLE V-1. PROBABILITY DISTRIBUTION OF q IN N = 50 OFFSPRING
FROM A PARENTAL POPULATION WITH q = 0.50*

					q				
	<0.35	0.35–0.40	0.40–0.45	0.45–0.50	0.50–0.55	0.55–0.60	0.60–0.65	0.65+	Total
Prob.......	0.002	0.021	0.136	0.341	0.341	0.136	0.021	0.002	1.000

*In a population where N = 50 individuals whose parental q = 0.50 the gene frequency will be expected to vary about the mean value q = 0.50 with a standard deviation of

$$\sigma_q = \sqrt{\left(\frac{0.5 \times 0.5}{2 \times 50}\right)} = 0.05.$$

The probabilities of the various values that q may take in the gene pool are shown in the table (approximation by normal curve).

Source: Reprinted from C. C. Li, *Population Genetics* (University of Chicago Press, 1963), p. 315, by permission of the publisher.

ability statement regarding the effects of sampling in a small population.[2] Each generation is regarded as a sample of gametes produced by the preceding generation. As the result of sampling each generation from a gene pool of a limited, small size, the gene frequencies fluctuate (drift) from generation to generation, because any new generation's gene pool may not be representative of the parental gene pool from which it was drawn. The probability that q (gene frequency) may take various values in the next generation is derived from the normal curve of distribution (Table V-1). It is obvious that the amount of variance is directly dependent on sample size; the smaller the sample, the greater the variance, and the greater the fluctuation of gene frequencies (see the Overview in Chapter I). The operation of genetic drift leads to a net loss of heterozygosity from the gene pool (the so-called *decay of variability*) at the rate of $\frac{1}{2} N_e$ (where N_e equals the effective population size) per generation. Theoretically, the ultimate result of the "decay" is homozygosity at each genetic locus through the fixation of one allele at 100 percent in the gene pool and the concomitant loss of the other allele(s). The loss occurs because as the allele frequencies drift from the central values ($p = q = 0.50$) that yield maximum heterozygosity, the magnitude of variation becomes smaller, and as the q value approaches $q = 1$ or $q = 0$, the chance factor operates with greater finality (Figure 1).

The calculations for drift assume the conditions defined by the Hardy-Weinberg formula, with the single exception that the population size is finite. The above formula is, therefore, applicable only to the "ideal" small

[2] Random drift per generation in a population may be calculated by the use of the formula $\sigma_q^2 = \dfrac{q(1 - q)}{2N_e}$, where σ_q^2 is variance due to drift, q is the gene frequency and N_e the effective breeding size (Li, 1963).

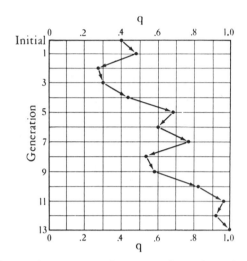

FIG. 1. Genetic drift. Random variation of q in a small population leads to ultimate extinction or fixation of a gene. The smaller the population, the more rapidly will it reach the end points. (Reprinted from C. C. Li, *Population Genetics* [University of Chicago Press, 1963], p. 317, by permission of the publisher.)

population. For this reason the breeding population (N) must be corrected to the effective breeding population (N_e), that is, the "ideal" small population (see the Overview in Chapter II). However, the definition of what size constitutes a "small population" within which drift may be effective depends on the interrelationships between the number of breeding individuals (N), the selective value of the allele (s), mutation pressure (μ), and gene flow (m). Wright (1931) later refined his original probability model to mathematically express the interrelationships between these variables. Although it is usually stated that drift is effective in "small" populations (that is, where N numbers in the hundreds), and is inoperative in "large" populations (where N numbers in the thousands), the statement is an oversimplification. It is true that the smaller the effective size of a population the greater is the random fluctuation in gene frequencies; yet at the same time the stronger the systematic pressure (s, μ, m) on any locus the less effective is the indeterminate random process. "Small population" is therefore a relative term, and should be defined quantitatively. According to Wright (1931), a population is "relatively small" (that is, drift predominates) when

$$N \leq \tfrac{1}{2}s, \ N \leq \tfrac{1}{2}\mu, \ N \leq \tfrac{1}{2}m.$$

Conversely, the systematic pressures predominate when

$$N \geq \tfrac{1}{2}s, \ N \geq \tfrac{1}{2}\mu, \ N \geq \tfrac{1}{2}m.$$

Both systematic pressures and drift interact in the overlapping ranges of N.

It is obvious that the measurement and demonstration of drift in any real population is an extremely difficult task. Furthermore, it is doubtful that any human population has ever approximated the ideal conditions requisite for the application of the mathematical models of drift. However, within the extensive methodological limitations imposed by actual populations, population geneticists have attempted to assess genetic drift as it may operate in human groups. The problem has been approached in the following three ways (see Roberts' first article below):

(1) Variations in gene frequency between local populations, and between offshoot and their parental populations are assumed to be attributable to drift.

(2) The size and structure of populations are analyzed to estimate the magnitude of drift that theoretically could occur.

(3) Mathematical models are applied to the genetic and demographic data from populations to test whether the observed variations in gene frequencies can be explained by the operation of drift alone.

Each of the three approaches is discussed in the following sections.

Comparison of Gene-Frequency Variations in Related Populations

Under the first—and most simplistic—approach, it is assumed by an investigator that if a single isolated population shows aberrant gene frequencies as compared to other genetically related populations, and the other forces of evolution can be excluded, the genetic differences are due to drift. For example, genetic studies on various groups of Eskimos indicate that smaller more isolated groups differ strikingly from the major larger Eskimo groups. In another study, involving Australian aborigines, the inhabitants of one island possessed no blood group A and a higher R_0 than found in any other aboriginal group, whereas the inhabitants of nearby island had low A and low R_0 (studies cited in Glass, 1954). In these and other similar cases, genetic drift has been given as the probable cause of certain aberrant gene-frequency distributions. However, to cite drift as an explanation and to demonstrate its actual operation are two different matters entirely, because only genetic data spanning many generations could provide conclusive evidence of drift in operation, and the data are lacking for human populations.

The geneticist Bentley Glass has attempted to use genetic data from three generations to document the operation of genetic drift in the Dunker religious isolate of Pennsylvania. The Dunker isolate fulfills the conditions under which it could be predicted that genetic drift would be operative. The isolate was established in 1719 by members of a religious sect who migrated from West Germany to Pennsylvania, where they were later joined by other migrants. In 1881, the sect split and only a minority remained in the isolate, where they continue to intermarry and retain certain distinctive cultural practices. Glass counted 90 parents in the isolate as constituting the breeding population (N).

In the original study (Glass, *et al.*, 1952), comparison of the isolate with populations of West Germany (representing the parental population) and the United States showed that the isolate differed from both populations in its ABO and MN blood-group frequencies. Glass attributed the observed differences to genetic drift. However, as he recognized in a later paper (Glass, 1956), the original analysis did not show a distinction between population differences due to genetic drift (that is, fluctuation of gene frequencies over several generations) and those due to the founder effect (a unique distribution of genotypes in the original progenitors of the isolate). Glass therefore subdivided his population by age and compared the gene frequencies of three generations. The ABO frequencies showed no generational differences, but the MN frequencies showed exaggerated differences between generations. For example, between Generations I and II, the frequency of L^M increased from 0.550 to 0.685, a difference of +0.135, which Glass concluded was due to genetic drift.

But a question arises as to whether Glass's generational data do, in fact, document drift. As Roberts (1965) has pointed out, the probability that such a magnitude of drift would occur over a single generation is very remote; the more likely explanation for the loss of the L^N allele may be the emigration from the isolate of families carrying the gene. Thus, the single case purporting to actually document the results of genetic drift having operated over several generations in a human population is a tenuous one. On the other hand, small isolates often do show aberrant gene frequencies, and in many cases it is hard to see, although impossible to prove, how any evolutionary process other than genetic drift could account for the observed situation.

The Opportunity for Drift

Under the second approach, Wright's mathematical models have been applied to various human populations to give rough approximations of the opportunity provided by a population for the operation of drift. Wright (1940, 1943) showed that gene differentiation due to drift depends on the product of the effective population size (N_e) and the migration rate (m) where m is the proportion of a population replaced in each generation by migrants. The product ($N_e m$) is a measure of the degree of reproductive isolation of a population, which in turn is indirectly a measure of the opportunity provided for genetic drift. If the product is 5 or less, the expected changes in gene frequency due to genetic drift are likely to be very marked; between 5 and 50 the changes will still be appreciable; above 50 the changes will be slight. The physical anthropologist Gabriel Lasker (1960) has calculated the product or the so-called "index of isolation" for selected isolated groups. He obtained indices ranging from less than 3 for Buzios

TABLE V–2. INDEX OF ISOLATION OF TWELVE SELECTED GROUPS

Place	Total Population	Migration Rates (m) percent	Index of Isolation (N_em)	Source
Buzios Island, Brazil	126	7*	2.9	Lasker, 1960
Ranches of Quiroga, Mexico	133	10	4.4	" "
Fox Indian Reservation, U.S.A.	342	7	8.0	" "
Tristan da Cunha Island	274	12	11.0	" "
Hopewell, N.C., parish	250–330	12	10.0–12.0	" "
Dunkers, Pennsylvania	298–350	10–22	9.9–25.7	" "
Ulithi Atoll, Pacific	421	15–20	21.0–28.1	" "
Xavante Villages, Brazil	691	16	49.0	Salzano, et al., below
Tzintzuntzan, Mexico	1,231	12	49.2	Lasker, 1960
Dinka village, Tir	375	50	55.0	Roberts, 1956
Dinka section, Abuya	1,300	35	133.0	" "
Dinka tribe, Ageir	13,400	11	430.0	" "

*In Lasker's figures the migration rate (m) is the percent of the total population born elsewhere; m for the Xavante and Dinka (Salzano and Roberts respectively) is the percent of the breeding population born elsewhere. Lasker's effective breeding size (N_e) is ⅓ of the total population, whereas the Xavante and the Dinka breeding populations have been corrected to the "ideal" N_e.

Island, Brazil to approximately 50 for the community of Tzintzuntzan in the state of Michoacan, Mexico (Table V-2). Applying Wright's formula to the demographic data from the three Xavante villages, the index obtained is approximately 49. In contrast, Roberts' figures from three exogamous divisions of the Dinka of North Africa give higher indices, ranging from 55 to 430. The low indices suggest that genetic drift may be an important force of evolution operating on the gene pools of the populations considered here, but the conclusion is inferential only.

The Application of Mathematical Models

The third approach involves applying models of population structure to demographic and genetic data to see if the population dynamics, as indicated by the models, could account for the observed genetic composition of the population(s). Presently, there are three major types of mathematical models which describe isolate structure:

(1) The *island model* developed by Sewall Wright (1943) assumes "island" populations of equal size randomly exchanging genes at an equal rate per generation (m) which is independent of the distance between islands.

(2) The *distance model*, also developed by Wright (1943, 1946), deals with "neighborhoods" offering uniform distribution and density along a line or on a surface. Marriages are determined by distance and the diffusion of individuals occurs in terms of some law, usually that of the normal curve of distribution. Migration of individuals is measured in terms of the distance between the birthplaces of parent and offspring.

(3) The *stepping-stone model* proposed by Kimura and Weiss (1964) assumes a discontinuous distribution of the population in "colonies," i steps apart (the "stepping-stone" concept), which exchange genes at a given rate. In the linear version each colony has two neighbors; its exchanges a fraction $(m/2)$ of its genes with each. In the two-dimensional version each colony has four neighbors with which it exchanges $m/4$ of its genes; in the three-dimensional version there are six neighbors.

The inadequacies of any one of the models for testing the operation of genetic drift in actual human populations are apparent. Wright's island model was devised to deal with laboratory populations. Its basic assumption—that the exchange of genes between populations is unaffected by distance—is far removed from the reality of human mating patterns. Studies in human populations have shown that the fact of residential propinquity is an important or often the decisive factor in determining one's choice of a mate. For example, in the city of Columbus, Ohio, of 281 couples applying for marriage licenses, slightly over one-half of the mates resided within 16 standard city blocks of one another (Clarke, 1952). The "distance" and the "stepping-stone" models, although more relevant to human populations, rest on certain mathematical assumptions describing the distance distribution of populations—parents and offspring—that do not necessarily hold true in human populations. Taking only one of these assumptions—that the curve of parent/offspring distances is a normal one—it has been shown in actual human populations that the curve is more leptokurtic (Cavalli-Sforza, 1958; Roberts, 1965). One cause of the leptokurtic distribution is the tendency of individuals to remain in the area in which they themselves were born. Finally, both models assume "neighborhoods" or "colonies" of uniform size, an assumption which observation alone will show is universally violated by human populations.

In general, all of the models described assume constants of size, time, and space; therefore, none cope with the fact that real human populations vary widely in geographic distribution, size, density, and mobility. The Italian population geneticist Luigi Cavalli-Sforza, in the research described in his article included here, found that the classical models were not applicable to the actual demographic parameters of the populations in the Parma Valley, Italy. His solution was to use the actual population data to construct an artifically subdivided population which simulated as closely as possible the demographic conditions and gene frequencies which had been recorded for the real population. The programmed data was then fed into a computer and the simulated populations allowed to reproduce over a number of generations. Comparison between the computated gene frequencies and the actual gene frequencies showed that the variation in blood-group frequencies between the simulated villages corresponded very closely to the variation in frequencies between the actual villages. On the basis of the close correspondence, Cavalli-Sforza concludes that a large part of the population heterogeneity observed in the upper Parma River Valley is the result of drift. The computer method as used in this study is a valuable tool to the population geneticist. For example, Cavalli-Sforza was able to cycle his simulated population through multiple generations, thereby obtaining an assessment of the probable operation of genetic drift over time, a parameter lacking for the majority of human populations.

As has been discussed, genetic drift involves the loss of heterozygosity in a gene pool, and it is within this context of "decay" that two special phenomena, the *founder effect* and *inbreeding*, must be considered. Although the two cannot be strictly equated to genetic drift per se, they are related in certain ways.

The Founder Effect

The term *founder effect*, or "bottleneck effect," refers in its most limited meaning to a unique drastic reduction in the size of a population (Roberts, below). However, founder effect is best defined as the total genetic impact experienced by a population which has undergone the combined events of (1) drastic *reduction* in population size, (2) rapid population *expansion*, and (3) relative *isolation* with little genetic admixture. If these three conditions are fulfilled, then, as Roberts states, "the effects of these size reductions persist and, in accumulation, shape the profile of ancestral contributions to the gene pool of the present population. . . . Thus, the genetic constitution of a population at any moment is regarded as comprising a pack of probable ancestral contributions which can be calculated." The present genetic composition of populations described in the following articles, the Tristan da Cunha islanders (Roberts) and the Amish demes (McKusick, *et al.*), are the results of the founder effect.

The initial event of the total founder effect, the unique drastic reduction in the size of the population, occurs primarily in two ways: (1) from the death of large numbers of people in a population from disease, famine, war, or other natural disasters (for example, the drastic population reductions described by Motulsky in Chapter IV as due to the high mortality exacted by epidemics); and (2) from migration. In the first case the breeding population remaining after mortality has taken its toll constitutes the "founder population" for future generations. In the second case, if there has been net mass migration so that only a small group remains, the latter is considered the "founder" population. On the other hand, if the migration consisted of only a very small number of people who, after leaving behind the majority, "found" a new community, the immigrants are considered the "founder" population.

On Tristan da Cunha from 1816 to 1961 the island population grew from 15 to 270, but the over-all growth was marked by two separate major population reductions. The first reduction was due to mass emigration from the island; the second was from the combined effects of a boating accident which took the lives of most of the adult males and the subsequent emigration of many of the widows. As a result of these events the total population was twice passed through a so-called "bottleneck," by which some individuals were extracted and others were left on the island. The individuals remaining each time constituted the "founder" population for succeeding generations. In fact, Roberts concludes that "the gene pool of the 1961 population derives in its major features principally from the reduction in population that occurred between 1884 and 1891. . . ."

The Amish presently live in a number of local endogamous communities (demes) which range in size from about 1000 to 9000 people. The separate demes were founded by successive small waves of immigrants who came to the United States from Europe between 1720 and 1850 and settled primarily in Pennsylvania, Ohio, and Indiana. Each modern deme is therefore descended from a very few immigrant founders, as shown by the unique set of relatively few family names which account for the majority of the people in each deme. For example, in three demes three different sets of eight names account for approximately 80 percent of the Amish families. The extensive demographic and genetic studies on the Amish demes by Victor McKusick and his group have shown that the circumscribed occurrence of certain rare recessive genetic syndromes (McKusick, et al., below) and the unique gene-frequency distribution of the blood groups (McKusick, et al., 1967) are best explained by the founder effect.

In addition to being "founded" by a few individuals, the populations described here also fulfill the two other conditions for a true founder effect, having undergone rapid expansion in relative isolation. The population of Tristan da Cunha grew from 15 in 1816 to 270 in 1961 in extreme geographical isolation, interrupted only rarely by an immigrant. The total Amish population has grown from approximately 8200 in 1905 to the present estimated size of 45,000. The Amish remain socially isolated from the surrounding non-Amish; marriage outside the group is forbidden and few outsiders enter it. Within the total Amish population the individual demes are endogamous, marriages being contracted within a rather small radius.

How does the founder effect relate to Wright's concept of genetic drift? In one respect, the two processes are identical in that both are sampling phenomena in which sampling error can act to alter gene frequencies. Just as each generation of a breeding population is a "sample" of the preceding generations' gene pool, the founder population is a "sample" drawn from its parental population. In each situation the small sample may be unrepresentative of the population from which it was drawn. On the other hand, the small sample created by size reduction is a unique event—a unique sampling error which occurs once and, depending on the sample drawn, may drastically alter the gene frequencies and reduce heterozygosity of a gene pool immediately. By contrast, genetic drift, as developed by Wright, is effective only through time; its theoretical results are achieved only after numerous successive generations.

Actually, the founder effect is best thought of as *creating* the conditions under which genetic drift may operate on a population's gene pool through succeeding generations. Because an expanded, isolated population is descended from a small "founder" group, its gene pool will consist of a large proportion of related gametes. Therefore, although a gene pool may appear "large" with respect to the total number of gametes contained therein (some of the Amish demes number in the thousands), the effective population (N_e), with its "ideal" gene pool of unrelated gametes may, in actuality, be very "small." In this way the founder effect creates gene-pool conditions under which genetic drift may be highly effective; thus, the genetic composition of such a population is the result of the interaction of the two phenomena. For example, in the two populations described, the accumulated effects of differential fertility—the result of a few families contributing a disproportionate number of gametes to the gene pool—certainly have had an impact on the genetic composition of the population. It appears that in the Amish demes certain distinctive families of high fertility have left significant marks on the gene pools in the form of rare recessive disorders and blood-group frequencies (McKusick, *et al.*, 1967). Differential fertility within the population, as translated into variance in family size, is, of course, random variation in the sampling process; that is, an example of Wright's concept of genetic drift.

Inbreeding

Inbreeding is the second phenomenon which is often incorrectly equated with genetic drift. The two processes are related only in that they generally coincide in a small population. If, for example, a new mutant in a population increases through drift, then it follows that all persons carrying that gene are necessarily related (assuming no recurrent mutation). The increase in the mutant-gene frequency and the increase of genetically related individuals in a population are, therefore, coincidental parts of the same sampling phenomenon. However, unlike genetic drift, inbreeding per se does not directly affect the gene-pool frequencies, although the results of inbreeding—morbidity and mortality—may. Rather, the effects of inbreeding are manifested at the phenotypic level.

Consanguineous matings, which constitute the *inbreeding system* of a population, may occur nonrandomly and/or randomly in human populations (cf. Allen, 1965). *Nonrandom consanguineous matings* are made by choice; they are actually assortative matings where the basis for the mating is biological relatedness through common ancestry. *Random consanguineous matings* occur in a finite population where the size of the isolate limits the number of mates available, thereby forcing consanguinity, although consanguineous marriages may not be the preferred type of mating.

Thus, frequency of consanguineous marriages in populations varies, depending on factors such as the size, distribution, and structure of the population, and cultural ideas regarding the desirability or nondesirability of such matings. Table V-3 shows the frequency of the most common type of consanguineous marriage, that of first cousins, in selected populations. The table includes populations in which first-cousin mar-

TABLE V-3. PROPORTION (%) OF FIRST-COUSIN MARRIAGES AMONG ALL MARRIAGES IN SELECTED POPULATIONS

Population	Period	Number of marriages	First cousins	Source
United States, Baltimore (urban)	±1935–1950	8,000	0.05	Stern, 1960
Netherlands (national statistics)	1948–1953	351,085	0.13	" "
Brazil, Rio de Janiero	1946–1956	1,172	0.42	" "
Portugal (national statistics)	1952–1955	276,800	1.40	" "
Xavante (3 villages)	1963–1964	257	3.50	Salzano, *et al.*, below
Spain, Las Hurdes (rural)	1951–1958	814	4.67	Stern, 1960
Japan, Nagasaki (urban)	1953	16,681	5.03	" "
Hopi, (3 mesas)	1962	388	5.90	Woolf and Dukepoo, below
Japan, Dainu (village)	1948–1949	323	7.12	Neel, *et al.*, 1949
India, Bombay (Caste of Parsees)	1950	512	12.9	Bonné, 1963
Japan, Nishi Nagashima (village)	1949	1,433	14.3	Neel, *et al.*, 1949
Kurdistan (nontribal village)	1951	46	17.4*	Bonné, 1963
Brazil, Parnamisini (village)	1950	179	19.5	" "
Fiji Islands (villages)	±1850–1895	448	29.7	" "
Samaritans, Israel and Jordan	1933–1960	51	43.1	" "
Kurdistan Village (tribal)	1951	21	57.1*	" "

*In the Kurdistani tribal village, in relation to first-cousin marriages, which are frequent, more distant consanguineous matings are low. In the nontribal Kurdistani village, where first-cousin marriages are less frequent, there is a high percentage of more distant consanguinity.

riages are preferred (nonrandom) and others in which they are not (random). In some human populations certain types of consanguineous marriages are considered desirable. In Japan, for example, where consanguineous matings are encouraged, in the cities of Hiroshima, Kure, and Nagasaki, an average of 7.45 percent of all marriages were consanguineous matings of some type; 4.30 percent of all marriages were first-cousin marriages. In the Japanese villages of Dainu and Nishi Nagashima, the percentage of first-cousin marriages is even higher: 7.12 and 14.3 percent, respectively (Neel, *et al.*, 1949). The table also indicates that first-cousin matings are more frequent in rural areas than in urban areas. The higher frequency may be due to preference in some cases, but in other cases it is probably due to the random consanguinity forced by a limited population size.

Random consanguinity has occurred in the Hopi population (Woolf and Dukepoo, below), the Amish populations (McKusick, *et al.*, below), and the Tristan da Cunha islanders (Roberts, below); yet consanguineous marriage is not a preferred type of mating in any of these groups. For example, in the Amish demes a large number of the individuals are genetically related, as is illustrated by the distribution of family names. It is not difficult to see that within each endogamous deme the number of unrelated individuals available to a person for marriage is severely limited. Because consanguineous marriage is not a preferred type of marriage among the Amish, first-cousin marriages are still uncommon, but the closed nature of the Amish demes is slowly forcing random consanguinity. Consequently, second-cousin marriages are rather frequent, and the average level of consanguinity of most matings in a deme may be the equivalent of third- or fourth-cousin relationships (McKusick, *et al.*, 1964). In his second included article, Roberts graphically describes a similar restricted-marriage situation on Tristan da Cunha. The early inhabitants of the island did not encourage consanguineous marriages; in fact, marriage records show that they did not occur at the frequencies expected under random mating. However, as the years progressed and their isolation continued uninterrupted, marriage with relatives was imposed on the islanders by the limited number of mates available:

Thus there was a conflict between the opposing tendencies on the one hand to outbreeding that their marriage system engendered, and on the other to inbreeding that the limited number or absence of non-relatives to be found in a small closed population imposed. The resultant of this conflict was the development of a slight degree of inbreeding (Roberts, below).

If one assumes that the original founders of an isolate were all unrelated, consanguineous matings are delayed to at least the third generation, when the mating of first cousins becomes possible. The fourth generation would therefore be the first generation potentially able to show the effects of inbreeding in the form of homozygous recessives. For example, on the island of Tristan da Cunha the first inbred individual was not born until forty years after the settlement had been founded. The time element of random consanguinity is also well illustrated in the Amish pedigree of "Strong Jacob" Yoder (see diagram, p. 367). Strong Jacob was one of the original founders of the Mifflin County deme, and probably carried the recessive gene for pyruvate kinase deficiency hemolytic anemia (PK deficiency). The first random consanguineous mating of individuals descended from Strong Jacob occurred in the fourth generation. Probably by chance, the first homozygous cases of PK deficiency did not occur until the seventh generation, although there were a number of intervening consanguineous matings potentially capable of producing a homozygous offspring. However once random consanguineous mating is initiated, as Roberts shows, as long as the population remains relatively isolated, the level of inbreeding, and therefore the number of homozygotes, increase through time.

Genetic drift results in the loss of heterozygosity in the gene pool by shifting gene frequencies from the central value of 0.50, whereas inbreeding results in the loss of heterozygosity in the individuals (phenotypes). In a population where inbreeding is high and the gene pool contains certain recessive genes, a number of individuals may be affected by recessive disorders which are rarely found in large populations. In fact, the presence of numbers of normally rare recessive homozygotes in a population, such as the albinos among the Hopi and the various recessive defects among the Amish, is in itself evidence of inbreeding, for a homozygous individual is often the sign of a previous consanguineous marriage. The reason that homozygosity indicates inbreeding is that a consanguineous mating increases the chance of homozygosity for rare genes in the offspring. However, inbreeding does not significantly increase the probability of homozygosity for common genes (Tables V–4 and V–5).

The chance of homozygosity is measured by the *coefficient of inbreeding* (α or F). The coefficient expresses the probability that both genes an individual possesses at a given locus are "identical by descent," that is, derived from a single allele present in a common ancestor. The F can be calculated for any individual by tracing his lines of descent back to the common ancestor of his parents.[3] This method is called the

---

[3] The formula for the coefficient of inbreeding F is $F = (\frac{1}{2})^N$, where N is the number of ancestors in the path of relationship from the individual through one of his parents and back through the other. The formula refers only to the probability of homozygosity if the common ancestor is himself not inbred ($F_a = 0$). If he is inbred ($F_a > 0$), then the inbreeding coefficient of his descendent is larger and is calculated by the formula $F = (\frac{1}{2})^N (1 + F_a)$.

TABLE V–4. FREQUENCIES OF HOMOZYGOUS RECESSIVE OFFSPRING FROM MARRIAGES OF UNRELATED PERSONS AND FROM FIRST COUSIN MARRIAGES FOR VARIOUS VALUES OF THE GENE FREQUENCY *

Frequency of Recessive Gene, q	Frequency of aa Offspring From Unrelated Parents, q^2	Frequency of aa Offspring From First Cousin Marriages, $q^2 + (1/16)pq$	Ratio of aa Offspring Related to Unrelated
0.5	0.25	0.2656	1.06
0.2	0.04	0.0500	1.25
0.1	0.01	0.0156	1.56
0.05	0.0025	0.00547	2.19
0.02	0.0004	0.00163	4.08
0.01	0.0001	0.000719	7.19
0.005	0.000025	0.000336	13.44
0.002	0.000004	0.000129	32.25
0.001	0.000001	0.000063	63.00

*If the frequency of q is large, there is very little difference in the frequency of *aa* among the offspring of unrelated or first-cousin parents. If q is very small the likelihood of homozygous *aa* offspring is much greater for children of first-cousin marriages.

Source: From *An Introduction to Human Genetics* by H. Eldon Sutton. Copyright © 1965 by Holt, Rinehart and Winston, Inc. Reprinted by permission of Holt, Rinehart and Winston, Inc.

TABLE V–5. PERCENTAGE OF AFFECTED INDIVIDUALS FROM COUSIN MARRIAGES*

Trait	Whites	Japanese
Albinism	18–24	37–59
Tay-Sachs disease	27–53	55–85
Ichthyosis congenita (skin disease)	30–40	67–93
Congenital total color blindness	11–21	39–51
Xeroderma pigmentosum	20–26	37–43

*The incidence of consanguinity among the parents of offspring affected by rare recessive diseases is high. For this reason, an increased frequency of consanguinity among parents of individuals affected with a given trait is regarded as important evidence for a recessive basis for the trait.

Source: Neel, *et al.*, 1949.

path coefficient method; it consists in summing the probabilities of transmission of the gene in question through each step (ancestor) in the relevant paths of descent. For the offspring of a first-cousin marriage, the total probability that both alleles of Generation IV are derived from a single allele in either ancestor of Generation I is $F = 1/16$ (Figure 2).[4]

If, instead of an individual, a whole population is considered, as Roberts has done for Tristan da Cunha and Woolf and Dukepoo for the Hopi, a *mean coefficient of inbreeding* ($\bar{F}$) can be calculated.

[4]The probability given by Roberts of 1/32 for homozygosity in the offspring of a first-cousin marriage is the probability that both alleles are derived from one specific ancestor in Generation I; that is, from the great-grandmother or the great-grandfather, rather than nonspecifically from either ancestor ($F = 1/16$).

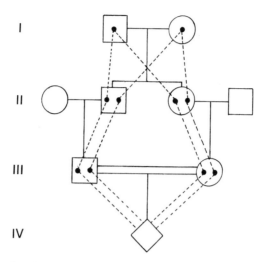

FIG. 2. Path Coefficient Method. Diagram showing
paths of relationship for the offspring (IV) of a first-
cousin marriage. Since therè are two common ancestors (I) of the parents (III), there are two paths of
descent. The coefficient of inbreeding of IV is therefore $F = (\frac{1}{2})^5 + (\frac{1}{2})^5 = \frac{1}{16}$.
Source: From *An Introduction to Human Genetics*
by H. Eldon Sutton. Copyright © 1965 by Holt, Rinehart and Winston, Inc. Reprinted by permission of
Holt, Rinehart and Winston, Inc.

This coefficient is the- average of the coefficients of inbreeding for
the offspring of unrelated parents ($F = 0$) and those of consanguineous matings of different degrees of relationship ($F > O$).[5] Roberts reports that the mean population coefficient of inbreeding for Tristan da
Cunha is $\bar{F} = 0.0403$. As he points out, this value is exceeded by some
other small populations, but as compared with most world populations
the figure is high (see Table V–6). It is obvious from the figures
given here that no human population ever approaches complete inbreeding ($F = 100$). The highest coefficient reported—$\bar{F} = 0.0434$, for
the Samaritans of Jordan and Israel (Bonne, 1963)—is considerably

[5] For example, if a population of 1000 contains 909 individuals of nonconsanguineous
parentage ($F = 0$), 1 individual from an uncle-niece marriage ($F = 1/8$), 30 from first-
cousin marriages ($F = 1/16$), and 60 from second-cousin marriages ($F = 1/64$), the mean
coefficient of inbreeding for the population would be:

$$\bar{F} = \frac{(909 \times 0) + (1 \times 1/8) + (30 \times 1/16) + (60 \times 1/64)}{1,000} = 0.00294$$

That is, the probability that the average individual in this population possesses two genes
identical by descent at a given locus is approximately 0.3 percent.

TABLE V-6. ESTIMATES OF THE MEAN COEFFICIENT OF INBREEDING ($\overline{F}$)
IN VARIOUS POPULATIONS

Population	Period	Number of Marriages	$\overline{F}$	Source
Nagasaki, Japan	1945–1949	842	.0004	Spuhler, 1959
Argentina	1954	23,000	.0005	" "
Rio de Janiero, Brazil	1954	1,272	.0008	" "
Puerto Rico	1954	6,013	.0013	" "
Hiroshima, Japan	1948–1949	10,547	.0029	" "
Kure, Japan	1948–1949	5,510	.0033	" "
Xavante (3 villages)	1963–1964	691	.0040	Salzano, et al., below
Waifu, Japan	1949	2,908	.0046	Spuhler, 1959
Alagoas, Brazil	1954	3,566	.0055	" "
Ramah Navaho, New Mexico	1948	1,118	.0066	" "
Hopi (3 mesas)	1962	388	.0080	Woolf and Dukepoo, below
Hasojima, Japan	1959	175*	.0241	Ishikuni, et al., 1960
Tristan da Cunha	1961	267	.0403	Roberts, below
Samaritans	1960	343	.0434	Bonné, 1963

*The $\overline{F}$ of offspring from a total of 45 marriages on the island was .03341.

below what might be considered a "high" level of inbreeding, as is achieved, for example, by inbred laboratory populations.

In any population in which the mean coefficient of inbreeding is "high," the frequency of a rare recessive genotype, such as albinism among the Hopi, is a function of both the frequency of the gene (q) in the gene pool and the amount of inbreeding ($\overline{F}$) in the population. The Hardy-Weinberg formula cannot be directly applied to the phenotypic data from an inbred population to calculate the frequency of a rare recessive gene because the formula assumes genetic equilibrium with $F = 0$. For this reason, the formula must be corrected by the inbreeding coefficient (in this case the symbol used is α), as Woolf and Dukepoo have done, to calculate the frequency of the recessive gene.[6]

Summary

The interrelatedness of genetic drift, founder effect, and inbreeding as they prevail in a small population may be summarized as follows: The specific genetic results of random mating in the first generation of a newly founded isolate depend on the initial sampling error deriving from the number and type of founders drawn from the source population. The gene pools of the first generation and all subsequent generations produced in the new isolate are subject to genetic drift and decrease in heterozygosity due to the small "ideal" breeding size (N_e) of the population. Random consanguineous mating is usually delayed until

[6]The formula used is $R = q^2 + \alpha pq$, where R is the frequency of the homozygotes (albinos) in the population.

the third generation, but once intiated, inbreeding increases through time as long as the population remains relatively isolated. Because a consanguineous mating increases the probability of homozygosity in the offspring, the number of rare recessive homozygotes also increases in frequency in the small population. In these ways the processes which occur in a small population can have significant effects on both the gene pool and the phenotypes of the population. As a result, small populations often show two characteristics that identify them as genetic isolates; namely, (1) gene frequencies significantly different from other genetically related populations, and (2) a larger number of rare homozygous recessives than would be expected from random mating in a large population.

Isolates therefore provide valuable information to the human geneticist in the form of aberrant gene frequencies, rare recessive diseases, and demographic data which is often excellent in its completeness and sometimes in its generational depth. Although unchallengeable proof of genetic drift operating in human populations is impossible to obtain, and the roles that genetic drift, founder effect, and consanguinity may have played in human evolution are difficult to determine, isolates continue to be the "laboratory" of the population geneticist (cf. Symposium, 1964).

Genetic Drift ("The Sewall Wright Effect")

See the following readings: "Fisher and Ford on 'The Sewall Wright Effect,' " by Sewall Wright, p. 321; and " 'Genetic Drift' in an Italian Population," by Luigi Luca Cavalli-Sforza, p. 330.

The Founder Effect

See the following readings: "Genetic Effects of Population Size Reduction," by Derek F. Roberts; and "The Distribution of Certain Genes in the Old Order Amish," by Victor A. McKusick, John A. Hostetler, Janice A. Egeland, and Roswell Eldridge.

Inbreeding

See the following readings: "The Development of Inbreeding in an Island Population," by Derek F. Roberts; and "Hopi Indians, Inbreeding, and Albinism," by Charles M. Woolf and Frank C. Dukepoo.

REFERENCES TO THE LITERATURE CITED IN THE OVERVIEW

1. ALLEN, G., 1965. Random and nonrandom inbreeding. *Eugen. Quart.*, 12: 181–198.
2. BONNE, B., 1963. The Samaritans: A demographic study. *Hum. Biol.*, 35: 61–89.
3. CAVALLI-SFORZA, L. L., 1958. Some data on the genetic structure of human populations. *Proc. Tenth Internat. Cong. Genetics*, 1: 338–407.
4. CLARKE, A. C., 1952. An examination of the operation of residential propinquity as a factor in mate selection. *Amer. Soc. Rev.*, 17: 17–22.
5. DAHLBERG, G., 1929. Inbreeding in man. *Genetics*, 14: 421–454.
6. GLASS, B., 1954. Genetic changes in human populations, especially those due to gene flow and genetic drift. *Advances in Genetics*, 6: 95–139.
7. _____, 1956. On the evidence of random genetic drift in human populations. *Amer. J. Phys. Anthrop.*, 14: 541–555.
8. _____ , M. S. SACKS, E. F. JOHNS, AND C. HESS, 1952. Genetic drift in a religious isolate; an analysis of the causes of variation in blood group and other gene frequencies in a small population. *Amer. Nat.*, 86: 145–159.
9. KIMURA, M. AND G. H. WEISS, 1964. The stepping stone model of population structure and the decrease of genetic correlation with distance. *Genetics*, 49: 561–576.
10. LASKER, G. W., 1960. Small isolated breeding populations and their significance for the process of racial differentiation. In A. F. C. Wallace (ed.), *Selected Papers of the Fifth International Congress of Anthropological and Ethnological Sciences*, Philadelphia, September 1–9, 1956, University of Pennsylvania Press, Philadelphia.
11. LI, C. C., 1963. *Population Genetics*. Chicago, University of Chicago Press.
12. MCKUSICK, V., J. A. HOSTETLER, AND J. A. EGELAND, 1964. Genetic studies of the Amish: Background and potentialities. *Bull. Johns Hopkins Hosp.*, 115: 203–222.
13. _____, *et al.*, 1967. Blood groups in two Amish demes. *Humangenetik*, 5: 36–41.
14. NEEL, J. V., M. KODANI, R. BREWER, AND R. C. ANDERSON, 1949. The incidence of consanguineous matings in Japan. *Amer. J. Hum. Genet.*, 1: 156–178.
15. ROBERTS, D. F., 1965. Assumption and fact in anthropological genetics. *J. Royal Anthro. Instit.*, 95:87–103.
16. SPUHLER, J. N., 1959. Physical anthropology and demography. In P. Hauser and O. Duncan (eds.), *The Study of Population*. Chicago, University of Chicago Press, pp. 728–758.
17. STERN, C. 1960, *Human Genetics*. 2nd. ed. San Francisco, W. H. Freeman.
18. SUTTON, H. E., 1965. *An Introduction to Human Genetics*. New York, Holt, Rinehart and Winston.
19. Symposium on *"Methodology of Isolates,"* 1964. Organized by the Problem Commission of Neurogenetics of the World Federation of Neurology. New York, S. Karger.

20. WAHLUND, S., 1928. Zusammensetzung von Population und Korrelationsersch einungen von Standpunkt der Vererbungslehre aus betrachtet. *Hereditas*, 11: 65–106.
21. WRIGHT, S., 1931. Evolution in Mendelian populations. *Genetics*, 16: 97–159.
22. ———, 1940. Breeding structure of populations in relation to speciation. *Am. Nat.*, 74: 232–248.
23. ———, 1943. Isolation by distance. *Genetics*, 28: 114–138.
24. ———, 1946. Isolation by distance under diverse systems of mating. *Genetics.*, 31: 39–59.

Fisher and Ford on "The Sewall Wright Effect"

SEWALL WRIGHT

In a paper in 1947, R. A. Fisher and E. B. Ford published data on the fluctuations in frequency of a certain color factor in a colony of a moth, *Panaxia dominula*.[8] They claimed to prove that these fluctuations were too great to have been due to accidents of sampling from generation to generation, and that the fluctuations must be supposed, by elimination, to have been due largely to variability in the direction and severity of selection. The authors held that their demonstration was fatal to the theory of evolution which they ascribe to me: that there is "a special evolutionary advantage to small isolated communities."

In a reply[22] I pointed out that their statement of the theory omitted its most essential features. As to the fatal effect of their demonstration, it was noted that a great many gene frequencies (possibly including all relating to genes with major differential effects) might fluctuate largely from selective variability without affecting significantly the role actually ascribed to accidents of sampling in partially isolated communities of a large species. While not relevant to this main point, it was also noted that the supposed demonstration of these investigators was highly questionable in its biological premises and in the statistical handling of the data. It seemed at least as probable from the published data that the observed fluctuations were actually due more to the small and highly variable numbers of parents as to variations in selection from year to year.

In a reply to this paper[9] Fisher and Ford do not discuss these points but reiterate their criticism of the theory. It may be noted here that the use of my name for the evolutionary effect of inbreeding is hardly appropriate. The first author to suggest that random differentiation among small isolated populations was something that must be taken into account seems to have been Gulick, in 1872.[11] I have ascribed only minor evolutionary importance to this process *by itself*. It is even more apparent than before that we are writing almost wholly at cross purposes with respect to what my theory actually is.

It is accordingly desirable to try to define the issue by listing certain points on which there is probably agreement: (1) Evolutionary transformation consists almost wholly in changes in frequencies in the system of chromosome patterns and Mendelian alleles of populations. (2) The course

Reproduced by permission of the publisher and Sewall Wright from American Scientist, *39:452–458 (1951).*

Sewall Wright, a geneticist, was born in Massachusetts, in 1889. He received his B.S. from Lombard College in 1911, his M.S. from the University of Illinois in 1912, and his Sc.D. in 1915 from Harvard University. He is Emeritus Professor of Genetics, University of Wisconsin, and Emeritus Professor of Zoology, University of Chicago. Along with Ronald Fisher and J. B. S. Haldane, Wright is one of the foremost mathematical geneticists of the twentieth century. Much of his research and writing is directed toward the theoretical mathematical formulation of evolutionary processes in populations. He has also conducted an extensive laboratory program on the physiological and developmental genetics of the guinea pig. Although he has done no field work himself in human populations, his mathematical models dealing with finite populations have been used extensively by people working with the genetics and dynamics of human populations. Probably best known for his concept of "genetic drift" ("the Sewall Wright effect"), he has received numerous honors and awards for his contributions to genetics, including nine honorary degrees, the National Medal of Science in 1966, and memberships in the National Academy of Sciences and the American Academy of Arts and Sciences. He has served as the president of several scientific associations, and in 1958 was president of the Tenth International Congress of Genetics, Montreal. He has recently completed two volumes of a treatise of three entitled Evolution and the Genetics of Population, *published by the University of Chicago Press (Vol. I.* Genetic and Biometric Foundations, *1968; Vol. II.* Theory of Gene Frequencies, *1969; Vol. III.* Experimental and Natural Populations, *forthcoming).*

of such transformation is controlled largely by selection. (Fisher and Ford do not seem to recognize the agreement here.) (3) Mutations merely furnish random raw material for evolution and rarely, if ever, determine the course of the process. (Polyploidy may in a sense be considered an exception.) (4) Mutations with slight effects are more likely to be fixed by natural selection than are those with large effects. (5) Fluctuations in gene frequencies in small completely isolated communities rarely if ever contribute to evolutionary advance, but merely to trivial differentiation, or in extreme cases to degeneration and extinction. (Fisher and Ford accept at least the first part of this statement but do not seem to recognize my agreement as stated in 1931 and in many later papers.)

Real disagreement seems to center on whether or not control by selection is so directed and complete as to preclude any significant role of random processes above the level of mutation. Fisher's belief in the virtual completeness of control by direct action of selection is illustrated in his theory of the prevailing dominance of type over deleterious mutations, which he advanced in 1928 and has recently reiterated.[5, 7] He pointed out that genetic factors modifying the appearance of a rare intermediate heterozygote, in the direction of type, tend to increase at a rate which he showed to be of the order of the mutation rate of the deleterious gene in question, or less, if multiple modifiers are required to make that latter completely recessive. As there are several hundred loci at which deleterious recessives occur in such a form as *Drosophila melanogaster*, the hypothesis postulates the occurrence of alleles at several hundred loci (or even thousands, if multiple modifiers are typical) at which the selective differential due to the modifying effect is only of the order of 10^{-6} or less per generation, and at which there is no differential effect in the absence of the mutation that takes precedence over this.[15, 19, 12, 14]

In the present paper,[9] the above belief seems to be involved in the contention that a demonstration that year-to-year fluctuations in the frequency of a particular gene are predominantly selective is fatal to the view that random fluctuations due to inbreeding may have a significant function for any genes.

I shall not attempt a full exposition of my views here, but refer to a recent nonmathematical summary.[24] The mathematical aspects, apart from the details of the theory of population structure, have also been summarized in a recent paper,[23] and the later aspect of the problem in another.[25] The central concept is that certain states of labile balance among the factors (mutation rate, inbreeding, cross breeding, and selection) give a more effective evolutionary mechanism than the operation of any single factor, even of selection, in excess.

The divergence in views probably traces ultimately to the view which I have held, and Fisher and Ford reject, that the effectiveness of selection is restricted in a random breeding population by the prevalence of pleiotropic effects of alleles and of nonadditive effects in combination. If such effects are unimportant in relation to adaptive value as a character, mass selection in a random breeding population guides the genetic system undeviatingly toward the single, most favorable complex possible from the genes that are present. If, however, such effects are important, there tend to be many distinct harmonious combinations (adaptive peaks). Mass selection tends merely to guide the genetic system toward the one that happens to have been most nearly approached for historical reasons, and this is not likely to be the highest.

Fisher's theory requires that mutations occur at an appreciable rate to provide raw material for the continuing operation of selection, but he does not ascribe special evolutionary advantage to the typical mutation.

Similarly, I have held that random changes in the genetic complexes of small local populations within a species are unlikely to be of evolutionary advantage to the populations individually, and that this is also true of the much larger deviations that accumulate from the interplay of such changes and mass selection. Nevertheless, it is held that in the whole array of such differentiated local populations there is the raw material for a different sort of selection in which the genetic complexes as wholes, rather than the mere net differential effects of the individual genes, are the units. This process (intergroup selection) depends on a certain balance between local isolation and cross breeding. There must be sufficient isolation to permit local differentiation, but there are no consequences of importance to the species as a whole, unless there is also sufficient possibility of excess emigration from centers in which superior genetic complexes have appeared and have been improved by local selection, to permit grading up of neighboring regions to the point at which selective improvement along the same line becomes autonomous in them. Once started such a process may spread through the species step by step. There may be spread from several centers coincidentally. Interaction and differentiation along the way insure continual shifting of the centers. The course of evolution of the species as a whole is, of course, guided predominantly by the genetic complexes of the population *sources* rather than of the population *sinks*. Haldane has supported this concept, with especial reference to the human species during the thousands of centuries of the Old Stone Age.[13] The idea that the variation in the frequencies of the blood groups is a relic of such a prehistoric condition of mankind[16, 17, 18] has been supported as at least one of the factors, by Boyd.[1]

While the author has stressed especially the role of accidents of sampling in causing random differentiation among neighborhoods, other random processes have also been considered important. Among these are effects of long-time fluctuations in the positions of the optima of quantitatively varying characters among local populations.[20] Other categories are discussed in a later paper.[22] In addition to these, selection related to differences in local conditions has been treated at length.[16] This, however, is not a random process. Its primary significance lies in current adaptability of the species, but it also has significance on the one hand in maintaining a store of potential variability within the species and, on the other, in leading occasionally to cleavage. It has, however, been noted as possible that an adaptation to special local conditions may occasionally prove to confer a general advantage, and thus contribute to intergroup selection in a way similar to random differentiation.

The crucial questions with respect to the place of the inbreeding effect in evolution are: (1) how far genetic complexes may have selective significance beyond the sum of the effects ascribable to the component genes, (2) to what extent favorable population structures exist in nature, (3) whether the selective differentials of alleles at many loci are sufficiently slight, and (4) whether the consequences brought about by intergroup selection on the basis of complexes of such genes may be important in evolution.

I do not pretend that it is easy to answer these questions from observations. The theory was developed from the attempt to decide on the most favorable conditions for evolution indicated by the mathematical theory on taking simultaneous account of all of the factors. These are not necessarily the conditions most likely to hold in a typical species at a given time. Other processes were indicated to be effective, but less than this.

With respect to (1), it is a commonplace that major genes nearly always have pleiotropic effects and that the effects of combinations are rarely if ever wholly predictable from the separate effects and are often utterly un-predictable. We do not, however, know much about pleiotropic and combi-nation effects of minor factors. With respect to (2), the mathematical anal-ysis of the consequences of restricted dispersal in continuous populations indicates that the conditions for a sufficiently fine-scaled structure, while severe in a species with uniform density, are by no means impossible and are probably often realized where density is not uniform.[21, 25] Realization is especially likely where there are numerous largely isolated colonies or "demes."[3] With respect to (3), it is not necessary to assume that many alleles have such exceedingly slight selective differentials as required by Fisher's theory of dominance. Differentials of the order of 10^{-3} or 10^{-4} are small enough and can easily occur in multifactorial systems, and presum-ably between isoalleles. As to (4), the theory is undoubtedly concerned most directly with the quantitative variability of the species and very little with conspicuous polymorphism.

It is probable that conspicuous polymorphism is usually a device for adaptation to diverse conditions encountered by the species. It may relate to adaptation to different seasonal conditions as shown by Dobzhansky in the case of chromosome patterns in *Drosphila pseudoobscura.*[4, 26] In other instances it may give a basis for adaptation to diverse microenvironments as discussed in the last reference. The observations of Cain and Sheppard on the frequencies of different color patterns in the land snail *Cepaea nemoralis* under different ecological conditions suggest this interpreta-tion.[2]

Polymorphism in which the advantage to the species comes from the mere fact of diversity, rather than in the special properties of alleles, does not come under this head. The occurrence of multiple self-incompatibility alleles of many plants is an example. The prevalance of antigenic polymor-phism in animals (e.g., blood groups of man) suggests that diversity may be of advantage in this case also. In such examples the differences among local populations may be due to stochastic processes either wholly or in part, even though the maintenance of the multiple alleles depends on selection. If wholly stochastic, the differences have, of course, only trivial importance in evolution. Let me emphasize that it is only as stochastic differentiation gives a basis for selection in the form of differential population growth and migration that it contributes significantly to evolution.

Judgments of the importance of quantitative variability in evolution vary. Goldschmidt, for example, gives it little weight.[10] I shall not attempt to go into the reasons for which I (and I think also Fisher and Ford) ascribe major importance to it.

It may be stated, however, that the processes of random differentiation and intergroup selection, operating on complexes of modifiers may be of great importance indirectly, and are perhaps necessary, for the establish-ment of just such major mutations as Goldschmidt would consider signifi-cant. The primary objection to the hypothesis of evolution by major mutations is the negligible chance of harmonious adjustment to the rest of the organism. Suppose, however, that one of these mutations offers the possibility of a major advance if only such an adjustment can come about. If the mutation recurs with ordinary frequency and is of such low penetrance as to withstand rapid elimination, it may be expected to appear in all local populations of a large species from time to time. Random drift can never carry the mutation beyond very low frequencies, however. There seems no appreciable chance of improving its adjustment and bringing it to fixation by selection of modifiers to rise above the threshold imposed by their effects in the absence of the mutation, If, however, the frequencies of numerous almost neutral modifiers are drifting largely at random in many local populations, there is a good chance that somewhere, at some time, a complex will be acquired which carries the adjustment of the major mutation, if it appears locally, beyond the threshold for successful adaptation, and it (and later the complex of modifiers) comes under direct favorable action of selection. The stage is then set for the spread of the whole complex through the species by the process previously described. Observations at any given time would be interpreted as indicating direct control by selection, yet the processes of random drift and intergroup selection have been of primary importance in the process.

The criticism stressed by Fisher and Ford as completely "fatal" to the author's theory, is that "it is not only small isolated populations, but also large populations that experience fluctuations in gene ratio," and this because of fluctuations due to variable selection. They correctly note that "it is only the random sampling fluctuation which is accentuated by the small size of an isolated population; other causes like selective survival varying from year to year will influence large populations equally. . . . This central criticism seems to have escaped Wright's attention, so that in a recent article in *Evolution* he has attributed to us opinions entirely contrary to those which we hold and clearly express in our paper. Thus on p. 291 he says: 'They hold that fluctuations of gene frequencies, of evolutionary significance, must be supposed to be due wholly to variations in selection (which they accept) or to accidents of sampling. This antithesis is to be rejected.' This passage constitutes a direct misstatement of our published views. There is nothing in our article even to suggest the antithesis which Wright ascribes to us. Not only do we presume throughout that accidents of sampling produce their calculable effects in causing fluctuation in gene ratios, but we take some care to evaluate them. An earlier and slightly different statement by Wright to the same effect occurs on p. 281. 'Thus Fisher and Ford insist on an either-or antithesis according to which one must either hold that the fluctuations of *all* gene frequencies, that are of any evolutionary significance, are due to accidents of sampling (attributed to me) or that they are *all* due to differences in selection, which they adopt.' "

I certainly had no intention of stating that Fisher and Ford had ignored *trivial* fluctuations due to accidents of sampling. I was considering only fluctuations of *evolutionary significance* and introduced this phrase into both sentences to guard against such misinterpretation. The first quotation is perhaps somewhat ambiguous out of context. The next two sentences were as follows: "The fluctuations of some genes are undoubtedly governed largely by violently shifting conditions of selection, but for others in the same population, accidents of sampling should be much more important, and for still others both may play significant roles. It is a question of the relative values of certain coefficients."

The antithesis which Fisher and Ford claim not to have suggested inheres in their use of the word "fatal" in the connection indicated above. A demonstration that the fluctuations of a particular gene frequency are largely selective could not be held by them to be fatal to my hypothesis unless they ascribed to me the view that the significant fluctuations of *all* gene frequencies are due to accidents of sampling, and they themselves adopted the antithetic view that *all* changes of possible evolutionary significance are selective. If, however, they will accept my phrase "of evolutionary significance" in place of their substitute "calculable," and rewrite one of their sentences quoted above as follows, "Not only do we presume throughout that accidents of sampling produce effects of evolutionary significance in causing fluctuations in gene ratios but we take some care to evaluate them," I will be glad to acknowledge that I misinterpreted them.

The authors go on to accuse me of misinterpreting my own earlier published views: "Nothing could be further from our actual criticism of the particular contribution to evolutionary theory which is due to Sewall Wright. He tells us that he now attaches importance of accidents of gene sampling only as one of many factors, and (p. 281) that he has always done so. This latter statement is hard to reconcile with his earlier writings. Thus in the *Statistical Theory of Evolution* he says of 'nonadaptive radiation' (p. 208): 'In short, this seems from statistical considerations to be the only mechanism which offers an adequate basis for a continuous and progressive evolutionary process.' He ends the same paper with the sentence: 'In particular, a state of subdivision of a sexually reproducing population into small, incompletely isolated groups provides the most favorable condition, not merely for branching of the species, but also for its evolution as a single group.' "

The sentences quoted above are again taken out of context. The two sentences in my paper that immediately preceded the former of these were as follows: "This process of *intergroup selection* may be very rapid as compared with mass selection of individuals among whom favorable combinations are broken up by the reduction-fertilization mechanism in the next generation after formation. With partial isolation and differentiation accompanying expansion of the successful subgroups, the process may go on indefinitely." The antecedent of "this" in the quoted sentence was thus "intergroup selection" instead of "nonadaptive radiation" as stated by Fisher and Ford. Intergroup selection depends on the balanced action of multiple processes as suggested in the rest of the quotation. In the case of the last sentence, the context also shows that the reference was to the process of intergroup selection, not to random drift *per se*.

I do not, of course, wish to maintain that my views on evolution have stood entirely still since 1931 when the above statements were written. The former of these is undoubtedly too extreme. Qualifications and additions have been made, beginning in 1932, and have continued up to the present time. The interplay of directed and random processes in populations of suitable structure has, however, continued to be the central theme.

REFERENCES

1. BOYD, W. C. *Genetics and the races of man.* Little, Brown & Co., Boston, 1950. 453 pp.
2. CAIN, A. J., and SHEPPARD, P.M. Selection in the polymorphic land snail Cepaea nemoralis. *Heredity*, *4*, 275–294, 1950.
3. CARTER, G. S. *Animal evolution.* Sidgwick and Jackson, Limited, London, 1951. 368 pp.
4. DOBZHANSKY, TH. Genetics of natural populations. IX. Temporal changes in the composition of populations of Drosophila pseudoobscura. *Genetics, 28,* 162–186, 1943.
5. FISHER, R. A. The possible modification of the response of the wild type to recurrent mutations. *Amer. Nat., 62,* 115–126, 1928.
6. _____. *The genetical theory of natural selection.* Oxford, Clarendon Press, 1930. 272 pp.
7. _____. *The theory of inbreeding.* Oliver & Boyd, Edinburgh, 1949. 120 pp.
8. FISHER, R. A., and FORD, E. B. The spread of a gene in natural conditions in a colony of the moth Panaxia dominula L. *Heredity, 1,* 143–174, 1947.
9. _____. The "Sewall Wright Effect." *Heredity, 4,* 117–119, 1950.
10. GOLDSCHMIDT, R. *The material basis of evolution.* Yale University Press, 1940. 436 pp.
11. GULICK, J. T. On the diversity of evolution under one set of external conditions. *Linnean Soc. Jour. Zool., 11,* 496–505, 1872.
12. HALDANE, J. B. S. A note on Fisher's theory of the origin of dominance; and on a correlation between dominance and linkage. *Amer. Nat., 64,* 87–90, 1930.
13. _____. Human evolution: past and future. Chapter 7, pp. 405–418, in: *Genetics, paleontology and evolution,* edited by G. L. Jepsen, G. G. Simpson, and E. Mayr. Princeton University Press, 1949.
14. LUSH, J. L. The theory of inbreeding (by Ronald A. Fisher). *Amer. Jour. Human Genetics, 2,* 97–100, 1950.
15. WRIGHT, S. Fisher's theory of dominance. *Amer. Nat., 63,* 274–279, 1929.
16. _____. Evolution in Mendelian populations. *Genetics, 16,* 97–159, 1931.

17. WRIGHT, S. Statistical theory of evolution. *Amer. Stat. Jour.* (March supplement), 201–208, 1931.

18. _____. The roles of mutation, inbreeding, crossbreeding and selection in evolution. *Proc. 6th Internat. Congress of Genetics, 1,* 356–366, 1932.

19. _____. Physiological and evolutionary theories of dominance. *Amer. Nat.,* 68, 25–53, 1934.

20. _____. Evolution in populations in approximate equilibrium. *Jour. Genetics, 30,* 257–266, 1935.

21. _____. Isolation by distance. *Genetics, 28,* 114–138, 1943.

22. _____. On the roles of directed and random changes in gene frequency in the genetics of populations. *Evolution, 2,* 279–294, 1948.

23. _____. Adaptation and selection. Chapter 20, pp. 365–389, in: *Genetics, paleontology and evolution,* edited by G. L. Jepsen, G. G. Simpson, and E. Mayr. Princeton University Press, 1949.

24. _____. Population structure in evolution. *Proc. Amer. Phil. Soc., 93,* 471–478, 1949.

25. _____. Genetical structure of populations.*Annals of Eugenics,* 1951 (in press).

26. WRIGHT, S., and DOBZHANSKY, TH. Genetics of natural populations. XII. Experimental reproduction of some of the changes caused by natural selection in certain populations of Drosophila pseudoobscura. *Genetics, 31,* 125–156, 1946.

"Genetic Drift" in an Italian Population

LUIGI LUCA CAVALLI-SFORZA

The variety of hereditary types in a human population originates with mutations in the genetic material. The survival and preferential multiplication of types better adapted to the environment (natural selection) is the basis of evolution. Into this process, however, enters another kind of variation that is so completely independent of natural selection that it can even promote the predominance of genes that oppose adaptation rather than favoring it. Called genetic drift, this type of variation is a random, statistical fluctuation in the frequency of a gene as it appears in a population from one generation to the next. Sometimes genetic drift seems to exert only a moderate influence, causing the frequency of a gene to fluctuate by 5 or 10 percent. At other times it may result in one gene overwhelming other genes responsible for the same characteristic.

How strong is the influence of genetic drift in evolution, and what factors control it? Together with my colleagues Franco Conterio and Antonio Moroni of the University of Parma, Italo Barrai and Gianna Zei of the University of Pavia and our collaborators at other institutions, I have for the past 15 years been investigating genetic drift in the populations of the cities and villages in the Parma Valley in Italy. We have examined parish books, studied marriage records in the Vatican archives, made surveys of blood types, developed mathematical theories and finally simulated some of the region's populations on a computer. We have found that genetic drift can affect evolution significantly, and we have been successful in identifying factors that control it.

Hypothetically genetic drift can happen in the following way. Suppose a small group of Europeans, perhaps 10 people, colonized an island (as the mutinous sailors of the *Bounty* and their Tahitian women did). Among 10 such randomly chosen people there might well be no one with blood of Type B or Type AB, because the genes for these blood types are respectively carried by only 15 percent and 5 percent of Europeans. Forty percent of Europeans have blood of Type O, and the same percentage have blood of Type A. In this small group, then, the frequency of the Type B gene might be zero rather than 15 percent because, in the nature of statiscal processes, it is absent and cannot reappear unless a rare mutation takes place. The gene may also be extinguished if only one or a few members of

Reproduced by permission of the publisher from Scientific American, 221:30–37 (1969). Copyright © 1969 by Scientific American, Inc.

Luigi L. Cavalli-Sforza, *a physician and population geneticist, was born in Genoa, Italy in 1922. He received his M.D. degree from the University of Pavia in 1944 and his M.A. from Cambridge University in 1950. At present he is Professor and Chairman, Department of Genetics, University of Pavia. He has done extensive work in bacterial genetics with special emphasis on the origin of resistance to antibodies. His interest in population genetics dates back to 1952 when he was lecturing in genetics at the University of Parma. Since that time he has devoted much of his research in populations to analyzing the role of chance in determining human evolution and developing methods of phylogenetic analysis for application to the problem of human origins and evolution. His field research in genetics has taken him to the Parma Valley in Italy and the central areas of Africa where he has worked among African pygmies. His more recent publications include* Analisi Statistica per Medici e Biologi e Analisi del Dosaggio Biologico *(Turin, 1961);* "Analysis of human evolution under random genetic drift," *with I. Barrai and A. W. F. Edwards,* Cold Spring Harbor Symposium on Quantitative Biology, *24:9–20 (1964);* "Population structure and human evolution," *Proceedings of the Royal Society, London,* 194:362–369 (1966); "Human diversity," *Proceedings of the XII International Congress of Genetics,* 3:405–416 (1969); "Studies on African Pygmies. I. A pilot investigation of Babinga Pygmies in the Central African Republic (with an analysis of genetic distances)," *American Journal of Human Genetics,* 2:252–274 (1969).

the group carry the Type B gene and they produce no descendants. Conversely, the frequency of a rare gene may sometimes increase until it becomes "fixed," or predominant. In remote valleys of the Alps, for instance, there is a relatively high frequency of such traits as albinism, mental deficiency and deaf-mutism, which are normally subject to negative selection.

This view of genetic drift suggests that two factors determine its strength: population size and migration. In the population of an alpine village, or among our hypothetical island colonists, a gene might vanish or become fixed in relatively few generations because the population is so small that even a slight change in the actual number of people carrying a gene causes a large change in the percentage of the population endowed with that trait. In a larger population a change in gene frequency would affect a smaller percentage of the people, and thus drift would be less pronounced. Migration can offset the movement of a gene toward predominance or extinction by increasing the frequency of rival genes. Like migration, natural selection may also restore equilibrium, by promoting adaptive combinations of genes, even after the situation has been profoundly disturbed by genetic drift.

Isolated observations, however, offer only glimpses of the significance of genetic drift, and speculating on such observations cannot provide us with a basis for measuring the phenomenon or identifying the factors that control it. Accordingly my colleagues and I began to search for an experimental group in which we could conduct tests that would clarify these issues. We

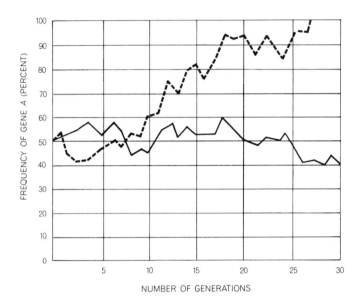

GENETIC DRIFT can cause the frequency of a gene to vary markedly from one population to another. In this calculation performed with random numbers Gene A appears in
50 percent of the members of two populations. Only 27 generations later Gene A has become "fixed" in one (broken line) and in the other its frequency fluctuates from 40 to 60
percent. [Redrawn.]

decided that the Parma Valley, which stretches for 90 kilometers to the
south of the city of Parma, would provide an ideal population.

The Parma Valley, located in north-central Italy, is named after a stream
that flows through it into the Po from the Apennines. The river has carved
out what in its upper reaches is a steepsided, inhospitable valley that gradually opens out into gentle hills and finally into a broad plain on which lies
the city of Parma. The very geology of the valley creates an almost
complete spectrum of the patterns of human habitation. In the highlands
the steep countryside encourages people to gather in small villages of about
200 to 300 inhabitants. Farther downstream the rural villages become bigger as the hills give way to the plain, and where the stream flows into the
Po stands the city of Parma [see illustration].

People have lived in the Parma Valley since prehistoric times. Because there have been no major immigrations since the seventh century
B.C. a certain demographic and genetic equilibrium has been
reached. The effect of natural factors such as migration, natural selection and genetic drift can therefore be studied under the simplest
conditions: when they are, or can be reasonably believed to be, in
equilibrium. On the plain and in the hills, however, immigration and

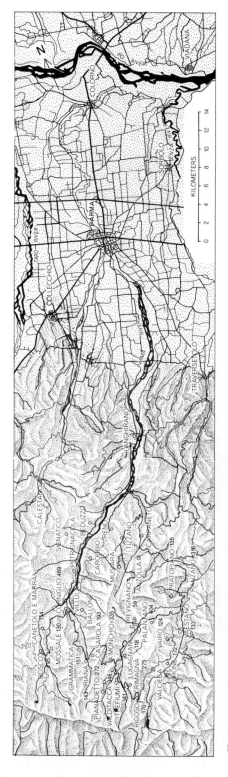

The Parma Valley stretches from the ridges behind the village of Rigoso at left to the plains of the Po River, 90 kilometers to the north. Because the settlement patterns of the Parma Valley include isolated villages in its steep-sided upper reaches, hill towns at lower altitudes and the city of Parma on the plain south of the Po, the valley constitutes an excellent natural laboratory for studying how genetic drift affects human evolution. In order to study genetic drift, blood samples were taken throughout the cities and towns of the valley, and those upland villages whose population sizes are indicated by numbers were simulated in a computer experiment.

333

migratory exchanges are more frequent than in the mountains. Important demographic information is supplied by the parish books of marriages, births and deaths in the area since the end of the 16th century. Accordingly the valley offers excellent opportunities for measuring the effects of population size and migration on drift.

One of the first hypotheses we decided to test was the one that drift should be more pronounced in a small, isolated population than in a large one, since a large population has a wide variety of gene types and is more susceptible to migration. If this proposition were true, we could expect to detect the strongest drift in small, isolated mountain villages in the uplands of the Parma Valley, less drift in the hill communities farther down the valley and the least drift on the plain.

The most convenient way to measure drift is suggested by the nature of the phenomenon itself. Under the influence of drift village populations will tend to become more and more different, even if at the beginning they were homogeneous in their composition of hereditary types. Taking a particular hereditary characteristic, say blood of Type A, individuals with blood of this type may become more frequent in one village and rare in other villages. What is needed is a measure of this kind of variation among villages. If we had only two villages, we could take

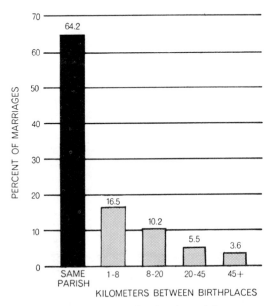

MIGRATION in upper Parma Valley has been infrequent, a conclusion drawn from the fact that most marriages recorded from 1650 to 1950 in parish books unite men and women who are from the same village. The number falls as the distance separating birthplaces increases. [Redrawn.]

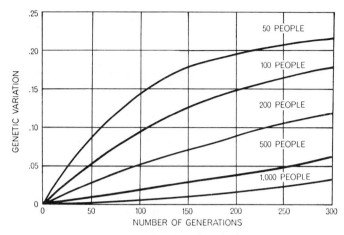

POPULATION AND DRIFT are closely related. The frequency of a particular gene in each population begins in the first generation at 50 percent, equivalent to zero on the vertical scale marked according to a measure of genetic drift called variance. After 300 generations the variation of the gene in the smallest population has increased to .22, almost as far as it can go, whereas in the largest population it has reached only .03. Genetic drift, then, is strongest in smaller populations and weakens as the population size increases. [Redrawn.]

the percentage of individuals with blood of Type A in one village and the percentage in the other village and compute the difference between the two percentages as a measure of the variation between the two villages. In examining many villages we might consider the differences between all possible pairs of villages and average them out. In actuality we use a somewhat different way of measuring the variation, but the principle is much the same. Here let us simply call the result "measure of genetic variation between villages." Estimates of the variation were obtained after we had grouped villages in somewhat larger local areas, from the highlands down through the hill towns to the city of Parma. The measure we used excludes the effects due to sampling because we used a fraction of the total population.

As predicted, the variation between villages declined as population size increased, from .03 in the high valley, where the population density was well under 50 people per square kilometer, to less than .01 in the hill country, where there are about 100 people per square kilometer, to almost nothing on the plain, where the density reaches 200 people per square kilometer [*see illustration*, p. 336].

It is possible that the variation between villages could be caused by adaptation to different environments rather than by genetic drift. As unlikely as it may seem, it is possible that the environmental conditions differ from village to village so that different genes are favored in each place. It may also be that people of diverse origins and therefore of

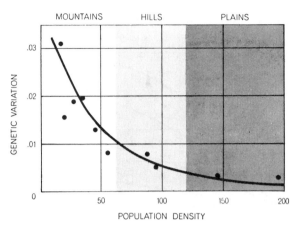

VARIATION in the frequency of a blood type between one village
and another was greatest, as predicted, in the isolated upland
hamlets, and declined as population density increased farther
down the valley in the hill towns, on the plain and in the city of
Parma. [Redrawn.]

diverse blood groups have settled in the more populous regions, and
that because of these historical accidents there has not yet been time to
reach an equilibrium.

If natural selection or historical accidents were responsible, the percentage of individuals possessing a certain gene would vary from village to village. The percentage would not necessarily vary in the same
way for all genes, since there is no reason why the selective factors or
historical accidents should operate with equal force on all genes. It
would, in fact, be a strange coincidence if they did. If the variations in
genes were caused by genetic drift, however, they would be the same,
on the average, for any gene. The reason is that genetic drift, being a
property of the population rather than of the gene, should affect all genes
in the same way. Our evidence shows that the variations between villages are indeed the same for any gene.

This first test of our ideas about genetic drift was convincing, but in
order to test our analysis more severely we wanted to make exact
forecasts of the amount of genetic variation caused by drift. Such an
exercise would require a precise quantitative prediction rather than
a simple qualitative statement. Unfortunately the classical mathematical theories of population genetics (put forward by Sewall Wright,
Motoo Kimura and Gustave Malécot) require that villages be of equal
size, and that migrations between them follow a highly homogeneous
pattern, simplifications that are rather far from reality.

To avoid this difficulty we developed other methods of predicting variations in the frequency of a gene on the basis of population size and migration. With the help of Walter Bodmer of Stanford University it was possible to devise a new theory that takes account of the actual observed migration pattern from village to village, however complex it may be. The model removes many of the oversimplifications of the classical theories but not all of them. We have therefore also developed a more general method that makes it possible on the basis of simple demographic information to predict the expected amount of drift with unlimited adherence to reality. This method consists in the use of artificial populations generated in a computer. Before I describe it, however, I should mention an apparently independent but in fact closely related approach, using a substantially different body of data, that we have followed in parallel with the study of genes.

This alternative approach we have followed is the study of relationships between individuals, which can be obtained from pedigrees or similar sources. One intuitively understands that the relationship between people must be associated with the similarities (or the differences) between the genes they carry. Both depend on common ancestry. Greater isolation between villages implies a lesser degree of common ancestry between the people of the villages, and therefore both a lesser degree of relationship between them and more differences between their genes. Thus the study of pedigrees, making it possible to estimate common ancestry, or degrees of relationship, should yield almost the same information as an analysis of the frequency of genes in the various villages. It has been shown that even data as simple as the identity of surnames can, in indicating common ancestry, supply information similar to what can be obtained from the direct study of genes.

It is the availability of parish books in the Parma Valley that makes it possible to carry on this investigation in parallel with the study of genes. Unfortunately the reconstruction of pedigrees from parish books is a laborious task, and it has not yet been completed. We do, however, have data on relationships from another source: records of consanguineous marriages. We found it particularly interesting to test the validity of this method as an alternative to the direct study of the effects of drift on genes. In the Parma Valley we could compare all these approaches. We could see, for instance, if we could predict drift from consanguinity or vice versa, or better still, predict both from a common source: simple demographic data.

Consanguineous marriages and genetic drift are similarly affected by common factors: population size and migration. A small population encourages consanguineous marriage because after a few generations most marriage partners would also be relatives. Migration, on the other hand, tends to decrease the frequency of consanguineous marriage by introducing new partners who are not relatives, or, if the flow is outward, by removing relatives who would otherwise be available.

The mathematical model with which the frequency of consanguineous marriage can be predicted is relatively simple. It is based on the idea that the population whose size critically affects the frequency of consanguineous marriage and genetic drift is somewhat diffuse, geographically speaking. As the Swedish geneticist Gunnar Dahlberg pointed out in 1938, this population basically consists of a group of people who are potential marriage partners for one another. This population is therefore not identical with the marriageable population because there are social barriers that reduce marriage choice. A village or a town might also be so large that not all the available partners would know one another. Such factors tend to make the population of eligible partners smaller than it is in a smaller village. The group can, however, extend across political boundaries, so that marriages are made between people living in different villages.

Since a simple census will not yield the size of the population of marriageable individuals, the population must be determined mathematically. By definition the population consists of a circle of N people available to one another for marriage. Assume now that an individual is not prohibited from marrying a blood relative (provided they are not so closely related that the marriage is forbidden by law). In this case the probability that he (or she) will marry a relative will be equal to the ratio between the number of eligible blood relatives, c, and the number of candidates who are not relatives. The probability of consanguineous marriage, m, will therefore equal c/N. This probability is also identical with the overall frequency of consanguineous marriage; all other factors being excluded, the frequency of consanguineous marriage would depend only on the number of available partners who are also kinsmen. Thus if there were 40 available partners and 20 of them were blood relatives, the frequency of consanguineous marriage would be one in two, or 50 percent. Knowing the number of blood relatives from simple calculations, and the frequency m of consanguineous marriages from ecclesiastical records, we can determine the size N of the population of eligible mates because it is a function of m and c. In other words, if $m = c/N$, then $N = c/m$.

Having determined the population size, it would be convenient at this point if we could simply complete our model by taking into account the effect of migration on the available supply of marriage partners. We could then predict the frequency of consanguineous marriage and therefore estimate the amount of genetic drift. Reality, however, forces us to make a circuitous detour. It appears that there are certain factors (such as the tendency of people to marry people of a similar age, biases for or against certain kinds of consanguineous marriage and the fact that degrees of kinship too remote are recorded incompletely or not at all) that would distort the prediction of genetic drift because they affect only the frequency of consanguineous marriage without influencing the rate of variation for a gene. In order to calculate genetic drift on the basis of consanguineous marriage we must identify and compensate for such factors.

These factors can be inferred from the study of consanguineous marriages. The Vatican archives contain records of 590,000 dispensations for consanguineous marriages granted from 1911 to 1964, the year in which our gathering of data temporarily ended. Only the dispensations for the more distantly consanguineous marriages that were granted directly by the bishops in certain areas such as Sicily and Sardinia and a few other remote dioceses are excluded. This material provided our investigators (led by Moroni, who had been a student of mine at the University of Parma) with a mass of valuable information. The records provide, among other data, the given name and the family name of the parents of the couple and the degree to which the marriage is consanguineous.

These records exist because the Roman custom and religious belief that prohibited marriage between blood relatives was inherited and diffused by the Catholic church. During the Middle Ages only the Pope could grant dispensations from the prohibition, and the dispensations (at least those known to us) were not numerous. The degree of consanguinity eligible for dispensation, however, has varied through the centuries, and a progressively more liberal trend can be detected. In the 16th century the Council of Trent recommended that a special dispensation be required for marriages up to "the fourth degree" (third cousins). Since 1917 dispensation has been required only for marriages between second cousins. The Vatican Council has recently pushed the liberalization one degree further, so that today only a marriage between first cousins, or between uncle and niece, require dispensation. Marriages between closer relatives are not and never have been eligible for dispensation.

The dispensation must be requested by the parish priest from the Curia before the marriage can be celebrated. In some cases the bishop can grant dispensations, but customarily he must forward a copy of the request to Rome (or, in other countries, to the representative of the Pope in that country). The Vatican will then reply to the bishop, and he will reply to the priest. The Vatican practically always grants dispensations in allowed cases, and therefore the dispensation request has constituted, or at least constitutes today, only a formal obstacle.

Of the factors that seem to alter the frequency of consanguineous marriage, one is the very closeness of the blood relationship. Apart from legal and religious restrictions, there is the widespread knowledge that consanguineous marriages may result in hereditary handicaps for the offspring.

Age can also have an important effect. As the archives show, marriages in which the consanguineous mates are a generation apart, such as those between uncle and niece and between first cousins once removed, are rarer than those between first cousins and between second cousins. We can explain this fact if we assume that in both consanguineous marriages and nonconsanguineous marriages age affects the choice of mates in the same way. In Italy, for instance, there is a mean age difference between husband and wife of about five years; the difference is smaller for young spouses and larger and more variable for older spouses. Therefore by considering the age differences among children of the same family, between parents and children and between normal spouses we can predict that marriages between uncle and niece will be only 3 percent of what would be expected on the basis of the frequency of this relationship, and marriages between aunt and nephew will be still rarer. By the same token one could expect a higher frequency of marriage between first cousins, because they tend to be of a similar age.

Migration also tends to reduce the frequency of consanguineous marriage. In places where the population is small and migration is low, blood relatives remain in contact. Hence we are not surprised to find most of the consanguineous marriages in rural areas whose populations have been rooted in the same soil for many generations. In industrial areas migration tends to disperse blood relatives so that they may not even meet, much less marry. Therefore consanguineous marriage tends to be diluted in frequency, or even to vanish, just as genetic drift does.

The effects of migration become more complex when we study specific types of genealogical trees. From data gathered in northern Emilia, a broad region including the Parma Valley, it appears that in the past century the number of consanguineous marriages diminishes when among the immediate ancestors of husband and wife there are more females than males. The reason is that men and women migrate in different patterns. In a largely rural area such as the one we were examining, where land is inherited by the male child, fewer sons emigrate than daughters. Moreover, a woman who marries someone from another village will emigrate in the process, and the distance between the villages will make her descendants a little less available for marriage with descendants remaining in the original place. The more women there are in the genealogical tree, the more significant this pattern is [see illustration].

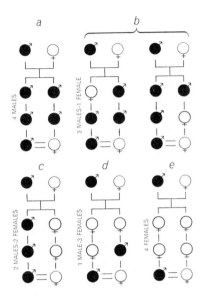

FAMILY TREES affect the frequency of consanguineous marriage. Different family trees associated with second-cousin marriages show that when spouses share only male ancestors (a), the number of consanguineous marriages reaches 774. As the number of female ancestors rises the number of marriages falls. Trees such as b (two varieties are shown) have produced 652 marriages, c 325 marriages, d 262 marriages and e 252 marriages, according to diocesan records from 1850 to 1950. Since land passes from father to son, men do not usually emigrate and marriages among relatives are therefore much more likely. There are 10 other trees for second-cousin marriages. [Redrawn.]

We have also isolated a factor of a sociological nature. It is scarcely important enough on the statistical level to merit consideration, but I shall cite it as a curiosity. It is the tendency of children of consanguineous mothers to intermarry. Probably this is because the mothers, when related, have a greater tendency to conserve bonds that favor marriage among the children.

With the help of Kimura (who is with the National Institute of Genetics in Japan) and Barrai, we incorporated these inhibiting factors and the effects of migration into a mathematical model that predicts the incidence of consanguineous marriage with some accuracy.

Even this carefully derived mathematical theory suffers, as all applications of mathematics in biology do, from the necessity of simplifying reality in order to make results calculable. The difficulty lies not only in solving complicated mathematical problems but also in successfully simplifying the terms of the problem to allow the use of appropriate mathematical instruments without losing essential features of the problem. Computers, however, make possible another technique for attacking these problems. If we have enough data available on the population under examination, we can reconstruct it in the computer and see what happens in experiments. The repetition of these experiments a sufficient number of times gives us a view of what we can expect in reality. In this way we can, without the use of higher mathematics and with the employment of real data, forecast the complicated effects of the genetic structure of a population on phenomena such as the frequency of consanguineous marriages or genetic drift.

Naturally we simplify the artificial population that we reconstruct in a computer as much as we can. Our computer "men" and "women" do not have hands and brains, only a number (0 or 1) that indicates sex, a number of several digits representing a name, and other numbers that identify the father and the mother of each individual, his distant ancestors and his descendants. If we study the effects of age, we must give our artificial subjects an age, and by the same token we can characterize their geographic location and social class [see illustration].

An artificial population so constructed marries, reproduces and dies according to certain probability tables drawn from reality. For reasons of

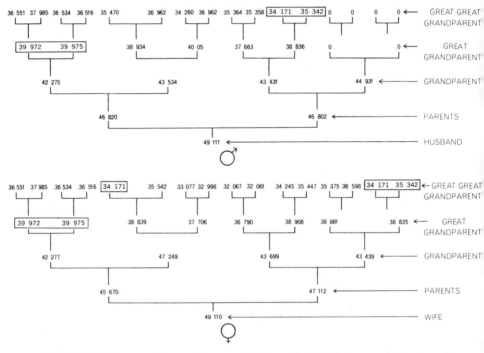

MAN AND WIFE are programmed for a computer run that simulates populations in order to determine the rate of genetic drift through the frequency of consanguineous marriage. The male's name is encoded in the digits 49111. Other numbers (not shown) can be programmed that indicate the person's sex, age, social class and genetic endowment. Shading indicates ancestors shared by man and wife. [Redrawn.]

economy we make the time advance crudely, in steps of 10 years. Thus if we have a 30-year-old person, we ask the probability of his dying before 40, and on the basis of tables of the real population we make him die according to a random procedure that has a probability equal to the real one. To determine whether a 30-year-old man dies before 40, for example, we calculate from the program a random number between 00 and 99. Such a number has equal probability of being one of the 100 numbers from 00 to 99. Since the probability of a man of this age dying is 12 percent, the individual dies if the chosen number lies between 00 and 11 and survives if it falls between 12 and 99. In the same way, using a random-number table that gives real probabilities, we decide if he marries and whom he marries by making the choice based on age, social class and geographical location. Finally, computer-generated marriages can be analyzed to determine the degree of consanguinity.

The results obtained by simulating the population of the Parma Valley confirm the impressions we had gained from our mathematical analysis of the actual population. That is, taking into account age, migration and the number of blood relatives of a given degree, the frequency of consanguineous marriages is about the same as what one would expect it to be if such marriages happened randomly. This is particularly true for the more remote degrees of consanguinity. Among first cousins, however, it seems that the frequency of consanguineous marriages is only half what it would have been if such marriages had been random.

Our final experiment consisted in simulating the populations of 22 villages in order to test our hypotheses of genetic drift. The 5,000 individuals in the test population were given genes of the three blood-group systems: ABO (governed by the gene types A_1, A_2, B and O), Rh (seven gene types) and MN (two gene types). At the beginning of the simulation the frequencies of these genes in each simulated village were the same as the average of the frequencies in the actual villages. With the passage of each simulated generation, however, the frequencies began to change by a factor that was approximately the same for each blood-group system. They finally leveled off when the number of people belonging to a blood group was two or three times the original value. This equilibrium was reached fairly soon (after about 15 generations) as a consequence of the establishment of a balance between drift, which tends to make villages different, and migration, which tends to make them more nearly the same.

We found that the variations among the simulated villages quite closely matched the variations predicted by Bodmer's theory, thus confirming its capacity to represent real data. We also found that the variations among the simulated villages matched, although less closely, the variations observed among the real villages. In both the simulated and the real villages no single gene type either vanished or became predominant; the drift was not strong enough to achieve this result. Fairly divergent proportions of genes could be found in the different villages. Which gene increased in which particular village was of course a matter of chance. We did not expect a real village to show the same proportion of a given gene as its artificial counterpart [*see illustration*].

It is clear that since the observed variation corresponds—within limits that we are now investigating—to the expected one it is not necessary to invoke explanations other than the action of genetic drift. The methods we have used in our study of the Parma Valley have now been applied to other populations as diverse as African pygmies, New Guinea tribesmen and the descendants of the Maya Indians. The results so far have confirmed the concept that genetic drift is the principal agent responsible for the variations among villages, tribes or clans. In fact, at the microgeographical and microevolutionary levels on which we worked the differences attributable to natural selection were not large (apart from some minor ones we intend to re-examine). Furthermore, there is evidence that drift can oper-

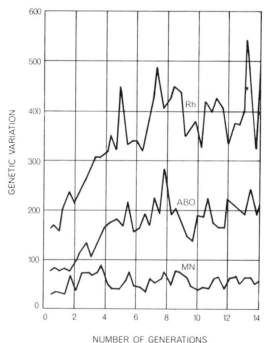

NUMBER OF GENERATIONS

GENE FREQUENCIES for the ABO, Rh, MN blood-group systems in the populations of the villages from the upper part of the Parma Valley vary with the passing of each generation. Differences in gene frequency between villages simulated in a computer program and measured by "chi square" were similar to the differences in the real villages. Therefore migration, population size and other demographic influences that control the simulated genetic drift are probably equivalent to the forces at work in the real villages. [Redrawn.]

ate on a macroevolutionary scale extending over millions of years. Recent comparisons of the sequences of amino acids in the proteins of separate species show differences that are at least in part caused by drift, although the evidence is still controversial.

In seeking to extend these conclusions one could study populations distributed over large areas. In that case, however, it would be easier to encounter disturbing factors. For example, in a large enough region one might encounter selective pressures whose diversifying effects would be added to the effects of drift on certain genes, whereas in another area or in the case of another gene natural selection might oppose drift and reduce the variations.

In any case, the discovery that genetic drift can affect evolution on a small scale over a short period of time gives the phenomenon a more important role in evolution than was once thought. It would be an error to assume, however, that evolution is almost entirely random. Only natural selection can bring about adaptation to the environment, and its importance must not be underestimated. The relative importance of drift and natural selection in determining the course of evolution remains to be assessed.

Genetic Effects of Population Size Reduction

DEREK F. ROBERTS

For evolutionary thought, the proof by Wright that nonadaptive differentiation can occur as the result of genetic accident appears in retrospect to be among the most important results of the application of mathematical analyses to the facts of genetics. Random processes, to whose results the term "genetic drift" has come to be applied, play a part in determining the character of populations. In theory, their role is particularly important in small and entirely isolated groups, whereas with a larger population distributed uninterruptedly over a wide territory their effects tend to be overshadowed by those of selection. The argument that random drift can occur is, however, different from establishing the fact that it has occurred in any given instance; the latter has proved unexpectedly difficult, even in human populations which according to Mayr[1] provide some of the best evidence for random fluctuations of gene frequency. Indeed, it has been argued (for example, refs. 1–3) that drift is of inconsequential influence in the evolution of populations.

Part of the difficulty lies in the way in which the term "random drift" is used to cover various interrelated genetic phenomena. More than two decades were required to appreciate their variety and elucidate their effects. The importance of random factors in the balance of those bringing about evolutionary advance was recognized by Wright[4] in 1932, though not fully. Random fluctuation in intensity or direction of selection had been touched on the year before[6], but was not stressed until 1935 (ref. 7). In 1938 Wright[5] published a method for dealing with several different types of random processes mathematically. The practically complete indeterminacy of events that are unique was not recognized until 1949 (ref. 8). Today "random drift" is taken to represent the cumulative effects of all the random processes affecting gene frequencies. These may be summarized as follows. (*a*) The random variation that occurs in the sampling process of passing on gametes from one generation to the next. This is the component to which the term "drift" is frequently restricted in common parlance. (*b*) Random variation in the magnitude and direction of selective and other systematic forces; these include random fluctuations in mutation rate, in selection pressures, and in rates of population movement. (*c*) Random unique events. These comprise category I 3 of Wright's[9] classification; here are included novel favourable mutations, unique hybridiza-

Reproduced by permission of the publisher and Derek F. Roberts from Nature, *220:1084–1088 (1968).*

DEREK F. ROBERTS, *physical anthropologist and population geneticist, was born in London in 1925. He received his M.A. from Cambridge University in 1952 and his D.Phil. from Oxford University in 1953. He is presently a Reader in Human Genetics and Director of the Laboratory of Human Genetics, University of Newcastle upon Tyne. His main research interests are in human population genetics, demography, and ecology, and he has published numerous articles in these areas (see References, this chapter). As a veteran field worker he has done genetic field work among various populations in Africa and Brazil. With G. A. Harrison he has coedited* Natural Selection in Human Populations *(New York, 1959).*

tions, swamping of a population by mass immigration, unique selective incidents, and unique reductions in numbers. The last of these is often termed the "founder" effect or the "bottleneck" effect, which represent the one phenomenon but at different points in a population's history in a given territory.

Studies of Drift

Discussion of drift as a mechanism for genetic differentiation in human groups has come from three different types of studies. (1) Variations in gene frequency between local populations, and between offshoot and parent populations, have been attributed to drift. For example, Birdsell's[10] survey in south-west central Australia of forty-two tribes showed sharp blood group gene frequency differences which he attributed to drift, recognizing, however, that it could "be no more than suggested in terms of broad probabilities." Glass et al.[11] showed in their Dunker study what seemed to be drift of MN gene frequencies. Giles et al.[12] demonstrated differences in several blood group systems between New Guinea villages of common history and environment, differences which they felt to be due to drift. In these studies, drift has merely been invoked, not proved or measured; there are at best only pointers to the type of random process involved. In the third study, it was felt that the founder principle, or population size reduction by epidemics, was the cause of the differences, though no numerical details were given. In the second study, the authors recognized the difficulty of distinguishing the founder effect from that of accumulated gamete sampling over several generations, and their subsequent attribution to the latter[13] of part of the MN frequency change was later questioned[14].

(2) The demographic and mating structures of particular societies have been examined to show the magnitude of drift that could occur (for example, refs. 15–19). Here the restriction was to random variation that occurs in gamete sampling from one generation to the next. The net outcome of these was to show that the mating structure of the societies examined

would only allow drift of very small degree to occur, though there were extremely segmented societies where the effect might be stronger.

(3) Mathematical models have been applied to populations for which
genetic, numerical and mating data are available to enquire whether the observed variations in gene frequencies are explicable by drift alone. For example, Cavalli-Sforza et al.[20] found that in North Italy they are greater
than might be expected by drift alone, and attributed this extra dispersion
to overestimation of the effective size of the population concerned. But because what was measured was the accumulated effects of intergenerational
gamete sampling, the finding could have reflected other components of
drift. None of these studies, however, has yet demonstrated conclusively
that drift has occurred; only the last has attempted to compare observed
variation with expected drift magnitude. None, moreover, gives sufficient
attention to the category (c) of drift processes, the random unique events.
These are almost completely indeterminate statistically, and are therefore
not adequately covered in any models so far applied, being treated as negligible components of δq, the random portion of gene frequency change per
generation, by Wright[9]. The genetic effects of these processes are illustrated and quantified in this article. It concerns the effects of accidents on
the genetic constitution of a population, and in particular how these effects
have been brought about through the resulting diminution of numbers, the
"bottleneck" phenomenon.

Population and Procedure

To identify the effects of a bottleneck in terms of the frequency of, say,
blood group genes, the investigator would have to be extremely fortunate.
He would have to know the gene frequencies in the population immediately
before, and immediately after, the bottleneck occurred. He would require
foresight to know that the size of the population is about to change, and
hindsight to know that it had finished changing. The same results, however, can be demonstrated if the genetic constitution of the population is specified in a way other than by the usual array of frequencies of
particular alleles and genotypes. This can be done if the population pedigree
is known.

A parent passes to his child half of his chromosomes and therefore virtually half of his genes. When the child himself reproduces he again passes
on half his genes, so a grandchild derives on average a quarter of his genes
from each grandparent. The probability that any autosomal allele present
in a grandparent is present in any one of his grandchildren is one quarter,
and the probable genetic contribution of any grandparent to the gene pool
of his grandchildren is a quarter. The genetic constitution of any generation of descendants can therefore be specified in terms of the contributions
to it from particular ancestors, and, by extension, the same method can be
applied to any group of individuals but covering more than one gener-

ation. Thus the genetic constitution of a population at any moment is regarded as comprising a pack of probable ancestral contributions which can
be calculated. They reflect the relative fertilities of the ancestors concerned and their descendants. Allowance can be made for the effects of
mortality and migration, though not for other mechanisms of gene frequency change, notably for the random variation that occurs in the sampling process of passing on gametes from one generation to the next, for
mutation, or for selection through differential mortality. Thus calculation
of these probable contributions at a series of points in time, tracing their
secular variation, shows how the genetic constitution of the population so
defined varies during the period. The calculations are simple, and multiple
lines of descent must be allowed for where the population is in any way
inbred.

This procedure has been applied to the population of Tristan da Cunha.
From the detailed data that are available for this population, from the
records kept by the earliest settlers, the observations of visitors and official
reports, the lists that at periods were kept of births, deaths, baptisms and
marriages, and the investigations of genealogies by Dr Woolley, Dr Loudon
and Dr Munch, it is possible to obtain the necessary data for the analysis.
First, a complete, albeit highly complex pedigree of the whole population
since its founding can be established. Second, the numerical evolution of
the population from its founding can be traced. Fig. 1 shows the size of the
population on December 31 of each year from 1816 to 1961. There have

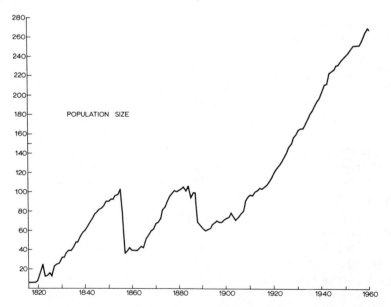

Fig. 1. The size of the population of Tristan da Cunha on December 31 of each
year from 1816 to 1960.

been three phases of increase in the population, one lasting until 1855, one until 1884 and one until October 1961, these three periods being separated by sudden and drastic population reductions—true bottlenecks.

Bottlenecks at Tristan da Cunha

In the first of these the numbers dropped from 103 at the end of 1855 to thirty-three in March 1857 (these figures include men temporarily absent from the island). This massive reduction seems to have been due to a combination of two principal factors, the death of one man and the presence of another. In 1851 the first missionary arrived, and in 1853 the founder, W. G., died. After his death the cohesion of the community appears to have relaxed, and twenty-five of his descendants left for America in 1856. The pastor who remained on the island until 1857 became increasingly convinced that emigration was necessary for the population. Whether this was true, or whether it was a projection of his discontent with his own lot there, will never be known, but he noted that there were "more than a dozen adult females here, with no prospect of a comfortable provision for life" and "it will be a happy day when this little lonely spot is once more left to those who probably always were . . . its only fit inhabitants—the wild birds of the ocean." Under his influence, when he departed, another forty-five islanders left with him for the Cape, settling at Mossel Bay and Riversdale. The presence of a pastor of this opinion at a time when the population was reorganizing itself after the death of its dominant character can only be regarded as a chance combination of chance occurrences. The growth curve of population size to this date shows no sign of flattening, and no indication that numbers were approaching the limit that the island could support.

The second bottleneck was neither quite so extreme nor quite so abrupt. It was triggered by a disaster. The island has no natural harbour, and any vessels that called stood offshore while the islanders put out to them in their small boats. Sometimes they also put out to board passing vessels, for trade. On November 28, 1885, a boat manned by fifteen adult males set off to intercept a passing vessel but, in full view of the watching remaining islanders, vanished beneath the waves. Not one man was saved. This disaster made Tristan an "island of widows" and depleted the population of its adult providers. It left on the island four adult men, of whom one was insane and two were very aged, to support the remaining population. Despite the considerable distress that immediately followed, by August of the next year the population as a whole had accommodated to the situation—for example, adolescent boys had taken over crew duties in the boat—and a petition, organized by those who had emigrated to South Africa, formally to evacuate the island was rejected. During the next few years, however, many of the widows and their offspring left the island of their own accord, and the population declined to a second minimum in 1891. The reduction in numbers this time was not by two-thirds but by a little less than half, from 106 to 59. Again, this reduction is a direct result of accident. The population thereafter began its most recent phase of increase.

Effects on the Gene Pool

The genetic constitution of the population in 1855 is shown in the second column of Table 1. Twenty ancestors had contributed genes, their respective probable contributions to the gene pool varying from 0·005 to 0·137; the two greatest contributions were from the two original settlers, W. G. and his wife, M. L., and the next three from two of the women from St Helena, S. W. and S. K., and from R. R., the earliest of the men to arrive after the founding of the settlement. These five individuals contributed more than half the genes in the gene pool of the population at the end of 1855.

TABLE 1. PROBABLE CONTRIBUTIONS TO THE GENE POOL OF THE POPULATION IN 1855, 1857, 1884, 1891 AND 1961

Ancestor		1855	1857	1884	1891	1961
W.G.	1	0.1275	0.0625	0.0548	0.0657	0.0691
M.L.	2	0.1373	0.0625	0.0548	0.0657	0.0691
T.S.	3	0.0662	0.1389	0.0943	0.1441	0.1339
S.W.	4	0.0907	0.1910	0.1392	0.1864	0.1602
F.M.W.	5	0.0245	0.0521	0.0448	0.0424	0.0263
R.R.	6	0.0637	—	—	—	—
S.K.	7	0.0858	—	—	—	—
T.R.	8	0.0049	0.0139	0.0177	0.0424	0.0382
M.W.	9	0.0637	0.1389	0.1380	0.0805	0.0424
A.C.	10	0.0637	0.1389	0.1285	0.0636	0.0424
P.G.	11	0.0490	0.1042	0.0896	0.0847	0.0526
C.T.	12	0.0490	—	—	—	—
P.M.	13	0.0441	—	—	—	—
W.D.	14	0.0588	—	—	—	—
M.F.	15	0.0049	—	—	—	—
G.	16	0.0049	—	—	—	—
A.H.	17	0.0245	0.0833	0.0684	0.0890	0.0365
F.R.C.	18	0.0049	0.0139	—	—	—
F.F.K.	19	0.0221	—	—	—	—
B.	20	0.0098	—	—	—	—
J.B.	21	—	—	0.0472	—	—
S.P.	22	—	—	0.0613	0.1186	0.1045
F.R.	23	—	—	0.0283	0.0170	—
R.A.B.	24	—	—	0.0142	—	—
M.J.	25	—	—	0.0142	—	—
F.S.G.	26	—	—	0.0047	—	—
A.R.	27	—	—	—	—	0.0625
E.S.	28	—	—	—	—	0.0445
G.L.	29	—	—	—	—	0.0435
G.C.	30	—	—	—	—	0.0126
A.S.	31	—	—	—	—	0.0543
R.L.	32	—	—	—	—	0.0037
F.P.S.	33	—	—	—	—	0.0019
J.B.	34	—	—	—	—	0.0019

The effect of the first reduction in population is shown by comparison of the third with the second column of Table 1. The primary effect was deprivation of the population of eight of its founder ancestors and a recent arrival whose total contributions in 1855 had been more than one-third—genes from only eleven were to be found in the new gene pool. Second, there was a change in their relative contributions. Genes from two of the principal contributors were among those that completely disappeared. The greatest contribution now was from S. W. (0·191), and the next three largest were from her husband, T. S., her sister, M. W., and her husband, A. C. (0·139 each)—the contributions from these three individuals were more than doubled. These four individuals together now contributed 60 per cent of the genes in the new population compared with their previous total of less than 29 per cent. The contributions of the first two settlers were halved, and that of a former minor contributor (T. R.) multiplied nearly threefold (from 0·005 to 0·014).

In the phase of increase that followed this first reduction, the contribution of one further ancestor was lost, but new contributions to the island's gene pool came from six more arrivals. The relative contributions from the original ancestors correspondingly declined with the exception of one (T.R.). By the end of 1884 the gene pool derived from sixteen individuals. It was relatively little affected by the boat disaster itself—the greatest changes were in the contributions of S. P. (0·061 to 0·071), W. G. and M. L. (0·055 to 0·047)— but the subsequent population size reduction had a much more pronounced effect, shown by comparison of columns 4 and 5. Again, there was loss of all the genes from several contributors, and again there was a rearrangement of the relative contributions of the remainder. The contributions (totaling some 8 per cent) lost were from four relatively recent arrivals, so that in 1891 the gene pool derived from only twelve individuals. The greatest contribution was still that from S. W., increasing from 0·139 to 0·186, the second largest from her husband T. S., increasing from 0·094 to 0·144. But the third largest was from a much later immigrant woman, S. P., which nearly doubled, from 0·061 to 0·119, while the former principal contributors, M. W. and A. C., dropped to fifth and eighth places, respectively, from 0·138 to 0·081 and from 0·129 to 0·064.

The Gene Pool in 1961

The final column of Table 1 shows the probable contributions to the gene pool in October 1961, when the island was evacuated. In the intervening period all the genes from one earlier arrival had been lost, and eight new arrivals had contributed 22.4 per cent of the total to the gene pool, which thus derived from nineteen ancestors. The principal changes were the diminution in the contributions of A. H. and M. W. by about half, and of P. G. and A. C. by about one-third, and the great contributions (sixth and

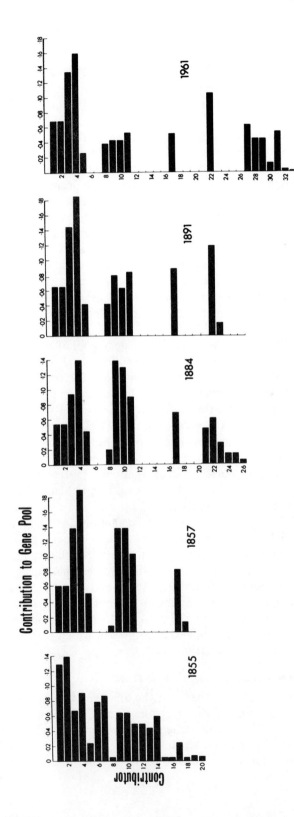

Fig. 2. Contributions of particular ancestors to the gene pool.

seventh in order of size) made by two of the new arrivals. But the three greatest contributions were made by the same individuals as in 1891. Indeed, there is a marked overall similarity between the figures for 1961 and those for 1891.

Tracing the contributions of particular ancestors (Fig. 2), it was the first exodus that halved the contributions of the first two settlers, and then, after a period of slight decline, the second size reduction increased them to nearly the 1961 figure. The predominance of the two principal contributors to the 1961 population is traceable to the first exodus that doubled their original contribution, and then, after a decline, to the second exodus that elevated them to the 1961 values. From the minor contributor, T. R. (No. 8), nearly 4 per cent of the 1961 gene pool derives, largely because the first exodus nearly trebled his contribution and the second exodus multiplied it by another 2·5 times. Two of the most prolific of the early settlers, M. W. and A. C. (Nos. 9 and 10), contribute only a little more than 4 per cent each to the 1961 gene pool; their contribution was doubled by the first exodus and then halved again by the second.

The overall effect is shown in Fig. 3. The gene pool of the population is taken as unity, and the proportion of it derived from each ancestor is shown to scale by the width of the band; time along the abscissa is not to scale. Both the exoduses exerted a considerable effect on the genetic constitution. This effect was two-fold: (a) the elimination of all genes derived from an appreciable proportion of the ancestors and (b) alteration of the actual and relative sizes of the contributions from other ancestors. Immigration in the periods 1857–84 and 1891–1961, although diminishing the contribution of each pre-existing ancestor, did not, it seems, appreciably change the size of these contributions in relation to each other, although further work is required to establish this point. But at the present state of the analysis it seems to be chiefly the two reductions in population size which bring this about. Furthermore, the effects of these size reductions persist, as comparison of the 1961 figures with those for each other year shows. The general similarity of the 1961 figures to those for 1891 implies that, apart from the contributions of the recent immigrants, the gene pool of the 1961 population derives in its major features principally from the reduction in population that occurred between 1884 and 1891, and the modification that occurred then acted on the gene pool the features of which had been chiefly derived from the effects of the earlier exodus.

Reduction of Population and Genetic Drift

I have documented here, perhaps for the first time in a human population, the effects of drastic population reduction on the genetic constitution of the population. In both of the bottlenecks there are two primary effects: (a) deprivation of the population of the genetic contribution of some founder ancestors, that is, a reduction in the number of contributing an-

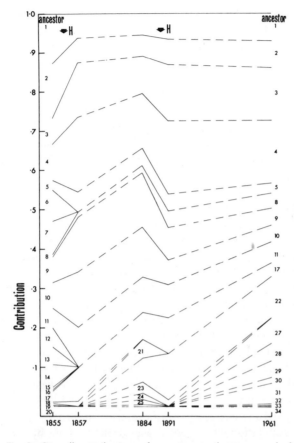

Fig. 3. Overall contributions of ancestors to the gene pool of the population in 1855, 1857, 1884 and 1961. **H** shows serious bottleneck.

cestors, and (*b*) change in the relative contributions of the remainder. The effects of these size reductions persist and, in accumulation, shape the profile of ancestral contributions to the gene pool of the present population; this in 1961 bears much more resemblance to the 1891 profile than to that of 1855. The present profile bears little relationship to the fertility of the ancestors, though further analysis is in progress to assess the relative importance of accumulated differential fertility effects. But there can no longer be any doubt that reduction in population size of the magnitude that occurs in nature must have a severe effect on the genetic constitution of a population.

There are, of course, drawbacks to this method of envisaging the gene pool as a pack of probable ancestral contributions. It obviously depends on the reliability of the pedigree. All possible steps were taken to ensure accuracy by comparison of the accounts of descent collected by different investigators, by checking these against the known movements of individuals to and away from the island, and by comparing all pedigrees with the information on the large number of genetic markers that is available for this population. In the few cases where paternity remained in doubt, it was possible to assign the most probable father on the basis of the known genotypes. No other population has been so intensively studied and the pedigree information now available can be regarded as the most reliable possible. The second drawback is that the work is couched in terms of probabilities. Although on average a child possesses a quarter of the genes of each grandparent, it is an extremely remote possibility that a given child actually carries no gene derived from a particular grandparent. While this would be quite unlikely taken over all loci, on the other hand, it is much more likely for a single locus. It is therefore interesting to compare observed numbers of genes with the probable numbers expected from the present work. The C_5 serum cholinesterase phenotype occurs in thirty-six individuals out of 213 tested[21]; there is little doubt that this gene was introduced by S. W., whose contribution to the 1961 population was 0·16; this gives an expected number of thirty-four heterozygous carriers in the population, instead of the observed thirty-six. The very slight gain of the C_5 gene is perhaps a result of intergenerational gamete sampling, or perhaps of selection. The pack method taken as a starting point for this analysis of the effects of population size reduction is obviously a powerful one for the elucidation of the effects of other processes on the gene pool.

This example illustrates the difficulty of assigning any given gene frequency change to any one category of evolutionary processes. Although the reduction in population was in both cases brought about by accident, there was a large non-accidental component to the reduction itself. In both cases family groups emigrated, so which genes and what proportions of them were lost from the population's gene pool were a partly random, partly non-random array. The families that departed obviously included individuals who felt they could no longer accept the conditions on the island or its prospects for them, so the non-random component in part may be identified as selective. The fact that man is a social animal, that individuals tend to move and often act as family units, brought about, in this population, what may be termed a "booster" effect, whereby, although the genes lost from the population by accident were in actuality a random sample, the loss of some of them was exaggerated by consequent deliberate emigration of families.

This study therefore shows that there is no doubt as to the importance of drastic reductions of population in shaping the genetic constitution of an isolated population. The gene frequency changes they bring about, however, can only be attributed to drift when drift is defined so as to incorporate all random unique events, for there may be an appreciable selective component and in man there is also an exaggeration of the effects of both these processes by family ties and relationships. In populations such as that discussed here, drift so defined cannot be regarded as of inconsequential influence on their genetic evolution, at least in the short term.

I thank Dr H. E. Lewis, the coordinator of the investigations of the Medical Research Council working party on Tristan da Cunha, for his help, and Dr Robson for her original pedigree diagrams, Dr Loudon and Dr Mourant for making available results of unpublished investigations, and Mrs Marjorie Smith for computational assistance.

REFERENCES

1. MAYR, E., *Animal Species and Evolution* (Harvard University Press, Cambridge, Massachusetts, 1963).
2. SHEPPARD, P. M., *Natural Selection and Heredity* (Hutchinson, London, 1958).
3. FORD, E. B., *Ecological Genetics* (Methuen, London, 1964).
4. WRIGHT, S., *Proc. Sixth Intern. Cong. Genet.*, **1**, 356 (1932).
5. WRIGHT, S., *Proc. US Nat. Acad. Sci.*, **24**, 253 (1938).
6. WRIGHT, S., *Genetics*, **16**, 97 (1931).
7. WRIGHT, S., *J. Genet.*, **30**, 257 (1935).
8. WRIGHT, S., *Proc. Amer. Phil. Soc.*, **93**, 471 (1949).
9. WRIGHT, S., *Cold Spring Harbor Symp. Quant. Biol.*, **20**, 16 (1955).
10. BIRDSELL, J. B., *Cold Spring Harbor Symp. Quant. Biol.*, **15**, 259 (1950).
11. GLASS, B., SACKS, M. S., JOHNS, E. F., AND HESS, C., *Amer. Nat.*, **86**, 145 (1952).
12. GILES, E., WALSH, R. J., AND BRADLEY, M. A., *Ann. NY Acad. Sci.*, **134**, 655 (1966).
13. GLASS, B., *Amer. J. Phys. Anth.*, **14**, 541 (1956).
14. ROBERTS, D. F., *J. Roy. Anth. Inst.*, **95**, 87 (1965).
15. LASKER, G. W., *Amer. Anth.*, **54**, 433 (1952).
16. ROBERTS, D. F., *Human Biol.*, **28**, 325 (1956).
17. ROBERTS, D. F., *Acta Genetica Stat. Med.*, **6**, 446 (1956).
18. ALSTROM, C. H., *Acta Genetica Stat. Med.*, **8**, 295 (1958).
19. BONNE, B., *Human Biol.*, **35**, 61 (1963).
20. CAVALLI-SFORZA, L. L., BARRAI, I., AND EDWARDS, A. W. F., *Cold Spring Harbor Symp. Quant. Biol.*, **29**, 9 (1964).
21. HARRIS, H. HOPKINSON, D. A., ROBSON, E. B., AND WHITTAKER, M., *Am. Human Genet.*, **26**, 359 (1913).

The Distribution of Certain Genes
in the Old Order Amish

VICTOR A. McKUSICK, JOHN A. HOSTETLER*,
JANICE A. EGELAND†, and
ROSWELL ELDRIDGE

Although perhaps not *well* known, the Old Order Amish are *widely* known for their adherence to old-fashioned social and technologic practices. Their closely prescribed manner of dress and use of the horse and buggy are familiar features. More profound sociologic characteristics, as well as the history of the group, are not so familiar. Revelant aspects will, therefore, be reviewed briefly (Bender, Smith, 1955; Hostetler, 1963).

Description

Most Amish are farmers, and all are rural-living. Amish society is theocratic. The unit is the church district in which lay clergy regulate all aspects of community life. Religious services are held in the home. The German Bible is used, and the south German dialect called Dutch (for *Deutsch*) is spoken within the group. Use of electricity and ownership of modern devices such as the automobile and telephone are forbidden. Resistance to consolidated schools and to education beyond the legal minimum, absolute pacifism, and, in general, separateness from "the world" are other cardinal features. Thus, a common religion and language and a "peculiar people" sense hold the group together.

From an estimated 8200 in 1905, the Amish population has grown to the present estimated 45,000 (Fig. 1). In this same period the population of the United States only doubled and part of the increase was due to immigration. Over 80% of Amish live in Pennsylvania, Ohio and Indiana (Fig. 2). Over 50% of Amish live in three counties: Lancaster Co. (Pa.), Holmes Co. (Ohio), and Lagrange Co. (Ind.). In Europe, the Amish culture was assimilated several decades ago. Except for the group in Ontario, all Amish live in the United States.

*Department of Sociology and Anthropology, Pennsylvania State University, Ogontz Campus, Abington, Penna.

†Department of Sociology and Anthropology, Franklin and Marshall College, Lancaster, Penna.

VICTOR A. McKUSICK, a physician and medical geneticist, was born in Maine in 1921. He received his M.D. from Johns Hopkins University School of Medicine in 1946. At present he is Professor of Medicine and Chief of the Division of Medical Genetics, Johns Hopkins University School of Medicine. His research interests are medical genetics, the mapping of the human X-chromosome, and the ethnicity of genetic disease in isolates. His major population-genetics research has been among the Amish isolates of Pennsylvania, about which he has published numerous articles. He is an associate editor of various medical and genetics journals, a member of several scientific research committees, including the Advisory Comittee for the 1970 United States Census, and a recipient of the Modern Medicine Distinguished Achievement Award for 1966. He is the author of several books on medical genetics including Heritable Disorders of Connective Tissue, (St. Louis, 3rd ed. 1966); Medical Genetics 1958–1960 (St. Louis, 1961); Medical Genetics 1961–1963 (New York, 1965); Mendelian Inheritance in Man (Baltimore, 2nd ed. 1968); The X Chromosome of Man, (Washington, 2nd ed. 1969); Human Genetics (New Jersey, 2nd ed. 1969).

History

The Amish sect originated in the Canton of Berne, Switzerland, in 1693 when Jacob Amman led a split from the older and more extensive Mennonite church. Converts were acquired in Alsace, Lorraine, the Palatinate, and neighboring areas of southern Germany and eastern France. Many of these converts were Swiss who had moved to these areas in the preceding century. This was a "movement within a movement."

Migration to eastern Pennsylvania began about 1720 and continued until about 1770. Most present-day Lancaster Co. Amish are descendents of pre-Revolutionary immigrants, who probably totalled no more than 200 persons. Waves of Amish immigration continued until about 1850, but the

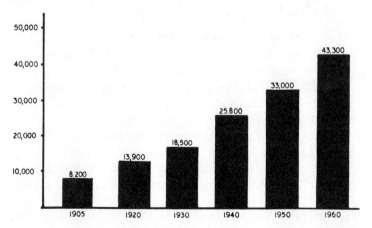

FIGURE 1. Growth of the Old Order Amish population. After data given by Hostetler (1963).

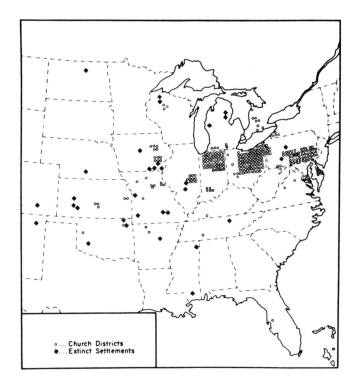

FIGURE 2. Location of Old Order Amish church districts. From Hostetler (1963).

later immigrants, finding the land taken up in eastern Pennsylvania, moved on to Ohio and Indiana. The migration patterns account for the peculiarities of distribution of family names and of certain genes in the old order Amish of the United States and Canada.

Prospects for Genetic Studies in the Amish

Several features of Amish society are favorable for genetic studies of certain types (McKusick, Hostetler, Egeland, 1964d).

1. It is a defined population, indeed a self-defined population. Although the Amish are often confused with other types of "plain people," the distinctions are clear to those familiar with the group.

2. It is a closed population. Although some leave, almost no new blood has entered the group since the immigrations.

3. Genealogic records are excellent. Almost all Lancaster County Amishmen can trace their complete ancestry to immigrants two centuries ago.

4. Undernutrition and infectious disease do not confuse interpretation of findings.

5. Standards of medical care are high. It is especially relevant that diagnostic standards are high.

6. Notable uniformity of socioeconomic circumstances reduces this source of variability.

7. The average level of consanguinity is high.

8. Families are large. As shown in Fig. 3, no reduction in family size occurred during the years of the last great depression.

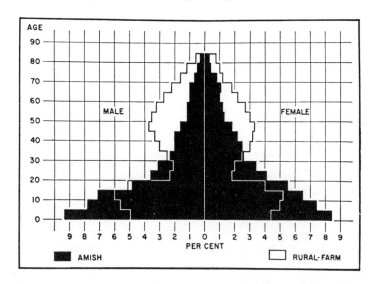

FIGURE 3. Age structure of Amish and non-Amish populations. From Hostetler (1963).

9. Because of their agrarian life, the Amish are immobile. Large kindreds are available for study in a limited geographic area.

10. The Amish are clannish and keep well informed of illness in groups throughout the country through the agency of *The Budget,* a weekly newspaper, and by other means.

Family Names as an Indication of Subisolate Formation

Differences in their background, as outlined briefly above, and presumably in the genetic constitution of Amish groups in six areas are reflected in the distribution of family names. (As will be described later, each of four of these six areas is known to have a relatively high frequency of a certain gene which is ordinarily rare.)

Table 1 presents the distribution of family names in Lancaster County, Pennsylvania, and Holmes County, Ohio. Eight names account, in each case, for about 80% of families, and no overlap is observed. Here is reflected the pre-Revolutionary and post-Revolutionary origins, respectively, of these groups, and the isolation which has been maintained since the immigrations.

Mifflin County, Pa. (Table 1), shows yet another distribution of family names. Although this settlement is also derived from pre-Revolutionary

TABLE 1. OLD ORDER AMISH FAMILY NAMES

Lancaster Co., Pa.		Holmes Co., O.		Mifflin Co., Pa.	
Stolzfus*	23%	Miller	26%	Yoder	28%
King	12%	Yoder	17%	Peachey	19%
Fisher	12%	Troyer	11%	Hostetler	13%
Beiler	12%	Hershberger	5%	Byler	6%
Lapp	7%	Raber	5%	Zook	6%
Zook	6%	Schlabach	5%	Speicher	5%
Esh**	6%	Weaver	4%	Kanagy	4%
Glick	3%	Mast	4%	Swarey	4%
	81%		77%		85%
Totals:					
1106 families, 1957		1611 families, 1960		238 families, 1951	

*Including Stolzfoos.
**Including Esch.

The Lancaster County families are those listed in the *Fisher Family History*, which, published privately in 1957, included all but 4 of the Old Order Amish families in that county. The Holmes County data are from the *Ohio Amish Directory* assembled and privately circulated in 1960 by Ervin Gingerich, Star Route, Millersburg, Ohio. The Mifflin County listing was given by John A. Hostetler, in "The Amish Family in Mifflin County, Pennsylvania" (M.A. thesis, Pennsylvania State College, 1961).

immigrants who settled first in Berks County (Pa.), it was founded by a small group who, for the most part, had names different from those who started the Lancaster County settlement. Only two names, Zook and Beiler (or Byler), appear with high frequency in both Mifflin and Lancaster Counties.

Although family censuses are not available, peculiarities of family name are noted in at least three other areas where discrete immigration in the first half of the last century and relative isolation since then are known to have occurred. These areas and the leading names in each are as follows:

1. Adams and Allen Counties, Indiana: Eicher, Girod, Hilty, Longacher, Neuenschwander, Schmidt, Schwartz, Steury, Wengerd, Wicky.

2. Daviess County, Indiana: Knepp, Stoll, Wagler, Witmer.

3. Perth and Waterloo Counties, Ontario: Albrecht, Jantze, Koepfer, Steckle.

Distribution of Ordinarily Rare Genes as an Indication of Subisolate Formation

Inquiries about various genetic and/or congenital disorders were sent to over 500 physicians practicing in Amish areas of Pennsylvania, Ohio, Indiana, and Ontario. Field trips were made to each of these areas and both physicians and Amish families were visited. Amish informants in communities throughout the country were contacted by one of us (J.A.H.). Although the methods of these surveys have shortcomings, conclusions about the distribution of certain genes are possible.

Four recessive genetic disorders have been found to have relatively high frequency, each in a different Amish group. These are:

a. Ellis-van Creveld syndrome in Lancaster County (Pa.) Amish.

b. Pyruvate kinase deficient hemolytic anemia in Mifflin County (Pa.) Amish.

c. Hemophilia B (Christmas disease), an X-linked recessive, in Holmes County (Ohio) Amish.

d. Limb-girdle muscular dystrophy in Amish of Adams and Allen Counties, Ind.

The Ellis-van Creveld syndrome (McKusick, Egeland, Eldridge, and Krusen, 1964b) is characterized by disproportionate dwarfism with relatively greater shortening in the distal part of the extremities, polydactyly, dysplasia of the fingernails and, in about half of cases, congenital malformation of the heart, most often single atrium. (See Fig. 4.)

Among the Amish, EvC is limited to those living in eastern Pennsylvania. Forty-three definite cases distributed in 26 sibships have been identified (Fig. 5). In all sibships except one, both parents are normal. Segregation analysis by the Lenz-Hogben method (Table 2) shows close agreement between the number of affected persons observed and the number expected on the basis of the autosomal recessive hypothesis.

Since EvC is ordinarily very rare, its natural history is known only on the basis of isolated cases and occasional families with more than one affected sib. Less than 50 cases have been reported (Ellis, Andrews, 1962). Basing descriptions of the natural history of rare recessive disorders on collections of individual case reports compounds the well recognized biases of hospital statistics. A useful aspect of the study of inbred groups is the opportunity to collect unusually large and homogeneous series of rare recessive disorders. From this study, it is clear that EvC is a cause of stillbirth (in at least 3 cases), and of neonatal and infant deaths. Of the 43

TABLE 2. THE ELLIS-VAN CREVELD SYNDROME

Sibship	Number of sibships	Number affected		Variance
		observed	expected	
2	2	2	2.2856	0.244
3	3	6	3.8919	0.789
4	6	7	8.7768	2.520
6	4	5	7.2992	3.104
7	2	6	4.0392	1.840
8	2	4	4.4450	2.344
9	1	3	2.4328	1.380
11	1	2	2.8710	1.805
12	1	2	3.0980	2.020
14	1	1	3.5630	2.446
TOTALS	23	39	42.7025	18.692
			SD $= \pm 4.323$	

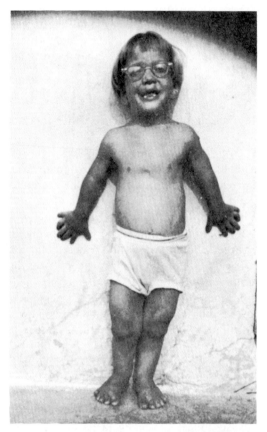

4A

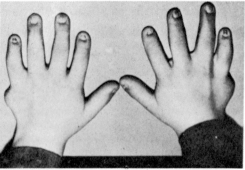

4B

FIGURE 4. The Ellis-van Creveld syndrome. A. 5-year-old
Amish boy showing short-limbed dwarfism (height 35 in.)
and polydactyly. B. Hands of 11-year-old Amish girl. The
extra fingers were amputated in infancy, by the crude
ligature method. Note the dysplastic fingernails.

THE ELLIS-VAN CREVELD SYNDROME

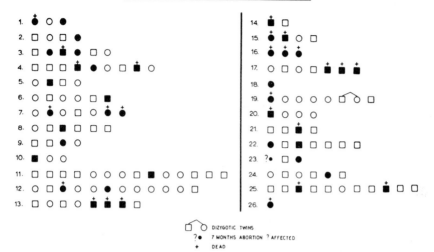

FIGURE 5. Twenty-six Amish sibships with at least one case of the Ellis-van Creveld syndrome. The father of sibship 22 is affected; all other sibships have unaffected parents. (Since submission of this paper, the number of sibships has been increased to 30 and the number of affected persons to 52. The information on sibship 24 has, furthermore, been revised. See McKusick et al., 1964b.)

cases, 16 died under 2 weeks of age and 23 died under 6 months of age. On the other hand, 8 of 17 living cases are adults. One adult is 58 years old and is married to a cousin who is presumed to be heterozygous for the EvC gene because they have two affected children (Fig. 6), one of whom is married to an unrelated person.

The EvC gene, ordinarily very rare, may be as frequent as 0.07 in the Lancaster Co. Amish.

Is there one couple who is ancestral to all 52 parents of the 26 affected sibships? Among the ancestors three names (Fig. 6) appear repeatedly: Samuel King (who immigrated in 1744), Nikolas Stoltzfus (who immigrated in 1767), and Christian Fisher (who was born in 1757). When all the sibships are traced back, it turns out that only Samuel King and his wife are ancestral to all parents. It is presumed, therefore, that one or the other carried the EvC gene. Samuel King left a large number of descendants. The unprecedentedly high frequency of the EvC gene in the Lancaster County Amish is an example of "founder effect," a form of genetic drift.

Pyruvate kinase deficient hemolytic anemia. Bowman and Procopio (1963) reported a form of hemolytic anemia due to deficiency of red cell pyruvate kinase (PK). The patients were from Old Order Amish families of Mifflin County, Pa. Although PK deficiency was first described by

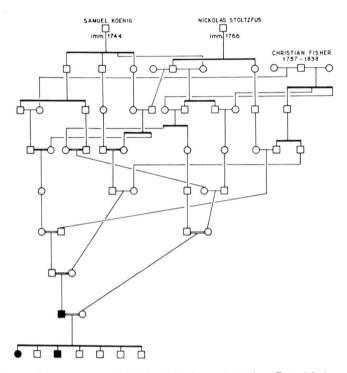

FIGURE 6. The ancestry of sibship 22 to demonstrate the affected father
(age 58 years) and the descent from the three ancestors who appear
most frequently in the pedigrees of Ellis-van Creveld cases.

Tanaka, Valentine and Miwa (1962), the disease in the Amish may be
distinctive from that observed by these workers and by others (Oski and
Diamond, 1963). In the Amish cases the anemia is much more profound,
leading to death in the first years of life if not treated, and splenectomy
has dramatic benefit, converting the fatal disorder into a compensated
anemia consistent with good health.

PK deficiency has not been found in Amish other than those living in
Mifflin County, Pennsylvania (Bowman and McKusick, 1965). All cases
can be traced to "Strong Jacob" Yoder, who immigrated in 1742 (Fig. 7).
In Mifflin County (Fig. 8), the Amish settlement, which started in 1792,
occupies a valley hemmed by mountains which have probably been a
factor in the isolation.

Christmas disease. Hemophilia B has been found in relatively high frequency in the Amish of Ohio, as indicated by the work of Wright, Doan,
Dodd, and Thomas (1948), Ratnoff (1958), and Wall and Doan (1963, personal communication), and corroborated by this survey. The cases may all
be descendants of two carrier sisters, Anna and Gertrude Hershberger,
who were born in the 1820's (Wall and Doan, 1963). There is no known

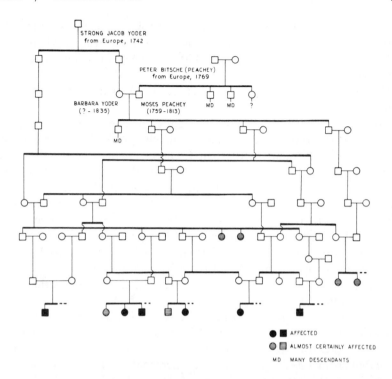

FIGURE 7. Pedigree of cases of pyruvate kinase deficiency hemolytic anemia to show descent of all cases from "Strong Jacob" Yoder. (New cases which also trace back to Jacob Yoder have been discovered since construction of the pedigree.)

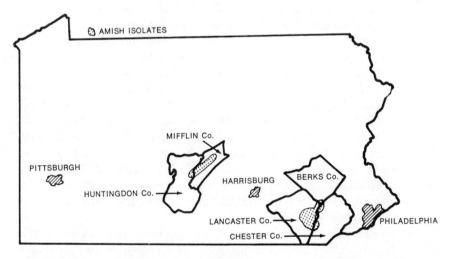

FIGURE 8. Location of Mifflin County Amish settlement where pyruvate kinase deficiency is found.

genealogic connection (Moor-Jankowski, 1964, personal communication) with the inhabitants of the Tenna district of Switzerland, where hemoophilia B has been present in rather high frequency for a long time (Moor-Jankowski, Huser, Rosin, Trvog, Schneeberger, and Geiger, 1958).

Hemophilia has not been identified in the Amish of Pennsylvania or Ontario.

Limb-girdle muscular dystrophy. Jackson and Carey (1961) have pointed out the rather high frequency of limb-girdle muscular dystrophy in the Amish of Adams and Allen Counties, Indiana. Cases have been found in the Amish of other areas but none in the Amish of eastern Pennsylvania. More than one immigrant probably carried the gene. Although a large number of the cases in Adams and Allen Counties and elsewhere can be traced to Johannes and Anna Schwartz, who immigrated in 1852, several others cannot be traced to them. Whereas the X-linked Duchenne type is the most frequent form of progressive muscular dystrophy in most parts of the world, the autosomal recessive limb-girdle type predominates in the Canton of Berne (Rossi, 1964 personal communication), from which the Amish originated.

Other genetic disorders. Other recessives identified in the Amish include the Swiss type of agammaglobulinemia, ataxia-telangiectasia, phenylketonuria (Martin, Davis, Askew, 1963), congenital deafness, plasma thromboplastin antecedent (Factor XI) deficiency, osteogenesis imperfecta congenita, metachromatic leukoencephalopaphy and a renal tubular defect of an atypical Fanconi type. Information on the Amish-wide distribution of these is, thus far, incomplete.

A few simple dominant disorders have been traced in Amish kindreds. Two dominant mutations have been demonstrated. One is the condition called symphalangism (Fig. 9) by neurosurgeon Harvey Cushing (1916), writing in the first issue of *genetics* (which also contained Bridges' pioneer paper on non-disjunction). Cushing's family was of Scottish descent. Symphalangism is of further genetico-historic interest, having been traced through fourteen generations of an English noble family (Drinkwater, 1917). Although the condition was first described in a French family (Mercier, 1838) and has been observed in Chinese (Hall, 1928), Japanese (Sugiura and Inagaki, 1960), Amerindians (Freud and Slobody, 1943), Peruvians (Velez, 1961) and Germans (Rochlin, 1928), few well-documented instances of new mutations have been reported. The low illegitimacy rate, the large families available for study, and the closed nature of the population permit greater certainty about new mutation in the Amish than is usually possible.

The second example of a new dominant mutation is spastic paraplegia (Fig. 10). Autosomal dominant and autosomal recessive types of spastic paraplegia are known (Bell and Carmichael, 1939) and at least one kindred with an X-linked recessive form has been reported (Johnston and Mc-

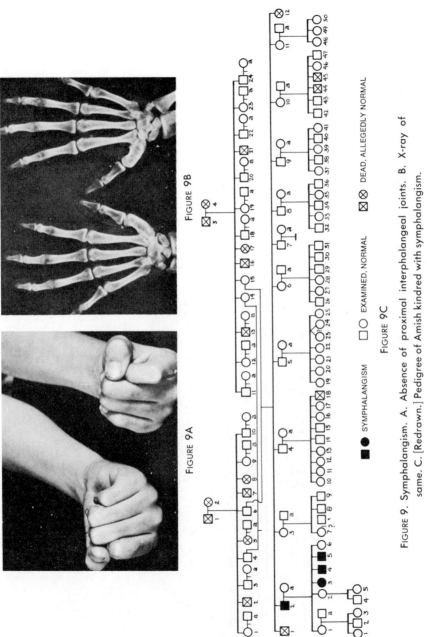

FIGURE 9A

FIGURE 9B

FIGURE 9C

FIGURE 9. Symphalangism. A. Absence of proximal interphalangeal joints. B. X-ray of same. C. [Redrawn.] Pedigree of Amish kindred with symphalangism.

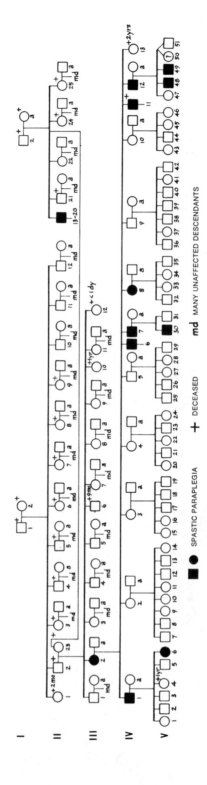

FIGURE 10. Pedigree of Amish kindred with new mutation for spastic paraplegia. (The symbol numbered II, 13–20 represents 8 unaffected brothers and sisters with progeny.)

■ SPASTIC PARAPLEGIA

+ DECEASED

md MANY UNAFFECTED DESCENDANTS

Kusick, 1962). Few families with dominantly inherited, early-onset, non-progressive spastic paraplegia of the type observed in this family have, however, been reported (Bell and Carmichael, 1939; Bayley, 1897; Spiller, 1902). In almost none of these, furthermore, has the original mutation been identified.

The Usefulness of Inbred Populations for the Detection of "New" Recessive Disorders

In the investigation of "achondroplastic dwarfs," who were said to be unusually numerous among the Amish, a newly recognized, recessively inherited disorder has been identified. Tentatively we are calling it *cartilage-hair hypoplasia* (CHH) (McKusick, Eldridge, Hostetler, Egeland and Ruangwit, 1964C). Clinically, short-limbed dwarfism and sparse, fine hair are the predominant features (Fig. 11). X-ray changes before closure of the epiphyses are characteristic (Fig. 12). Biopsy at the growth line shows cartilage hypoplasia; microscopically the hair has an abnormally small caliber (Fig. 13).

Forty-eight sibships containing 70 affected persons have been identified (Fig. 14). Five affected men married to normal women have 15 normal children. An affected man and woman have three children, all affected. Segregation analysis shows a deficiency of affected as compared with the number according to the recessive hypothesis. The preferred explanation is reduced penetrance, i.e., some homozygotes may be so mildly affected that they are not considered dwarfs. A disturbed sex ratio (41 F:29 M) suggests that sex influences penetrance. The penetrance is probably about 50%, when dwarfism, as interpreted by the Amish, is taken as the phenotype. The frequency of the CHH gene is probably about 0.05 when all Amish are considered. Although the disorder is widespread in the Amish, being found in all major communities, some lack of homogeneity is evident: Holmes and Wayne Counties, Ohio, have the largest number of cases. Adams and Allen Counties, Indiana, have none. None was found in the original Ontario settlement, although one case was found in a child of a family which migrated to Ontario from Holmes County in 1961. Genealogic data show that the largest proportion of the cases trace their ancestry to one John Miller, who was wounded by the Indians in 1757 in an encounter in which others of the noncombatant Amish were killed. From the initial settlement in Berks County, Pennsylvania, descendants of John Miller spread to other areas of Pennsylvania and to Ohio, Indiana, Illinois, Iowa, Kansas, Oregon and Ontario, where Amish cases of CHH are now found. However, since some of the 100 unaffected but presumably heterozygous parents of CHH cases cannot be traced to John Miller, several immigrants probably carried the gene.

The wide geographic distribution of the CHH gene differs, then, from the distribution of the other recessive genes which have been studied. It is noteworthy that the surname of Samuel King, who immigrated in 1744 and

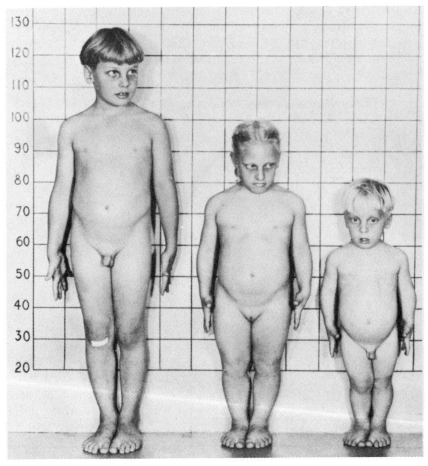

FIGURE 11. Cartilage-hair hypoplasia. Three Amish sibs (left to right): Normal boy, Marlin, age 7 years; Orpha, affected, age 9 1/2 years; Floyd, affected, age 5 1/2 years. The head hair in the dwarfed children is sparse, fine and light-colored. Eyebrows and eyelashes are similarly affected.

probably carried the gene for the Ellis-van Creveld syndrome, is largely limited to Lancaster County, Pa., whereas the name of Miller is much more generally distributed among the Amish.

This entity, cartilage-hair hypoplasia, illustrates the usefulness of inbred populations for the identification and study of "new" genetic disorders, especially those which are recessively inherited. Since attention was called to CHH in preliminary reports in *Lancet* (McKusick, 1964a) and in *Presse Médicale* (McKusick, 1964b), cases of the syndrome have been reported to us: affected brothers in a Finnish family and a single case in an Italian family (L. O. Langer, Jr. 1964, personal communication); three affected sibs in a French family (Maroteaux, Savart, Lefebvre and Royer,

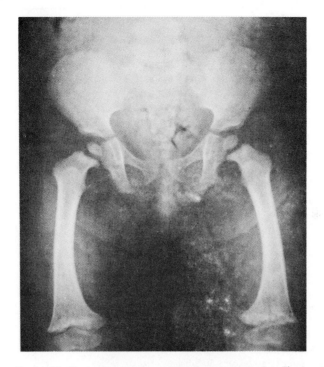

FIGURE 12. X-ray changes in cartilage-hair hypoplasia. The irregularity and other changes at the lower end of the shaft of the femur are typical.

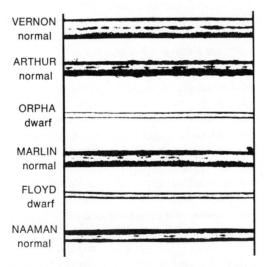

FIGURE 13. C. Microscopic appearance of hair in six sibs of ages varying from 15 (top) to 3 (bottom) years. (The third, fourth, and fifth come from the three sibs shown in Fig. 11.) The hair of the two dwarfed sibs is of abnormally small caliber and lacks pigment.

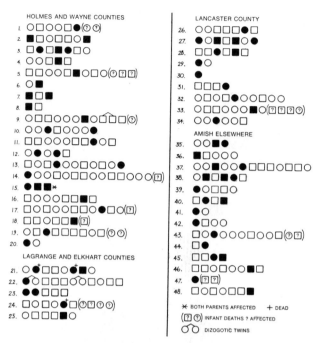

FIGURE 14. Amish sibships with one or more cases of CHH. All parents
are unaffected except in sibships 15, of which both parents are
affected.

1963; Maroteaux, personal communication 1964). Previously the cases were
too few to impress persons with the fact that hair and cartilage changes
are associated in a hereditary syndrome.

A Catalog of Rare Recessive Phenotypes in Man

There are, of course, other reasons than merely the nosology of genetic
disease for identifying recessive diseases in inbred populations. As an aid
to such identification, we have assembled a catalog of rare recessive
phenotypes in man. Confining consideration to *rare* phenotypes relieves us
of the knotty question of the genetics of conditions such as diabetes mel-
litus for which a recessive hypothesis has been advanced in the past. Speaking of *recessive* phenotypes and carefully defining the phenotype help
avoid the embarrassing situations with those conditions in which some
expression is certain or likely in the heterozygote. Maintaining two
classes of disorders—those in which recessive inheritance is quite certain,
and those in which it is possible or likely, but not yet proven—permits us
to enjoy the heuristic values of a complete enumeration, yet avoid the intellectual "sloppiness" of including conditions for which evidence of recessive inheritance is not complete. The catalog has been "computerized"
for ease of corrections, additions, deletions, translocations, and indexing.
Over 200 rare phenotypes in man can, in our judgement, be considered

recessive, but there must be many more. Studies in inbred populations can uncover as yet undescribed additions for the catalog.

The Usefulness of Inbred Populations for Identifying Recessive Inheritance of Rare Conditions Not Previously Considered Genetic

Related to the use of inbred groups for the identification of "new" diseases is their use to determine the genetics of a condition not previously considered to be simply inherited. This aspect can be illustrated, in the Amish, with a congenital malformation called *hydrometrocolpos* (McKusick et al., 1964a). Through a combination of transverse vaginal septum and excessive cervical secretions in response to maternal hormone, the newborn female infant has an abdominal tumor which consists of greatly dilated vagina and uterus (Fig. 15A). Two sibships, each containing two well-studied cases, have been identified. In each case the parents are consanguineous and all 4 parents can be traced to one ancestral pair (Fig. 15B). The findings suggest that at least one form of this very rare malformation is recessively inherited. It is, of course, female-limited. At least three useful features of Amish society are illustrated by this experience: The Amish use the diagnostic facilities of neighboring medical centers. Genealogic information is extensive and accurate. Amish are well informed about other Amish in all parts of the country; one affected family lived in Lancaster Co. (Pa.) and the other in Geauga Co. (Ohio); the first learned about the second following a detailed account of the ailment in *The Budget*.

Evidence for the autosomal recessive inheritance of endocardial fibroelastosis (Newton, 1963, personal communication), one form of pituitary dwarfism (unpublished observations) and one form of hydrocephalus (unpublished observations) is provided by Amish families ascertained in this study.

Pitfalls, of course, await him who uses this approach indiscriminately. Generally frequent conditions, e.g., mongolism, might be considered recessive on the basis of pedigree findings in such inbred groups.

Dwarfism and a Model for Quantitative Characters in Inbred Populations

In other connections reference was made earlier to two forms of dwarfism which are relatively frequent in the Amish. Some of the results from a systematic study of dwarfism in the Amish throughout the country were presented. In principle the methods of the study were simple. They consisted of (1) identifying all Amish dwarfs, (2) studying each case to categorize it, and (3) studying the genetics of each category. The two most frequent forms of high grade dwarfism in the Amish have already been described: the Ellis-van Creveld syndrome and cartilage-hair hypoplasia. In these the inheritance is autosomal recessive. In the ascertainment of cases, which was done mainly through physicians practicing in Amish areas

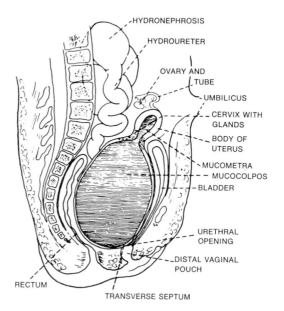

FIGURE 15A

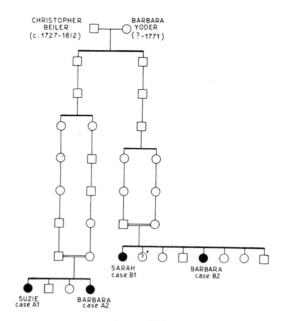

FIGURE 15B

FIGURE 15. Hydrometrocolpos. A. Artist's conception of the
malformation. B. Pedigree.

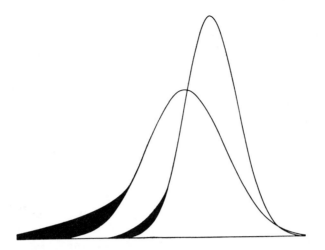

FIGURE 16. A model for a multifactorial trait such as stature. The distribution curve to the right is intended to represent a sizeable randomly mating population. The blacked-in area on the lower tail represents single gene entities and chromosomal aberrations. The curve to the left is intended to represent the distribution for an inbred population. The blacked-in part of the lower tail is larger because of the increased number of cases due to homozygosity for rare recessive genes. See text for a discussion of the justification for the lower mean and wider variance in the inbred population.

or through Amish acquaintances in various communities, cases of chromosomal aberration, i.e., cases of the Turner syndrome, cases of mongolism, were presented to us as dwarfs. Some cases were brought to our attention which, on study, we concluded are polygenic dwarfs. They have (1) normal proportions, (2) by X ray no abnormality except small size, (3) no associated non-skeletal anomaly, (4) no impairment of general health, and (5) relatives who are short with, however, a grading of stature and no distinct pedigree pattern.

The model schematized in Fig. 16 is suggested to account for these findings. It is applicable to other quantitative characters, such as intelligence. Indeed, the model is precisely that proposed by Fraser Roberts (1952) to explain the genetics of mental deficiency. Stature in "normals" (the right-hand curve) is distributed more or less symmetrically around a mean value with a resulting "bell-shaped" curve. The position of a given individual on the curve is multifactorially determined, i.e., multiple factors both genetic and environmental are involved, and the genetic factors are themselves multiple, or polygenic. In the lower "tail" of the "normal" curve a certain number of dwarfs will be found; the number depends on the arbitrary definition of dwarfism chosen for the study. Also in the lower tail there will be instances of dwarfism due to single major genes (such as the Ellis-van Creveld syndrome) and to chromosomal aberrations (such as the Turner syndrome). These are represented by the blacked-in part (Fig. 16).

In an inbred population the distribution may show changes with regard to the mean, the variance, and the size of the accretion on the lower tail. In the Amish, mean stature may be lower than in other Americans of western European origin. Preliminary data on Amish and non-Amish school children in Lancaster County suggest that such is the case. Some evidence of an effect of inbreeding to reduce stature is provided by the study of Mange (1938) in the Hutterites, and in Switzerland Hulse (1957) found increased stature with outbreeding.

As Newcombe (1964) pointed out, one expects in an inbred population a greater variance in a polygenically determined character. For stature one would expect an increased proportion of short persons and of tall persons. Tall Amish persons may be less conspicuous in comparison with an outbred population in which the mean height is higher. The proportion of Amish persons who have polygenic dwarfism is probably increased.

When an inbred population of homogeneous origin is compared with an outbred cosmopolitan population, the greater variance of the inbred population may be cancelled out by the heterogeneity of origin of the outbred group. Such seems to have been found by Neel and colleagues (1964) in a study in which an inbred group of Brazilian Indians was compared with a sample of the population of Hamburg, Germany.

Summary

The distribution of five recessive disorders in the Old Order Amish has been described and explained on the basis of migration patterns, founder effect, and isolation. These conditions are Ellis-van Creveld syndrome, pyruvate kinase deficiency, hemophilia B, limb-girdle muscular dystrophy, and cartilage-hair hypoplasia. Two dominant mutations—one for symphalangism and one for spastic paraplegia—have been identified.

Effectively isolated from the general population by religion, the Amish exist in several subgroups, each of which has a distinct history. The subgroups are semi-isolated from each other by distance and, to some extent, by social barriers arising from their different histories.

The usefulness of inbred populations for detecting "new" recessively inherited disorders is illustrated by cartilage-hair hypoplasia. The usefulness of inbred populations of demonstrating inheritance of rare malformations not considered genetic is illustrated by hydrometrocolpos. We have assembled a catalog of rare recessive phenotypes in man for use in the study of inbred populations and the nosology of genetic disease. The usefulness of inbred populations for assembling large homogeneous series of ordinarily rare recessive conditions whose natural history is not yet fully known is illustrated by the Ellis-van Creveld syndrome; almost as many cases have been identified in the Lancaster County Amish as are reported in the entire medical literature.

A study of dwarfism in the Amish suggests a model for the frequency distribution of quantitative traits in inbred populations.

ACKNOWLEDGMENTS

The work reported here was supported in part by Grant GM10189 from the National Institutes of Health and in part by a grant from the Kennedy Foundation.

REFERENCES

BAYLEY, W. D. 1897. Hereditary spastic paraplegia. J. Nerv. and Mental Dis. *26:* 697–701.

BELL, J., and E. A. CARMICHAEL. 1939. On hereditary ataxia and spastic paraplegia. Treasury of Human Inheritance *4* (Part III): 141–281.

BENDER, H. S., and C. H. SMITH, ed. 1955. The Mennonite Encyclopedia. Mennonite Publishing House, 4v. Scottdale, Pa.

BOWMAN, H. S., and V. A. McKUSICK, 1965. Pyruvate kinase deficient hemolytic anemia in an Amish isolate. Am. J. Human Genet., in press.

BOWMAN, H. S., and F. PROCOPIO. 1963. Hereditary non-spherocytic hemolytic anemia of the pyruvate-kinase deficient type. Ann. Int. Med. *58:* 567.

CUSHING, H. 1916. Hereditary anchylosis of the proximal interphalangeal joints (symphalangism). Genetics *1:* 90–106.

DRINKWATER, H. 1917. Phalangeal anarthrosis (synostosis, ankylosis) transmitted through fourteen generations. Proc. Roy. Soc. Med. *10:* 60–68.

ELLIS, R. W. B., and J. D. ANDREWS. 1962. Chondroectodermal dysplasia. J. Bone and Joint Surg. *44B:* 626–636.

FREUD, P., and L. SLOBODY. 1943. Symphalangism, a familial malformation. Am. J. Dis. Child. *65:* 550–557.

HALL, G. A. M. 1938. Hereditary brachydactylism and interphalangeal ankylosis. Ann. Eugenics *3:* 265–268.

HOSTETLER, J. A. 1963. Amish Society. Johns Hopkins Press, Baltimore, Md.

HULSE, F. S. 1957. Exogamie et heterosis. Arch. Suisses Anthropol. Gen. *22:* 103–125.

JACKSON, C. E., and J. N. CAREY. 1961. Progressive muscular dystrophy: autosomal recessive type. Pediatrics *28:* 77–84.

JOHNSTON, A. W., and V. A. McKUSICK. 1962. A sex-linked recessive form of spastic paraplegia. Am. J. Human Genet. *14:* 83–94.

MANGE, A. P. 1963. The Population Structure of a Human Isolate. Ph.D Thesis. Univ. of Wisconsin.

MAROTEAUX, P., P. SAVART, J. LEFEBVRE, and P. ROYER, 1963. Les formes partielles de la dysostose métaphysaire. Presse méd. *71:* 1523–1526.

MARTIN, P. H., L. DAVIS, and D. ASKEW. 1963. High incidence of phenyl-ketonuria in an isolated Indiana community. J. Indiana Med. Assoc. *56:* 997–999.

McKusick, V. A. 1964a. Dysotose métaphysaire et modifications des cheveux. Presse Méd. *72:* 907–908.

———— . 1964b. Metaphyseal dysostosis and thin hair: a "new" recessively inherited syndrome? (Letter) Lancet *i;* 832–833.

McKusick, V. A., R. L. Bauer, C. E. Koop, and R. B. Scott. 1964a. Hydrometrocolpos as a simply inherited malformation. J. Am. Med. Assoc. *189:* 813–816.

McKusick, V. A., J. A. Egeland, R. Eldridge, and D. E. Krusen. 1964b. Dwarfism in the Amish. I. The Ellis-van Creveld syndrome. Bull. Johns Hopkins Hosp., *115:* 306–336.

McKusick, V. A., R. Eldridge, J. A. Hostetler, J. A. Egeland, and U. Ruangwit. 1964c. Dwarfism in the Amish. II. Cartilage-hair hypoplasia. Bull. Johns Hopkins Hosp., in press.

McKusick, V. A., J. A. Hostetler, and J. A. Egeland. 1964d. Genetic studies of the Amish: background and potentialities. Bull. Johns Hopkins Hosp., *115:* 203–222.

Mercier, L.-A. 1838. Absence héréditaire d'une phalange aux doigts et aux orteils. Bull. Soc. Anat. Paris *13:* 35–36.

Moor-Jankowski, J. K., H. J. Huser, S. Rosin, G. Trvog, M. Schneeberger, and M. Geiger. 1958. Hemophilia B: Genetics, Hematology and Clinical Aspects. S. Karger, Basel.

Neel, J. V., F. M. Salzano, P. C. Junqueira, F. Keiter, and D. Maybury-Lewis. 1964. Studies on the Xavante Indians of the Brazilian Mato Grosso. Am. J. Human Genet. *16:* 52–140.

Newcombe, H. B. 1964. *In* Proc. 2nd Int. Conf. on Congenital Malformations N.Y., 1963. International Medical Congress. New York.

Oski, F. A., and L. K. Diamond. 1963. Erythrocyte pyruvate kinase deficiency resulting in congenital nonspherocytic hemolytic anemia. New England J. Med. *269:* 763–770.

Ratnoff, O. D. 1958. Hereditary defects in clotting mechanisms. Adv. Int. Med. *9:* 107–179.

Roberts, J. A. Fraser. 1952. The genetics of mental deficiency. Eugenics Rev. *44:* 71–83.

Rochlin, D. G. 1928. Uber die hereditare symmetrische Gelenkhypoplasie. Ztschr. f. Konst.-Lehre *13:* 654–663.

Spiller, W. G. 1902. Fourteen cases of spastic spinal paralysis occurring in one family. Phila. Med. J. *9:* 1129–1131.

Sugiura, Y., and Y. Inagaki. 1960. Symphalangism with carpal and tarsal bone fusions. Jap. J. Human Genet. *5:* 117.

Tanaka, K. R., W. N. Valentine, and S. Miwa. 1962. Pyruvate kinase (PK) deficiency hereditary non-spherocytic hemolytic anemia. Blood *19:* 267–295.

Vélez, C. 1961. p. 474. *In* McKusick, V. A. et al., Medical genetics, 1958–1960, C. V. Mosby Co., St. Louis.

Wright, C.-S., C. A. Doan, V. A. Dodd, and J. D. Thomas. 1948. Hemophilia; current theories and successful management in traumatic and surgical crises. J. Lab. and Clin. Med. *33:* 708–720.

The Development of Inbreeding in an Island Population

DEREK F. ROBERTS

Rather than add yet one more to the steadily mounting number of theoretical discussions of the quantification of breeding structure, it seems more useful for the purposes of this meeting to describe from human studies an actual example of the development of inbreeding in an isolated population. This it is hoped will illustrate some of the points and problems that may not be immediately apparent in an abstract discussion; it will once again show the important contribution that human genetics can make to the subject as a whole, for this study could not be matched in any free-living population of any other species; and it will show a practical application of population genetics to suggest a resolution to a human problem.

The population of Tristan da Cunha is practically unique in that one is able to trace its development from its very foundation. Thanks to the detailed records that were kept by the earliest settlers, the genealogical information collected by a variety of observers and in particular Dr. Woolley, a former Governor, the writings of the several missionaries on the island and of casual visitors, the official reports of captains of ships visiting the island, the records of births, deaths, baptisms and marriages that at periods were kept, and particularly the painstaking investigations of the social anthropologists Dr. Loudon and Dr. Munch, it is possible to trace with remarkable precision the development of this population. One can form a clear picture of the reproductive histories of the inhabitants, of the arrivals and departures, and so arrive at an almost year by year demographic recital. The present article concerns the way in which the isolation of the population led to the development of a degree of inbreeding.

This study was undertaken as part of an intensive investigation under the auspices of the Medical Research Council on the arrival of almost the whole population in the United Kingdom in 1961, when the facilities and services of modern medicine were made available to them. For whereas, from reports of the occasional visits by ships' surgeons and from the full-scale Norwegian expedition to the island in 1937/38 (Christopherson 1946), the impression was current that the islanders were remarkably healthy, full clinical examination failed to substantiate this impression (Lewis, 1963, Black et al 1963). A particularly striking feature of their

Reproduced by permission of the publisher and Derek F. Roberts from Ciência e Cultura, *19:78–84 (1967).*

health pattern was the high incidence of congenital abnormality and diseases with some genetic component, and the first interpretation of this was to attribute it to the inbreeding that must have occurred. It was obviously necessary to determine how extensive is the inbreeding, before one could attribute to it responsibility for the high incidence of these congenital disorders, and this was the object of the present study.

The History of the Present Population

With the exile of Napoleon on the remote island of St. Helena, the powers of Europe hoped there would be an end to the turmoil that had for so long convulsed the continent. To prevent any attempts at his rescue, a British military garrison from the Cape occupied the neighbouring though distant island of Tristan da Cunha in 1816, but from the Admiralty there came the opinion that Tristan was of no importance to Napoleon's security and so the garrison was withdrawn a few months later in 1817. A Scots corporal, William Glass, asked permission to remain with his wife, infant son and newborn daughter. Other men stayed with them, and they were joined by various others, who formed a firm to exploit the resources of the island. Most of the others who joined them stayed for only short periods, but two English seamen stayed permanently, Alexander Cotton, a naval veteran who arrived in 1821, and Thomas Swain, another naval man, in 1826. To balance the sex ratio some ladies arrived from St. Helena. A Dutch sailor, Peter Green, was shipwrecked and joined the community in 1836; an American sailor, Thomas Rogers, came in 1836 but departed after two years, having married one of Glass's daughters and left a son. Then there was Andrew Hagan, the skipper of an American whaler who after an unsuccessful antarctic trip decided to leave his ship and remain on Tristan. Two Italians were shipwrecked in 1892, another St. Helena woman came in 1863 and two white women in 1908. According to the literature, these 15 individuals are the ancestors of the present population of nearly 270, and these 15 included two pairs of sisters and one mother-daughter pair. The genes with which they endowed the population thus derived from a variety of sources—north-west continental Europe, the British Isles, the Mediterranean and Africa. There were several other men and women who came for a time, and then departed without leaving any descendants among the present population. Unfortunately, the story is not quite so simple. Although these 15 individuals provided the majority of the genes in the gene-pool of the present population, there were also a small number of "invisible" ancestors, for there were occasional illegitimate births, though many of these were due to antenuptial conceptions and were subsequently legitimised. Thanks to the completeness of the information now available, it is possible to identify these.

Living was by no means easy for this little community. The island consists almost entirely of the inhospitable volcanic slopes and cliffs, and though its total area is about 38 square miles, only three square miles are sufficiently level for cultivation. During the days of sail Tristan, in the middle of the Atlantic, 1800 miles from Cape Town, 2,000 from South America, was a valuable watering place for passenger ships, cargo vessels and whalers, and the inhabitants were able to carry on a trade of potatoes and fresh meat in exchange for flour, sugar, wood, paint, hardware, etc. But, with the coming of steam, boats no longer called; the islanders were just able to support themselves by their vegetable plots, their animals, and their fishing, but the standard of life seems to have slowly declined. A number of the community emigrated; either as individuals seeking a living elsewhere perhaps dissatisfied with the difficulties of existence, or as a group, as for example after a particularly severe tragedy in 1885 when 15 adult males were drowned as a result of an accident at sea, and the departure of their widows and children in the years that followed reduced the population to 50, whereas in 1884 it had been 106. With increasing isolation the islanders depended increasingly on charity supplied through missionaries. On three occasions (1904, 1907, 1932) evacuation to the Cape was officially suggested, but the islanders always refused.

A new era dawned for them in 1940 when a British service garrison was stationed there, for the islanders became acquainted with money, employment and the goods money could buy: the garrison was withdrawn after the war but then the Tristan da Cunha Development Company was established to fish for crayfish, freeze them in a factory on the island, then market them via Cape Town, thus providing contact with the outside world through the company's two fishing boats. Success came to the enterprise, enabling funds to be provided for the installation in 1960 of running water in each house, modern sanitation, and the prospect that once again the island could become selfsupporting.

But then on 10th October, 1961, with very little warning, the volcano erupted just behind the settlement, and soon the lava destroyed the factory and landing beach. But the folk had left the settlement almost immediately and they took temporary refuge on nearby Nightingale Island. The whole population of 267 was subsequently transported to Cape Town and then to England, where they remained until the island was reported safe for their return.

The Calculation of Inbreeding

From the sources mentioned at the outset, it is possible to draw up a complete, albeit highly complex, pedigree of the whole population since its founding. In those cases where a birth was the outcome of an irregular union, it has been possible, thanks to the completeness of the information now available, to identify in nearly all cases, and with a high degree of probability in many, the actual fathers. These identifications in recent generations have been checked against the remarkably full information on blood groups and other genetic markers in the Tristan population now available, as a result of the investigations of Dr. Mourant, Dr. Harris and others.

First it is necessary to distinguish between two uses of the term "inbreeding." There is the loose social use, equatable with endogamy, mating within a community. Secondly, there is the biological term, referring to mating between two individuals who have derived some of their genes from the same common ancestor so that their offspring can be homozygous for identical genes derived from one he or she possessed. The genetic consequences of endogamy and biological inbreeding may be quite different and a small endogamous population is not necessarily biologically highly inbred. Certainly the former applied to the Tristan population, but it remained to be seen how extensive was the latter.

Biological inbreeding can be quantified in terms of the inbreeding coefficient, developed by Sewall Wright, which measures the probability that both the genes an individual possesses at a given autosomal locus are replicates of a single gene that was present in the common ancestor, one replicate descending to him through his mother, the other replicate through his father's line. Thus, in the case of the offspring of a first cousin marriage, the probability that an ancestor transmitted a given gene to his son is $\frac{1}{2}$; the probability that the ancestor transmitted a given gene to his son's son is $\frac{1}{2} \times \frac{1}{2}$ is $\frac{1}{4}$, and to his great-grandson is $\frac{1}{8}$; but there are similar probabilities that the same gene was passed from the ancestor via his daughter and her daughter to the great-grandson; so the probability that the great-grandson is homozygous for this identical gene is $\frac{1}{8} \times \frac{1}{8}$, equals $1/64$. But the ancestor could also have passed on the allele of the given gene, and again the probability that the great-grandson is homozygous for this is also $1/64$. Hence the probability that the great-grandson is an identical homozygote for either of the alleles of the ancestor is $1/64 + 1/64 = 1/32$. To calculate the inbreeding coefficient for an individual then, all that one has to do is to sum these probabilities for all the common ancestors. This has been done for all members of the Tristan population assuming that none of the founders were themselves inbred, and that the two pairs of sisters among them were full sibs i.e. the parents of the founder members are the base population which provides the point of reference for the calculated values of F. In the few cases where it has not been possible to assign paternity definitely to one of several men, in the present analysis I have taken the lowest of the several possible values of their inbreeding coefficients.

Results

1. The Present Inbreeding Level

The results of the analysis of inbreeding coefficients in the Tristan population of the island on the 10th October, 1961 was .0403 which is to say that the probability that the average individual possesses two identical genes at a given locus [is] approximately 4%. This figure is less than two-thirds that of the offspring of a marriage between two first cousins, and by comparison with what obtains in some domesticated animals is not at all high. It is moreover very little above the probability (.033) of identical homozygosity in any random mating population derived from fifteen founder ancestors; the exact figure taking into account differential fertility is in process of calculation. It is exceeded in other small human populations, for instance the Samaritans of Israel and Jordan (.0434 calculated over a comparable number of generations, Bonne 1963) and probably the Japanese of Hosojima, (Ishikuni et al 1960) for their observed value .0334, was calculated over fewer generations. By comparison with the mean inbreeding coefficients that are published for human populations, the Tristan value appears to be high, falling at the upper end of the range. But in this context it should be remembered that—

(a) in none of the many human populations with an inbreeding mating system has the inbreeding coefficient yet been calculated

(b) it is not necessarily meaningful to compare an inbreeding coefficient x_1 in one population with a coefficient x_2 in another population even though they are both calculated over the same number of generations for they have as their point of reference different base populations; thus it is highly unlikely that the Samaritan population of the early 19th century contained no genes identical by descent or that its inbreeding coefficient was 0, whereas these are reasonable assumptions for the forebears of the Tristan population.

On balance it seems that in the Tristan population inbreeding so far is not excessive and so there should be no undue increase in homozygosity. This is borne out by the evidence of those blood groups and other genetic markers in which the heterozygote can be distinguished, namely the MN and Rhesus systems (Table 3); neither of these shows any tendency to increase in homozygosity over that expected in a random panmictic situation, but rather the converse, and indeed the difficulty is to account for the very considerable excess of heterozygotes.

2. The increase in inbreeding

Secondly the value of the inbreeding coefficient appears to be steadily increasing with the passage of time. Table 1 shows the breakdown by ten year age groups within the present population (figure 1). It is of interest to put this age trend within the present population into the perspective of the development of inbreeding since the inception of the population. Table 2 shows the values of the mean inbreeding coefficient and the number of inbred individuals by decade of birth until the present day, for all founder members in the ancestry and all members of the population born on the island. It is only in those born since 1910 that the mean inbreeding

TABLE 1. — INBREEDING BY AGE

Age group	F	n
0 — 9	.04854	46
10 — 19	.03855	50
20 — 29	.04745	46
30 — 39	.04399	40
40 — 49	.03900	31
50 — 59	.02344	24
60 and over	.02797	34

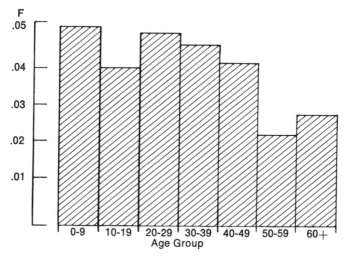

FIGURE 1. Inbreeding coefficients in the present population by age groups (1961). [Redrawn.]

TABLE 2 — INBREEDING BY DECADE OF BIRTH

Decade of Birth	Number of Individuals	Mean F.	N.° inbred.	Percentage Inbred
Before 1830	20	0	0	0
1830–39	24	0	0	0
1840–49	14	0	0	0
1850–59	19	.01316	5	26.3
1860–69	26	.01442	8	30.8
1870–79	39	.01402	11	28.2
1880–89	26	.02885	10	38.5
1890–99	20	.02110	10	50
1900–09	32	.03076	19	59.4
1910–19	28	.04011	21	75
1920–29	51	.04339	46	90.2
1930–39	51	.04952	51	100
1940–49	58	.03941	57	98.3
1950–59	48	.04863	47	97.9

TABLE 3 — FREQUENCY OF BLOOD GROUP PHENOTYPES LOCI AT WHICH THE
HETEROZYGOTE IS DISTINGUISHABLE

	N.⁹ Observed	Number Expected	
		random mating*	present degree of inbreeding
MM	47	54.79	57.14
MN	141	125.42	120.72
NN	64	71.79	74.14
CC	17	20.00	21.92
Cc	108	101.99	98.17
cc	127	130.01	131.91
EE	35	39.29	41.54
Ee	129	120.43	115.91
ee	88	92.28	94.54

*Not allowing for the effect of small population size.

coefficient may exceed that expected under random mating (figure 2); the low values in the earlier members of the population reflect the outbreeding mating practices. The few instances of very close inbreeding on Tristan ($F \geq .25$) have little effect on the average inbreeding coefficient for the population. The general increase during the history of the population is of course to be expected, since in a closed population of limited size, the later the generation an individual belongs to the more common ancestors are available to him. A similar increase showing the same pattern of an absence of inbreeding in the earliest generations followed by a regular rise, was documented in the Ramah Navaho isolate in the American southwest (Spuhler and Kluckhohn 1953). If the Tristan population remains closed, then this increase can be expected to continue. This will produce ever increasing homozygosity leading first to an increase in incidence of

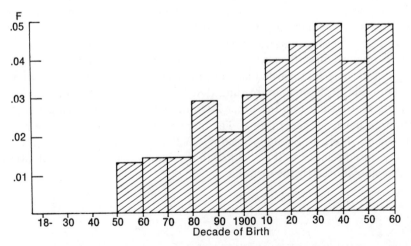

FIGURE 2. Mean inbreeding coefficients by decade of birth. [Redrawn.]

those defects which are manifestations of recessive genes and eventually to gene fixation.

This process, the loss of heterozygosis and increasing homozygosis in a finite group, is not due to the occasional very close inbreeding that occurs by chance in random mating for it continues steadily even with the maximum avoidance of inbreeding in each generation. It is due to two phenomena, one the closed nature of the population, the other its limited size. Its rate depends upon the mating pattern that is practised. It can, however, be avoided if the population is not totally closed, for an equilibrium can be reached and a steady degree of inbreeding maintained if there is some immigration into the population from outside at a regular rate; the rate required to maintain the balance depends upon the mating pattern practised. For a population of the size, structure and mating pattern of Tristan an immigration rate of about 0.1 is necessary to maintain the degree of inbreeding at is present level. There is a suggestion from figure 2 that the mean values of the inbreeding coefficient are attaining a plateau; whether they do so ultimately or not depends on the effect of the enforced recent migration on the marriage patterns of those who have returned to the island. The present plateau is partly due to incoming gene flow in illegitimate births by some visiting outsider, though these have not been sufficiently frequent to account totally for it; partly it is due to the rapid increase in population size and it may also reflect an effect of differential fertility.

Secular change in inbreeding may be further examined through analysis of the rates at which it increases per generation. This it is difficult to do in the present material on account of overlapping generations and the fact that the one individual may occur in two or three different generations according to which path of descent is traced. However, this difficulty can be partly overcome by tracing down the descent from each original ancestor separately and making the appropriate weightings for multiple appearances. The results are approximations but none the less appear to give some useful information. The inbreeding coefficient for a given generation may be regarded as comprising two parts, one the increment attributable to the new inbreeding that has occurred in the matings producing that generation, the other attributable to the inbreeding that has occurred in earlier generations. The increment, ΔF, is the rate of inbreeding, i.e. the increase of the inbreeding coefficient in one generation; it can be measured relative to the distance that remains before the population is completely inbred by the formula

$$\Delta F = \frac{F_t - F_{t-1}}{1 - F_{t-1}}$$

where F_t is the inbreeding coefficient in one generation and F_{t-1} that in the preceding generation. In the generations into which the descendants of Wm. Glass and Maria Leanders can be arranged, increments were zero until the third generation; from generations 3 to 7 the increments are positive; there was an especially large increment between the fourth and fifth generations which may reflect the population reduction consequent upon the boat disaster. Over all seven generations the mean increment is .0072 per generation and over the last four generations the mean is .0127. This rate of increase is about three times that observed in the Navaho both over the total duration of the Ramah isolate (.0026), and over the last four generations (.0045).

Discussion

In considering the development of inbreeding in this population, it is important to remember the nature and character of the early settlers. In the early days on this island there was none of the violence and lust that characterised for example the population of Pitcairn Island, founded by the Bounty mutineers. No-one who has worked through the early documents of the Tristan settlement can fail to be impressed by the business-like way in which affairs were handled. The original document setting up the co-partnership on the island in 1817 is a model of clarity of expression, properly and legally witnessed, and so is the 1821 revision of the constitution and direction of island affairs, both embodying clearly the concept of equality and co-operation. Returns of all stores and forage used were kept, and we know exactly what stores, provisions and effects were in the possession of the original founders, down to the detail of an old chest of drawers and one corn mill, out of repair. Glass was a God-fearing man who entered all details of family additions in the family bible; as a non-commissioned officer with long service in the British Army he was accustomed to discipline and the ordering of affairs, and so too were the naval men who elected to join him on completion of their service. They would receive no deserters or run-aways as settlers amongst themselves. They were very concerned with the education of their children. Glass sent his two eldest children to school in England, in 1830 a new arrival took charge of educating the children on the island, and upon his departure the three senior settlers advertised in a Cape paper for a teacher for the children, offering in return his keep and one-tenth of the proceeds of the sale of all their produce. They time and time again exposed themselves to considerable danger and inconvenience to rescue survivors from shipwrecks. These then were no undisciplined rabble of adventurers, but serious minded literate men who endeavoured to retain in their remoteness the values of their society, not least of which was observation of the prohibited degrees of marriage; it is only in extremely few instances that these have been transgressed in irregular matings. These do not occur at the frequencies expected under true random mating, so the situation in the early generations was in fact one of outbreeding. Marriage with relatives was imposed on them by the limited choice that their isolation and small numbers offered. Thus there was a conflict between the opposing tendencies on the one hand to outbreeding that their marriage system engendered, and on the other to inbreeding that the limited number or absence of non-relatives to be found in a small closed population imposed. The resultant of this conflict was the development of a slight degree of inbreeding.

The first inbred individual was not born until 40 years after the settlement had been founded, in the third generation. When his father, aged 22, wanted to marry in 1854, the only women of marriageable age, excluding his own sisters, were the five Cotton girls aged 24, 22, 21, 19 and 17, two Glass sisters aged 22 and 18, and one other girl about to leave the island; the Cotton girls were his first cousins and he chose the second oldest of them who was his own age. When the father of the second inbred individual wanted to marry in 1856, then aged 18, there were then nine unmarried women of marriageable age on the island, the same two Glass girls, aged 24 and 20, the three Cotton girls, now aged 23, 21 and 19, and the four Swain girls aged 21, 20, 18 and 17. The Glass girls were older than he, and he was thus left with the Cotton daughters (his first cousins one removed), and the Swain girls who were his mother's half-sisters. He chose the Cotton girl nearest to his own age. In these the earliest of the consanguineous unions there was some freedom of choice, but as Table 4 shows this rapidly disappeared; indeed throughout the marriages in this population the choice of spouse was conspicuously limited. A very few men brought in wives from outside, but for the last few decades anyone marrying another islander was in fact marrying a relative. Indeed, the only outbred individuals born since 1930 have been illegitimate by some visiting outsider. Were it not for the two ship-wrecked Italians who arrived

TABLE 4—THE REASON FOR CONSANGUINEOUS UNIONS

Consanguineous marriage	Date	N.⁹ of other available women	N.⁹ of non-relatives
1	1854	7	3
2	1856	8	2
3	1871	0	0
4	1876	0	0
5	1884	6	1
6	1888	7	0
7	1893	2	0
8	1898	0	0

Column 3 shows the total number of women of marriageable age (16 and over) whom the man in consanguineous unions might have chosen to marry instead of his wife (excluding his own sisters).

Column 4 shows the number of these who were not related to him.

in 1892 and the two Anglo-Irish women in 1908, the inbreeding coefficients would have been considerably higher. Unions with non Tristanians in the last decade or so will have the same effect in the next few generations, checking the rate of increase. It has been pointed out above what course is necessary to prevent any further increase.

Conclusion

The pedigree of the population of Tristan da Cunha, traced since its foundation, shows that some inbreeding has developed. This is attributable to the closed nature of this isolated population and to its small size, which severely restricted the number of potential mates from whom any individual could choose. The tendency to inbreeding that the latter engendered was in the opposite direction to the tendency to outbreeding that the marriage pattern of the islanders encouraged, and the result of this conflict was a relatively slight degree of inbreeding. For the present level is not excessive, and does not appear to have so far led to an undue increase in homozygosity. However, there has been a consistent increase in the inbreeding level over the years. To prevent the continuance of this trend an immigration rate of 0.1 per generation is required if the Tristan population retains its present size, structure and mating pattern.

ACKNOWLEDGMENT

Acknowledgement is gratefully made to Dr. H. E. Lewis for his enthusiasm and constant support; to Dr. Robson for her original pedigree diagrams; to Dr. Loudon for making available the results of his unpublished investigations and his helpful discussions; to Dr. Mourant for the blood group data; and to Mrs. Marjorie Smith for her assistance with the analysis.

REFERENCES

BLACK, J. A., THACKER, C. K., LEWIS, H. E., and THOULD, A. K. 1963—Tristan da Cunha: general medical investigations. Brit. Med. J. ii 1018–1024.

BONNE, B. 1963—The Samaritans: a demographic study. Human Biol. 35:61–89.

CHRISTOPHERSON, E. 1946—Results of the Norwegian scientific expedition to Tristan da Cunha 1937–8. Oslo. Norske Videnskaps-Akademi.

ISHIKUNT, N., NEMOTO, H., NEEL, J. V., DREW, A. L. and YANASE, T. 1960– Hosojima. Amer. J. Human Genetics **12**: 67–75.

LEWIS, H. E., 1963—The Tristan Islanders: a medical study of isolation. New Scientist **20**: 720–22.

SPUHLER, J. N. and KLUCKHOHN, C. 1953—Inbreeding coefficients of the Ramah Navaho population. Human Biol. **25**:295–317.

Hopi Indians, Inbreeding, and Albinism

CHARLES M. WOOLF and FRANK C. DUKEPOO

Albinism results from an inborn error of metabolism that involves the conversion of tyrosine to a brown or black insoluble polymer. Melanogenesis takes place in melanocytes, a group of specialized cells. The first step, the hydroxylation of tyrosine to dihydroxyphenylalanine (DOPA), is catalyzed by tyrosinase. The second step, the oxidation of DOPA to the quinone, is also catalyzed by tyrosinase. The rest of the pathway leading to the production of the polymer (melanin) may be nonenzymatic (*1*).

Although albinism in man is most often caused by a recessive autosomal gene, genetic heterogeneity may be present. The Trevor-Roper pedigree (*2*) of two albino parents producing normally pigmented offspring is suggestive evidence for at least two different types of recessive albinism in man. Supportive evidence comes from the experiments of Witkop, Van Scott, and Jacoby (*3*) who placed unfixed albino hair bulbs into a solution of L-tyrosine (50 milligrams per 100 milliliters, pH 6.8). They observed that the hair from some albinos formed pigment, but the hair from others did not. Genetic heterogeneity is not unexpected since a mutation in the structural gene, a controlling gene involved in the production of tyrosinase, or at some other locus resulting in the presence in the melanocyte of some enzyme-binding substance, would all disrupt melanogenesis.

Albinism occurs at different frequencies in various human populations. In Europe the overall frequency is about 1 in 20,000, with estimates ranging from 1 in 10,000 in Norway to about 1 in 30,000, or less, in southern Europe (*4*). In marked contrast are Indian populations in Central and North America. The Cuna Indians of San Blas Province, Lower Panama, show a frequency of about 1 in 200 (*5*). Frequency values of a similar magnitude characterize the Hopi Indians of Arizona and the Jemez and Zuni Indians of New Mexico (*6*). These North American Indians are not well known as having a high frequency of albinism, mainly because of the reduced population. A frequency of about 1 in 200 results in a conspicuous number of albinos in a total population of about 20,000 Cuna Indians; the impact of this frequency is not as great in populations of less than 6000 individuals.

The detrimental nature of albinism has been well established. Albinos are prone to skin cancer. Myopia and lateral nystagmus are usually part of the genetic syndrome, and albinos are extremely sensitive to sunlight. Neel *et al.* (*7*) conclude that, in present-day European and Japanese populations, the relative reproductive (Darwinian) fitness (f) of albinos is about 0.7 to

Reproduced by permission of the publisher, Charles M. Woolf, and Frank Dukepoo from Science, 164:30–37 (April 4, 1959).

CHARLES M. WOOLF, *a human geneticist, was born in Utah in 1925. He received his B.S. in 1948 and his M.S. in 1949 from the University of Utah. He received his Ph.D. from the University of California in 1954. At present he is Professor of Zoology at Arizona State University. His main research interests and publications deal with the genetics of carcinoma in humans and tumorous head strains of Drosophila. His work in human populations has included a study of consanguineous marriages among the Mormons in the United States and work on albinism in the Indians of Arizona and New Mexico, one report of which is included here. He is author of the book* Principles of Biometry *(Princeton, N.J., 1968).*

0.8, but in past generations it was more like 0.4 to 0.5. A relative fitness of 0.4 implies that albinos leave 40 functional offspring for every 100 left by nonalbinos.

When a deleterious trait is due to an autosomal recessive gene (c), and when carriers (Cc) and noncarriers (CC) have the same relative fitness value ($f = 1$), the frequency of affected individuals (cc) in a large population is a function of both the mutation rate (μ) and the selection coefficient ($s = 1 - f$) against affected individuals. The frequency at equilibrium is $R = \mu/s$. The mutation rate at the albino locus is unknown, but if a value of 1/40,000 is assumed (that is, there is one gamete containing the mutant gene among 40,000 produced), and the relative fitness of albinos is given the value $f = 0.4$, the frequency of albinos at equilibrium in the population would be $R = \mu/s = (1/40,000)/(0.6) = 1/24,000$. For this mutation rate, which is an acceptable value for man, the frequency of albinos at equilibrium would be relatively low, even if selection against them were less severe. For example, if the fitness of albinos were 0.9, then the frequency of albinos at equilibrium would be $R = (1/40,000)/(0.1) = 1/4,000$. The occurrence of Indian populations with an albino frequency of 1 in 200 indicates that some force other than mutation and selection against albinos has been operating in past generations.

Hopi Indians and Albinism

During the spring of 1961, one of us (C.M.W.) became interested in albinism among the Hopi Indians because an acquaintance remarked that she had observed an albino child in one of the Hopi villages. A trader then told her that albinos are common on the reservation, and the Hopis consider it a "good luck charm" to have one or more residing in their village. Hrdlička recounted a visit to the Hopis in 1900 and described 11 albinos from ten sibships (*8*). Based on this number of albinos and an estimated population size of 2000, the frequency in 1900 was 1 in 182.

The Hopi reservation is located in northeastern Arizona, where it is completely surrounded by the Navajo reservation (Fig. 1). The Hopi villages with one exception are located on the top or at the base of three mesas, known as First Mesa, Second Mesa, and Third Mesa, which are

FRANK DUKEPOO, *a graduate student in zoology, was born in Arizona in 1943. He received his B.S. in 1966 and his M.S. in 1968 from Arizona State University. He is a Hopi Indian, and his participation in the field research done on the Hopis made possible the collection of essential family data and contributed largely to the success of the study.*

fingers off Black Mesa (Fig. 2). The villages on the mesas are adjacent or separated by only a few kilometers. The villages of Hano, Sichomovi, and Walpi are on First Mesa. . . .[1] Polacca, which is part of the First Mesa population, is located at the base of the mesa. About 19 kilometers westward is Second Mesa with the villages of Mishongnovi, Shipaulovi, and Shongopovi. . . . Another 16 kilometers westward is Third Mesa with Bacavi, Hotevilla, and Oraibi on the top, and New Oraibi at the base. Oraibi. . . , occupied since about A.D. 1150, is the oldest continuously inhabited community in the continental United States (9). Moencopi, founded originally by members of Oraibi, is located 64 kilometers farther west on the western range of the Navajo reservation. The Hopi inhabitants of Moencopi are included in the Third Mesa population.

The village of Hano on First Mesa was established by Tewa emigrants from the Rio Grande area in New Mexico. Seeking refuge after the Spanish revolt, they arrived in about 1700. They were allowed to remain on First Mesa on the condition that they help protect the Hopis from hostile nomads (10). The Hopis and Tewas on First Mesa have intermarried ex-

FIG. 1. Locations of Hopi and Navajo reservations in Arizona.

[1]Figures 3, 4, and 5 have been deleted.

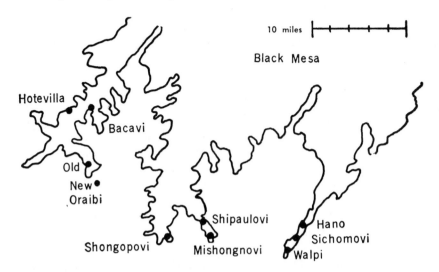

FIG. 2. Location of the Hopi villages on or at the base of First Mesa, Second Mesa, and Third Mesa.

tensively in recent generations. In this paper, unless specified otherwise, the Tewas are considered as part of the Hopi population. Accurate census data are not available for the Hopi population. The Hopi Indian agency, Bureau of Indian Affairs, Keams Canyon, Arizona, estimates (1968) that there are about 6000 Hopi Indians living on or off the reservation. The Hopi-Tewa population on First Mesa numbers about 1000.

Objectives of Study

A study was initiated with the following objectives: (i) to determine the frequency of albinos in the Hopi population; (ii) to construct pedigrees using living albinos and those described by Hrdlička as reference points; (iii) to estimate the mean population coefficient of inbreeding for the various village and mesa populations in order to determine the influence of inbreeding on the frequency of albinos; and (iv) to determine, if possible, what forces account for the high frequency of the detrimental gene in this population.

A pilot study (*11*) verified Hrdlička's conclusion that albinos are common in the Hopi population. The observed frequency in 1962 was 1 in 227. It was soon noted that generalized albinism occurs only in the Second Mesa and Third Mesa populations (*6*). Two male siblings with piebaldness (partial albinism) and subtotal nerve deafness are from First Mesa. These males are striking because of their uniformity in depigmented pattern even though they differ in age by 4 years (*12*). The gene responsible for this syndrome is different from the one causing albinism in the Second and Third Mesa populations.

It was a formidable and sometimes impossible task to construct Hopi pedigrees. Hopis tend to be suspicious of anyone, especially a white person, who attempts to obtain information about their families or culture. They fear he is going to "write a book" and expose Hopi secrets. The inhabitants of Hotevilla and Bacavi are especially suspicious. There are no written records, and traditional Hopis wish not to mention the name of deceased relatives. "Don't speak his name, let him rest in peace." It is not uncommon to find a person who has never heard the given name of a deceased grandparent. Among traditional Hopis, pre- and extramarital relations are accepted as part of the Hopi way of life. Divorce is easy to obtain, and new arrangements are accepted by the villagers. These activities introduce errors into pedigree studies; however, Hopis talk about these activities and are often aware of true paternities. The matrilineal clan system and occurrence of ceremonial and adopted relationships present a problem of communication in a genetics study. Hopis inherit membership in the clan of their mother. Men and women of the same clan and same generation are considered as brothers and sisters. When a young Hopi is initiated into a society he acquires a ceremonial father. Furthermore, an old man who befriends a Hopi in a special way may become an adopted father. In a conversation with a white person, a Hopi may not make a distinction between a biological, clan, ceremonial, or adopted relationship. He may refer to several different women as his mother, name distant relatives as brothers and sisters, or state that a given man is his father, then later refer to the same man as his older brother or son (13). Relationships are bewildering to white persons and even to young Hopis, as evidenced by an account in the autobiography of Sun Chief (14).

So this old man came in the morning for peaches and collected peach stones to plant. He would walk past me with his stick and say, "Good morning and good luck to you, my father." I did not like to be called father (ina'a) and showed that I was offended. One day my mother said to me, "Don't treat your son that way, Chuka. He used to have a father who belonged to our Sun Clan. His father was your great-great uncle; that makes me his aunt and you his father. Try to treat your son better." After a while I grew used to having him call me father.

Definitive progress was made in the collection of pedigrees when one of us (F.C.D., a Hopi Indian from First Mesa) became associated with the study. He obtained pedigree data (15) which would have otherwise been most inaccessible.

In this matrilineal society, all property is owned by females, and even though exceptions do occur, it is customary for an individual to identify with the place of residence of his mother, which is usually his place of birth and rearing. Even if a Hopi is born and reared off the reservation, he will identify with his mother's mesa. If his mother is not a Hopi, an individual may be adopted into a clan of his choosing and may identify with the place of residence of his father. With these few exceptions, the matrilineal system was used in our study to denote village and mesa membership.

In the total population of 6000, there are 26 albinos, or 1 in 231. As in 1962, all cases of albinism observed occur at Second Mesa and Third Mesa. Apart from First Mesa, the frequency is 26 in 5000, or 1 in 192. The albinos are affiliated with the following villages: Mishongnovi, nine; Shipaulovi, four; Shoṅgopovi, one; Bacavi, one; Hotevilla, five; New Oraibi, one; and Moencopi, five. Three albinos living in 1900 were not described by Hrdlička—two from Oraibi and one from Shipaulovi.

Abbreviated pedigrees showing living and related albinos are given in Figs. 6–10. Also shown in these pedigrees are the 11 albinos observed by Hrdlička in 1900. These latter albinos are designated by the letter *H*. A photograph of a young Hopi albino girl is shown in Fig. 11.

Inbreeding in the Hopi Population

The frequency of a recessive trait in a large population is a function of the frequency (q) of the recessive gene and the degree of inbreeding among the members of that population. This is expressed by Wright's equilibrium law (*16*) where the frequencies of *CC*, *Cc*, and *cc* individuals in a population at equilibrium are given by $D = p^2 + \alpha pq$, $H = 2pq(1 - \alpha)$, and $R = q^2 + \alpha pq$, respectively. The mean population coefficient of inbreeding symbolized by α, can range, theoretically, from 0 (random mating) to 1 (complete inbreeding). The Hardy-Weinberg law is a special case of Wright's equilibrium law, where $\alpha = 0$. When, for example, the frequency of a recessive gene is $q = 0.001$, the frequency of those homozygous for the gene ($R = q^2 + \alpha pq$) would be 20.98 times more frequent in a population where $\alpha = 0.02$ than in a random mating population. Inbreeding has relatively little effect on the frequency of individuals homozygous for a common recessive gene. If $q = 0.1$, the frequency of individuals homozygous for the gene would only be 1.18 times more frequent when $\alpha = 0.02$ than when $\alpha = 0$.

The history of the three mesa populations suggests that First Mesa should be the least inbred and should have the lowest frequency of the recessive gene for albinism. An obvious reason is the mixing of the Hopi and Tewa gene pools. Furthermore, First Mesa is the closest to Keams Canyon, which during recent times has been a center (boarding school, trading posts, Bureau of Indian Affairs government offices) where Hopis have encountered white Americans and members of other Indian tribes, mainly Navajo and New Mexico Pueblo Indians. On the basis of blood group studies, Brown *et al.* (*17*) concluded that the population of First Mesa shows considerable mixture of Hopi, Tewa, and Navajo genes.

Inbreeding in the village and mesa populations was measured by estimating mean population coefficients of inbreeding. Married couples were selected randomly in the various villages. For each couple, an in-

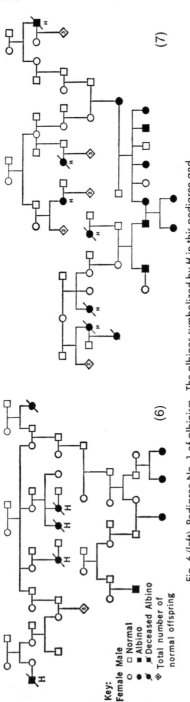

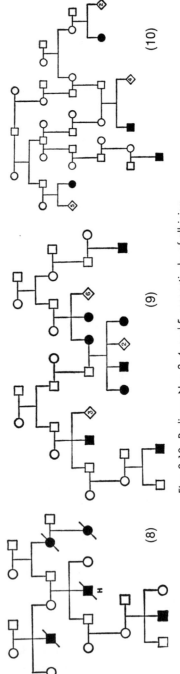

Key:
Female Male
○ □ Normal
● ■ Albino
● ■ Deceased Albino
◇ Total number of
   normal offspring

Fig. 6 (left). Pedigree No. 1 of albinism. The albinos symbolized by H in this pedigree and
other pedigrees were observed by Hrdlička in 1900(8). Fig. 7 (right). Pedigree No. 2 of albinism.

Figs. 8-10. Pedigree Nos. 3, 4, and 5, respectively, of albinism.

Fig. 11. Hopi albino girl. Photograph was taken during the period 1897–1900.
[Field Museum of Natural History, Chicago, Illinois]

breeding coefficient was calculated from the formula $F = (\frac{1}{2})^N[1 + F_A]$, where N is the number of individuals along the chain from one individual to another and F_A is the inbreeding coefficient of a common ancestor (*18*). The F represents the inbreeding coefficient of a child born to each couple; therefore, this method assumes that each married couple has the same reproductive potential. The number of married couples sampled was 388 (First Mesa, 131; Second Mesa, 133; and Third Mesa, 124). The F values were averaged to obtain an estimate of the mean population coefficient of inbreeding. Moencopi was not studied because it is not on the Hopi reservation, and it was considered more informative, for the time available, to concentrate on the villages geographically associated with the three mesas.

The data show that Second Mesa is the most inbred ($\alpha = 0.01365$), Third Mesa is intermediate ($\alpha = 0.00702$), and First Mesa, as expected, is the least inbred ($\alpha = 0.00250$). The mean population coefficient of inbreeding based on all 388 F values is 0.00797. These estimates are lower than the true values since it was difficult to trace pedigrees back for more than about six generations. Furthermore, even a Hopi Indian from another mesa is unable to obtain cooperation from all inhabitants of Hotevilla and Bacavi. Consultants from these villages helped collect data, and even they had difficulty in obtaining cooperation from their fellow villagers. Refusal to cooperate introduces additional bias in the estimates of α for these villages.

The Hopis are relatively inbred when compared with various European, white American, and Japanese populations (*19*). However, the Hopis are perhaps no more inbred than other southwestern Indians. Spuhler and Kluckhohn (*20*) obtained an α value of 0.0080 for the Ramah Navajos in New Mexico, a value almost identical to the overall value (0.00797) obtained for the Hopis.

Before the coming of roads and modern means of transportation, the mesa populations were semi-isolates. The distances involved, the rigors of a trip during the wintertime, and the threat of attack by Navajos and Utes during certain periods of time discouraged the casual exchange of visitors among mesas. These factors promoted mesa endogamy in the biological sense. Even today cultural activities are strongly mesa-oriented, resulting in a certain degree of rivalry among the mesas. Although the physical location of the villages has strongly favored biological endogamy, cultural regulations forbidding certain types of consanguineous matings have reduced the inbreeding potential. Closely related clans are grouped into a single phratry, a brotherhood or fraternity of clans. An individual is discouraged from marrying or having a sexual relationship with a member of his own clan or phratry, or with a member of his father's clan or phratry. Such restrictions eliminate certain relatives as spouses. However, it is of interest that this cultural regulation allows for marriage between half-sibs when they have the same father and their mothers belong to unrelated clans. Such marriages occur.

The frequency of the albino gene may be used as an indicator of the "purity" of the Hopi gene pool. Second Mesans tend to boast that they marry non-Hopis less frequently, and that they are the "most Hopi" of all. They also boast the largest number of albinos. No Hopi now living recalls the birth of an albino at First Mesa; however, the albino gene is present in this population. Two males from First Mesa married to females from Mishongnovi and Moencopi are the fathers of albino children. Each male had grandparents from Second Mesa. The mixing of the First Mesa gene pool has greatly reduced the frequency of the albino gene. Mixing is also found in a section of the Third Mesa population.

At the turn of the century, a serious dispute arose among the inhabitants of Oraibi, then the only village on Third Mesa. Some (the "friendlies") wished to support the government by enrolling their children in schools. Others (the "hostiles") wished to maintain the traditional Hopi way of life. As a result, the hostiles moved from Oraibi in 1906 and established the village of Hotevilla, and later the village of Bacavi. New Oraibi was established at the foot of the mesa by friendlies so that they could live in close proximity to the government school. Some of the friendlies from New Oraibi selected spouses who were not Hopi. The split has led almost to the total decline of Oraibi. Only a few families live there. The residents of Hotevilla and Bacavi soon found themselves in the dilemma of having too

TABLE 1. INBREEDING COEFFICIENTS IN DIFFERENT SEGMENTS
OF THE HOPI POPULATION

Village	Number of marriages	1/8	1/16	1/32	1/64	1/128	Other	α	
First Mesa									
Hano	50	0	0	1	1	1	0	0.00109	
Sichomovi	51	0	1	0	0	1	0	0.00137	
Walpi	30	1	0	1	0	0	1	0.00677	0.00250
Second Mesa									
Mishongnovi	43	1	2	2	2	2	3	0.01226	
Shipaulovi	28	0	5	3	3	1	3	0.02176	
Shongopovi	62	0	6	5	5	4	5	0.01222	0.01365
Third Mesa									
New (Old) Oraibi	59	0	2	1	2	1	1	0.00463	
Hotevilla	37	0	4	2	1	3	1	0.00950	
Bacavi	28	0	3	1	1	1	0	0.00864	0.00702
Total	388	2	23	16	15	14	14	0.00797	
		(0.5%)	(5.9%)	(4.1%)	(3.9%)	(3.6%)	(3.6%)		

few clans. They sought marriage partners from Moencopi rather than from Oraibi or Second Mesa. Moencopi, located 64 kilometers away on the Navajo reservation, consists of Hopi and Navajo inhabitants. It is divided into an upper and lower village. The lower village consists mostly of traditional Hopis; the upper village is composed of Hopis and Navajos. Mixing of Hopi and Navajo genes has occurred at Moencopi, especially in recent years. These marriage patterns are reflected in the data on inbreeding (Table 1) and the distribution of albinos. The 26 albinos tend to be of pure Hopi lineage. The five albinos from Moencopi are from the lower village. A major exception occurs in the village of Shipaulovi on Second Mesa. The mother of two albinos is a Navajo. The albino children have been adopted into a Hopi clan, identify with the village of their father, and are considered Hopis by the Hopis. Albinism is not uncommon to the Navajo, especially in the vicinity of the Hopi reservation (6). This is best explained by the Moencopi influence and the fact that for centuries the Navajos have been raiding the Hopi villages for the purpose of obtaining food, women, livestock, and other booty.

Estimate of Gene Frequency

The frequency of albinos and the inbreeding coefficient can be used to estimate the gene frequency in the Second and Third Mesa population. Based on an estimated population size of 5000, the frequency of albinos becomes $R = 26/5000 = 0.0052$. If $\alpha = 0.01088$ for Second and Third Mesa, and if we assume equilibrium, and solve for q in the equation $R = q^2 + \alpha pq$, we obtain a gene frequency of $q = 0.067$. The frequency of heterozygotes in this population becomes

$$H = 2pq(1 - \alpha) = 0.124.$$

The high frequency (12.4 percent) of carriers of the gene for albinism is reflected in the pedigrees shown in Figs. 6 to 10. The frequency of the gene for albinism is so high that even if the population resorted to random mating, the frequency of albinos would be reduced only slightly ($R = 1/223$ as compared with the present value of $R = 1/192$). The intriguing problem is to account for the high gene frequency in this population, as well as the Cuna, Jemez, and Zuni Indians.

Selection for the Heterozygote

A deleterious recessive gene is maintained at a relatively high frequency in a population when the heterozygote has a reproductive advantage. For example, if the relative reproductive fitness values for heterozygotes and albinos are set at 1.00 and 0.5, respectively, it can be shown that the relative fitness of homozygous persons would have to be only 0.962 in order to maintain an albino frequency of $R = 1/200$ in a large population. This slight advantage of the heterozygote, which would be difficult to measure in a practical situation, might manifest itself by increased fertility, vigor, maturation rate, or resistance to a disease. The implication is that, in the presence of a specific homozygous background produced by genetic drift and many generations of inbreeding, heterozygosity for certain mutant genes increases reproductive fitness. The corollary is that the advantage ceases with the advent of genetic variability. The presence of albinism at a high frequency in some populations and not others would imply that the latter lacked the proper genetic background or the allele had not been introduced by migration or mutation. This form of selection cannot be ruled out as the explanation for the high frequency of albinism in various Indian populations.

Genetic Drift

Genetic drift implies that a gene could be common in a population if one or more of the few early settlers carried the gene (founder principle), or because of chance segregation and recombination during a time when the population was small. Indian populations in the southwest have been plagued by epidemics, periodic starvation conditions, and raids by hostile groups. These populations have often consisted of a small number of individuals. If the albino gene were present in one of these small populations, the frequency would fluctuate and might rise by chance to a relatively high value. Furthermore, it is likely that many, if not all, of the present southwest Indian populations were founded by a small group of migrants. The frequency of the albino gene could be high in a present population if one or more of the founders were heterozygous for the gene.

If the high frequency of the albino gene in the Hopi population is accounted for by a chance process occurring in past generations, what is the origin of the mutant gene? The albino gene in the Hopi, Jemez, and Zuni Indians is likely "identical by descent" since gene flow occurred among the Pueblo Indians of Arizona and New Mexico in historic and pre-historic times. The presence of albinism in the remote Cuna Indians suggests further that the mutant gene is not of recent origin, and draws attention to Central America as the possible site of origin. There were many albinos in Montezuma's royal household at the time of the conquest of Mexico. Albinos along with malformed individuals were considered by the Aztecs as suitable wards of the state (6). The extent of this practice in Central America before the coming of Cortez is unknown. The protection of albinos in early American cultures would have favored the propagation of the albino gene in certain segments of the population. Groups migrating northward could then have carried the albino gene to the southwestern part of the United States. Climate would influence greatly the destiny of the albino gene acquired at a high frequency by migration. Natural selection against the gene would be more effective in the hot desert populations of southern Arizona than in the more temperate climates of northern Arizona and New Mexico. Pueblo dwellings which afford protection from the sun would lessen the effectiveness of natural selection.

The hypothesis of genetic drift may be questioned because of the existence of albinism at a high frequency in three different southwest Indian populations. It is not likely that a deleterious gene would reach a high frequency in all three populations by only a chance process. Some selective mechanism has likely been operating in one or more of these populations.

Role of Cultural Selection

A high frequency of a gene in a population is determined by processes operating in previous generations. We believe that the most acceptable hypothesis to explain albinism in the Hopis is the acquisition of the gene by migration (or mutation) and maintenance at a high frequency by cultural selection. Albinos have been protected in the Hopi society.

Traditional Hopis believe that albinism, along with other congenital defects, is the result of some specific action in the life of a parent or close relative. Typical retrospective explanations are as follows.

When _____ was a young man, he owned a white donkey. Everywhere he went he rode this donkey. He liked it so much two of his granddaughters were born white.

_____ liked to portray Eototo (white kachina). Therefore [his daughter] was born white.

_____ worked with white sand while his son was being formed.

His mother [father] slept with a white man [woman].

There is no recognition of the heredity phenomenon. An albino child born to an albino parent is explained merely by the existence of "whiteness" in the parent. Because of this logic it is assumed that albino parents may produce albino children.

Younger Hopis who have learned that albinism is a genetic defect may show a certain degree of discrimination against albinos; however, comments made by traditional Hopis reflect only a positive attitude toward them. Albinos are accepted completely in the Hopi society. They perform in ceremonies along with other Hopis. Several have been influential chiefs and priests. One, who was chief at Oraibi about 100 years ago, is legendary as a rainmaker. "During a plague of prairie dogs one fall he is said to have called for a Masau Katcina dance, which caused a veritable cloudburst that drowned all the prairie dogs" (*13*). Comments by traditional Hopis illustrate the acceptance of albinos in the Hopi society.

Albinos are smart, clean, nice and pretty. There is nothing wrong with them.

Albino girls are very pretty. I like their nice color. I would like to marry one, but they probably wouldn't have me.

I would like to have an albino baby. There were albinos when I was a child, so they are not new to me. In fact, I am related to one. He's my brother [clan affiliate].

I know lots of them, and they are good Hopis.

Here [on Second Mesa] we have lots of them and we are very proud of them. There are more here than at the other mesas. We take good care of them.

It is lucky to have albinos [in the village]. They are very special.

Albinos are considered as part of the Hopi heritage on Second and Third Mesa. This was evident in the revealing answer to the question asked of a member of Third Mesa as to why there are no albinos on First Mesa. "Oh, they are Tewas."

Marriage is important in the life of a Hopi. It is extremely rare to find an adult, especially a woman, who has never been married. For this reason, the fact that many albinos have remained unmarried contradicts the attitude of complete acceptance in the traditional Hopi society. One albino male stated he was unable to work in the fields and therefore could not grow corn for a wife and family. Although the inability to provide adequately for a family is the excuse usually given for males, the reason is not obvious for females. Answers to questions as to why specific albino females did not marry usually implied withdrawal of the female. The difficulty of being an albino in a dark-skinned population seems to have such a repressive action that suitors are rejected. This point needs further investigation. The acceptance by Hopis of this Hopi heritage apparently stops short of marriage. Females should reject males who cannot work in the fields. Titiev (*13*) quotes a discriminating comment regarding the albino chief from Oraibi who had the reputation of being a good rainmaker: "Nakwaiyamptiwa never married, 'because in those days ladies did not like albinos.'"

However, any real or subconscious bias towards marrying an albino apparently does not manifest itself in the selection of a sex partner. For this reason albino males in past years had an advantage in regard to sexual activity. Hopis for centuries have made their living by farming. Men and boys would leave the mesa tops and go to the fields, often many kilometers from the mesas. It is commonly known, among the inhabitants of Second and Third Mesa, that albino males were not expected to labor in the fields. They remained in the villages and performed tasks such as weaving. In the villages they were protected from the sun and hostile nomads, and most importantly, they had ample time and opportunity to engage in sexual activity. One albino is expecially legendary for this activity.

> I knew that old man. Some say he had about twelve kids; others say about fifteen. He never married.
> They say he was always around trading with the ladies. He would make babies with them.
> He was real funny, and knew a lot of good stories. Everyone liked him.

In a small population, a sexually active male will have an imposing effect on gene frequencies of the next generation.

The acquisition of the albino gene by migration (or mutation) and a form of cultural selection whereby albino males are given a slight sexual advantage would explain the frequency of the albino gene in the Hopi populations. Whether a similar hypothesis is acceptable for the high gene frequency among the Cuna, Jemez, and Zuni Indians remains to be determined. Indeed there is no reason to conclude that the high frequency in all four of these Indian populations is due to the same force or forces. There appears to be cultural selection against albinos in the Cuna population (5).

Albinos are not considered as "good luck charms" for a village. The word lucky is used frequently in the English vocabulary of a Hopi. The meanings of the expressions "He is lucky with corn" or "He is lucky with girls," are apparent. However, the statement "It is lucky to have albinos," can be given the wrong interpretation by an outsider. The statement implies that it is difficult to produce albinos because they do not appear in all families or in all villages. There is no evidence that any special religious significance was ever attached to albinism that would predispose to the present positive attitude toward them. This point was investigated thoroughly because Hopi tradition would allow for a religious symbol being attached to albinism.

Hopis share the legend permeating Central American Indian cultures that someday a white leader will return and lead them to a better life. Just as Cortez was welcomed by the Aztecs as their white leader, Pedro de Tovar, the first white man to visit the Hopis, was welcomed by the Hopis as their "lost white brother" (21). Tovar and his men quickly subdued the Hopis. The lost white brother has yet to arrive. If the birth of an albino was ever considered as a symbol or reminder of their lost white brother, it is unknown to any living Hopi. The intriguing problem remains, however, as to why albinos, especially albino children, are considered so affectionately in all villages. The admiration of their whiteness is clearly not an identification with white Americans; it represents instead an association of whiteness with cleanliness, goodness, and purity; attributes honored by traditional Hopis. Rain is essential for existence in the barren mesa country. White clouds symbolize rain and are a constant reminder of the necessity of living a pure life. "If Hopis live a clean, good and harmonious life, there will be plenty of rain and an abundance of food for the children to eat."

Conclusion

Although selection for the heterozygote or genetic drift may account for the high frequency of the albino gene in the Hopi population, the most apparent explanation is acquisition of the gene by migration and cultural selection of the type described here. A study of the Hopi people also indicates that time will soon erase albinism as a Hopi heritage on Second and Third Mesas. Paved roads now link the Hopi mesas so that only minutes are required to travel from one mesa to another by truck or automobile. These modern transportation facilities, schools on and off the reservation where members of different ethnic groups associate, and a growing population forcing young people to seek employment off the reservation all promote outbreeding, reduce the frequency of the albino gene, and decrease the probability of homozygosity for this gene. The eclipse of agriculture as a way of life negates any reproductive advantage held by albino males in past generations. The frequency of albinism among the Hopis will decrease rapidly with the decline of their culture.

REFERENCES AND NOTES

1. T. B. FITZPATRICK, in *The Metabolic Basis of Inherited Disease*, J. B. Stanbury, J. B. Wyngaarden, D. S. Frederickson, Eds. (McGraw-Hill, New York, 1960).
2. P. D. TREVOR-ROPER, *Brit. J. Ophthalmol.* **36,** 107 (1952).
3. C. J. WITKOP, JR., E. J. VAN SCOTT, JR., G. A. JACOBY, *Excerpta Med. Sect. I Anat. Anthropol. Embryol. Histol.* **F-1969,** Abstr. 381 (1961).
4. K. PEARSON, E. NETTLESHIP, C. H. USHER, *A Monograph on Albinism* (Draper Company Research Memoirs, Biometric Series, Dolan, London, 1911–1913), vols. 6 and 9.
5. C. E. KEELER, *J. Hered.* **55,** 115 (1964).
6. C. M. WOOLF, *Amer. J. Hum. Genet.* **17,** 23 (1965).
7. J. V. NEEL, M. KODANI, R. BREWER, R. C. ANDERSON, *ibid.* **1,** 156 (1949).
8. A. HRDLIČKA, "Physiological and medical observations among the Indians of southwestern United States and Northern Mexico," *Bull. Bur. Amer. Ethnol. No.* **34** (1908).
9. L. L. HARGRAVE, *Mus. N. Ariz. Mus. Notes* **4,** No. 7 (1932).

10. E. P. DOZIER, "The Hopi-Tewa of Arizona" *Univ. Calif. Publ. Amer. Archeol. Ethnol.* **44,** No. 3 (1954).
11. C. M. WOOLF AND R. B. GRANT, *Amer. J. Hum. Genet.* **14,** 391 (1962).
12. C. M. WOOLF, D. A. DOLOWITZ, H. E. ALDOUS, *Arch. Otolaryngol.* **82,** 244 (1965).
13. M. TITIEV, *Old Oraibi, A Study of the Hopi Indians of Third Mesa* (Peabody Museum Papers, Harvard Univ., Cambridge, Mass., 1944), vol. 22, No. 1.
14. SUN CHIEF, *The Autobiography of a Hopi Indian*, L. W. Simmons, Ed. (Yale Univ. Press, New Haven, 1942), p. 68.
15. F. C. DUKEPOO, thesis, Arizona State University, Tempe (1968).
16. CHING-TS'UNG LI, *Population Genetics* (Univ. of Chicago Press, Chicago, Ill., 1955).
17. K. S. BROWN, B. L. HANNA, A. A. DAHLBERG, H. H. STRANDSKOV, *Amer. J. Hum. Genet.* **10,** 175 (1958).
18. C. STERN, *Principles of Human Genetics* (Freeman, San Francisco, ed. 2, 1960).
19. C. M. WOOLF, F. E. STEPHENS, D. D. MULAIK, R. E. GILBERT, *Amer. J. Hum. Genet.* **8,** 236 (1956).
20. J. N. SPUHLER AND C. KLUCKHOHN, *Hum. Biol.* **25,** 295 (1953).
21. F. WATERS, *Book of the Hopi* (Viking, New York, 1963).
22. Supported by AEC contracts AT(11–1)-1415 and AT(11–1)-2013-01 to Arizona State University.

CHAPTER **VI**

Gene Flow

Overview

Three of the four evolutionary forces—mutation, natural selection, and genetic drift—act primarily to differentiate populations. The fourth force of evolution, *gene flow*, or genetic admixture, pushes in the opposite direction and makes populations genetically more similar. By gene flow the population geneticist refers to the process by which genes are introduced from the gene pool of one breeding population into the gene pool of another by reproductive sexual matings. Such "cross-matings" create an intermediate population which genetically resembles the parental populations in proportion to their respective genetic contributions.

Gene flow is sometimes equated with population migration, but in fact, the two processes may be independent. It is true that without initial migration enabling the individuals of two different breeding populations to come into contact with each other there can be no gene flow, but migration per se is not always accompanied by genetic admixture. A group may migrate to an area unoccupied by other human populations, as did the first groups of Asiatic immigrants who crossed the Bering Strait to settle a New World unpopulated by any other human group. Or, theoretically, a migrant group may move into an inhabited area and still maintain total genetic isolation by not mating with individuals from other populations. With respect to the latter situation, as was previously discussed in Chapter II, it is quite unlikely that reproductive barriers are ever complete between groups living in propinquity. Even the Amish, who wish to remain socially isolated, have incorporated into their demes some non-Amish individuals who have married Amish and reproduced, thereby introducing new genes into a deme's gene pool. Because genes are exchanged across population

409

boundaries only through reproductive "cross-matings," gene flow is better equated with *exogamy* than with migration.

Removal of the geographical and/or cultural barriers which maintain breeding isolates results in the influx of genes from other breeding populations and the consequent breakdown of genetic isolation. Although it is true that the breakdown of isolates has been accelerated in the modern world by rapid improvements in communication and transportation, the process has undoubtedly occurred continually throughout the history of man. If one accepts the inferences about the hunting-gathering groups of the Pleistocene to be made from the population dynamics of modern hunting-gathering groups, the living evidence indicates that even the former groups were not totally isolated genetically. Individuals of modern hunting-gathering cultures often mate outside their own breeding population; for example, among Australian aborigines, an average of 15 percent of all marriages occurs across tribal boundaries, so that language, cultural items, and genes are spread throughout the various aboriginal populations of Australia (Tindale, 1953).

However, the contact and intermixture of genetically very diverse populations from widely separated geographical areas are more probably post-Pleistocene phenomena dating back to the Neolithic, approximately 8,000 years ago. At that time the Agricultural Revolution triggered population explosions and subsequent large-scale migrations, as people sought new and distant lands and opportunities for their ever expanding numbers. This trend of population growth and expansion was further accelerated by colonialism and the Industrial Revolution. However, there is no doubt that the twentieth century especially has witnessed the accelerated disappearance of isolated primitive groups in many remote areas and the breakdown of rural isolation in the industrialized countries.

Effects of Gene Flow

The effects of gene flow on populations and their gene pools are manifold. Some of the effects have been investigated and documented in actual human populations, but the majority remain effects to be expected on theoretical grounds alone, uninvestigated or unproven in fact. The difficulties involved in investigating gene flow are the same as those generally pertaining to research on evolution in human populations discussed in previous chapters (see the conclusions by Siniscalco in Chapter IV). Given these limitations, what can be said about the operation and effects of gene flow in human populations?

Homogenizing Populations

First, from a strictly unidimensional point of view, gene flow may conceivably be regarded as a force diametrically opposing the effects of genetic isolation. The subdivision of a population into genetically iso-

TABLE VI-1. EXPECTED POPULATION VARIANCE UNDER
DIFFERENT PROPORTIONS OF IMMIGRANTS*

	Villages					
	C_1	C_2	C_3	C_4	C_5	C_6
Effective size (N_e)	125	125	125	1250	1250	1250
Proportion p_{io} of immigrants	0.05	0.10	0.20	0.05	0.10	0.20
Proportion p_{ii} of natives	0.95	0.90	0.80	0.95	0.90	0.80
Newly arising variance*	0.001	0.001	0.001	0.0001	0.0001	0.0001
$1/(1 - p_{ii})^2$	11.43	5.26	2.78	11.43	5.26	2.78
Total variance σ^2_{ii}	0.01143	0.00526	0.00278	0.00114	0.00052	0.00028

*The table shows six hypothetical villages with migration from the outside world into the villages but with no migration between villages. The gene frequency q in the outside world is 0.5 and the gene frequencies in the villages are not expected to deviate too greatly from this, that is, $\bar{q} = 0.5$. The variance newly arising each generation ($\bar{q}(1 - \bar{q})/2N_{e_i}$) is multiplied by $1/(1 - p_{ii})^2$ to show to what extent it accumulates over successive generations.

Source: Modified from Smith, 1969.

lated groups, each pursuing its own path of evolution under the pressure of forces unique to it, separates and diversifies a formerly homogeneous population. Gene flow, on the other hand, by combining gene pools, builds new, larger gene pools and progressively homogenizes formerly diverse isolated populations.

Table VI-1 shows this homogenizing effect as measured by the accumulated variance over successive generations in populations receiving different proportions of immigrants. With increased gene flow (represented by "proportion of immigrants"), the total accumulated variance (σ^2_{ii}) is reduced. As the total accumulated variance in each population decreases, the constituent populations come to resemble one another more and more. Reduced variance is synonymous with reduced genetic drift as drift is defined in Chapter V.

However, throughout the history of man population divergencies and convergencies have probably acted closely in conjunction with one another rather than in diametric opposition. This simultaneous interdependence and interaction of gene flow and population isolation are, of course, crucial to Sewall Wright's thesis concerning the evolutionary significance of genetic drift. He argues that only by the joint interaction of drift and gene flow can adaptive genetic complexes which are developed in semi-isolation be made available to the total species (see Sewall Wright, Chapter V). But the interaction and relative genetic impact of drift and gene flow are impossible to measure in human populations; therefore the evidence and arguments bearing on the question are of an *a posteriori* nature. A case in point is the situation described by the physical anthropologist J. B. Birdsell. He has suggested that the observed distribution of the phenotype "tawny hair" in the semi-isolated aboriginal tribes of Australia is indicative of the past outward diffusion from a single center of origin of a new allele with a probable selective advantage (Birdsell, 1950). The same line of reasoning can be applied to explain the distribution of any genetic trait throughout a series of populations, but such an explanation, based as it is on undocumented past dynamics, must remain purely speculative.

Reduction of Inbreeding

A second effect of gene flow is that, as the inevitable result of exogamous matings, it reduces the potential level of random inbreeding and its effects, which may be attained in an endogamous population. This effect of exogamy is shown on Tristan da Cunha, where, "[w]ere it not for the two ship-wrecked Italians who arrived in 1892 and the two Anglo-Irish women in 1908, the inbreeding coefficients would have been considerably higher" (Roberts, Chapter V). Exogamy, with its reduction of inbreeding, is a process which has rapidly accelerated in the twentieth century. The trend has been well-documented in France, where the true average coefficients of consanguinity ($F \times 10^5$) decreased from 85.6 to 22.4 in the period 1926 to 1958. In fact, with the exception of a few areas which remain strongly endogamous, endogamy almost disappeared throughout the country during the years 1926 to 1958 (Sutter and Gaux, 1964).

The increase in exogamy and the accompanying reduction in inbreeding has considerable genetic impact on the genotypes in a population. With a greater number of unrelated gametes available for mating, the level of homozygosity in an admixed population is lowered, as reflected in the decrease in the number of individuals affected by rare homozygous traits (the Wahlund effect). In fact, one generation of random mating between two or more formerly isolated populations will eliminate at a single locus all the accumulated genotypic effects of previous inbreeding in the populations involved. One would predict, therefore, that if the Amish demes were suddenly to engage in totally random mating with the surrounding populations of Pennsylvania, in all probability the offspring of panmixis would not exhibit any of the recessive disorders described by McKusick (Chapter V). A similar outcome might be predicted for the frequency of albinism in the Hopi. Woolf and Dukepoo (Chapter V) predict that, as more and more Hopi are exposed to the factors that promote mating outside their population, "[t]he frequency of albinism among the Hopis will decrease rapidly with the decline of their culture."

As the number of homozygous individuals decreases, the number of heterozygotes for both rare and common genes increases. This increase in heterozygotes may act to upset equilibrium in a population's gene pool. Theoretically, this situation occurs if a state of equilibrium had been reached between mutation and selection in the unmixed population by the operation of selection against lethal or sublethal homozygous recessives. With admixture the balance of the forces acting on the gene pool is disturbed, and the gene frequencies as maintained by these forces are also disturbed. The recessive-allele frequencies will then be expected to increase in the gene pool through a combination of mutation and segregation because they are not "hidden" from selection in the heterozygote. Eventually a new equilibrium point, as determined by the lower inbreeding rate, is established. However, the new equilibrium is reached only after hundreds of generations, and it is probable that in modern large exogamous human populations the recessive gene frequencies are not in equilibrium with the selective forces.

Heterosis and Recombination

With increased heterozygosity, the frequency of any deleterious or lethal recessive gene in the gene pool will be even greater if the heterozygote is more adaptively fit than either homozygotes (see Chapter IV). The increased total genotypic fitness of individuals and populations through heterozygosity—that is, the idea of *hybrid vigor*, or *heterosis*— is frequently suggested as a genetic consequence of admixture. Hybrid vigor, as measured by increased viability and/or fertility, has been observed in numerous laboratory cross-breeding experiments and in many classes of plants and animals where higher production is obtained from the progeny of crosses between lines or strains than from individuals of genetically isolated populations or lines. Yet, however convincing such data are for the hybrid-vigor concept, the biological reasons for the advantage of heterozygosity are imperfectly understood. One explanation has been offered by the geneticist F. Michael Lerner (1954), in the concept of *genetic homeostasis*. Genetic homeostasis refers to "the property of the population to equilibrate its genetic composition and to resist sudden changes" (Lerner, 1954). In Lerner's view, heterozygosity at the individual or population level is advantageous in that it provides each with a greater buffering ability to withstand environmental changes and therefore insures a greater chance of survival and reproduction. Homozygotes lack this favorable effect of alternative alleles, and are therefore less able to adjust to systematic or nonsystematic changes in the environment.

In contrast to heterosis, which supposedly increases fitness, an opposite effect of admixture has been suggested, the *recombination effect*. Theoretically, this effect of cross-mating reduces the fitness of the hybrid population. The assumption here is that the parental populations, which have undergone numerous generations of selection for favorable combinations of genes, are "homeostatic"; that is, in genetic balance with their environments. Cross-breeding produces new genotypes and populations which, because they have not been subjected through time to the stabilizing forces of natural selection, are less fit than the parental genotypes. Theoretically, then, depending on the differing magnitudes of heterozygosity, recombination, and selection, hybrid individuals or populations may be either more or less fit than were their parents. Whether hybridization is beneficial or detrimental is a difficult question to resolve, either in theory or in practice.

Studies investigating the concept of hybrid vigor in human populations have been primarily directed toward (1) demonstrating the "depressive" effects of inbreeding on individual offspring (see the discussion of "genetic load" in the Overview in Chapter III) or within endogamous isolates and/or (2) demonstrating the reverse in the offspring of exogamous matings or within hybrid populations.

Using the first approach, Schull and Neel in Japan (1965) and Roberts in Tristan da Cunha (1967) have suggested that mental capabilities may be susceptible to the depressive effects of inbreeding. The Japanese data showed that with increased inbreeding there was a consistently and significantly lower performance by children on seven indicators of school attainment. On Tristan da Cunha there was a significant association between a lower IQ and a higher inbreeding coefficient. Other investigators have postulated a correlation between the breakdown of endogamous isolates and the consequent increase of heterozygosity with the secular increase in body size in modern populations. Combining both approaches, the physical anthropologist Frederick Hulse (1958) has compared the stature and cephalic index of adults whose parents came from different villages in a Swiss canton with those whose parents come from the same village. He found that those of exogamous matings were taller and less brachycephalic than those of endogamous matings and concluded that hybrid vigor was responsible for the observed differences. Similar evidence comes from France, where a significant negative correlation has been shown between stature and the coefficient of inbreeding, results which have been interpreted as in agreement with the heterosis hypothesis (Schreider, 1967).

The results of these types of studies have been criticized as relatively inconclusive in that it is impossible to separate the effects of heterosis, recombination, natural selection, inbreeding, the dissimilar environments which the parental and hybrid populations usually occupy, and the possible genetic and nongenetic differences between endogamous and exogamous individuals (cf. Morton, Chung, and Mi, 1967). In an attempt to resolve some of these problems, the geneticists N. Morton, C. Chung, and M. P. Mi (1967) undertook a carefully planned investigation of the racially mixed population of Hawaii. Their study represents the most extensive effort to date designed to investigate the possible effects of heterosis or recombination on first-generation hybrids in a large admixed human population. It involved sex and twinning ratios, mortality rates, anthropometrics, and congenital malformation data, ABO and Rh blood typing, pertinent physical data on the mothers, and such sociological factors as might bear on the clinical material on 179,327 births and fetal deaths from 1948 to 1958. The study showed that first-generation hybrids are intermediate in size, mortality, and morbidity between the parental groups, an effect which has also been demonstrated in other groups of hybrid composition (J. C. Trevor, 1953 as cited in Morton, *et al.*). But, after detailed statistical analyses of the data, the investigators concluded that they could find no evidence to support either the hypothesis of heterosis or that of recombination effects on man. This is where the matter rests at the present.

Genetic Clines

The majority of the studies on gene flow and admixture deal not so much with the postulated biological effects of hybridization as with the quantitative measurement of the impact of gene flow on a population's gene pool, as inferred from gene frequency distributions. Substantial effort has been directed toward mapping gene-frequency distributions for purposes of detecting *genetic clines* which supposedly are the result of gene flow (see Glass, 1954). Populations distributed along a geographic line often show a consistent gradation in phenotypic or allelic frequencies, and the line connecting such a gradated series of populations is referred to as a *cline*. The assumption is that a cline reflects the number of migrants who have penetrated a region and mated with the indigenous populations. The indigenous populations, to the extent that they restrict cross-matings, act as genetic barriers to gene flow. The number of migrants and the frequencies of the introduced genes are, therefore, inversely proportional to the "distance" from their source population. The distance can be measured in terms of the numbers and distribution of the isolates in the area experiencing migration. This is the concept of *genetic space*; that is, the space measured in terms of the number of genetic barriers (isolates) penetrated (Birdsell, 1950).

Many clines in the human blood-group alleles have been demonstrated. One of the classic examples is the gradual decrease in the frequency of blood-group-allele I^B from Central Asia to Western Europe. Proceeding westward, the I^B frequency diminishes from 25-30 percent in Central Asia to a frequency approaching zero in areas of Western Europe. It has been suggested that the gradient is the result of the Mongolian invaders carrying blood-group B who pushed westward from 500 A.D. to 1500 A.D. in numbers which decreased in proportion to their distance from Central Asia (Candela, 1942).

Theoretically, a cline reflects past *continuous migration* with gene flow through a series of populations, showing a graded genetic similarity. A different situation occurs when *discontinuous migration* results in the sudden admixture of genetically very diverse populations. The contact and admixture of very different gene pools usually occurs when a group migrates over great distances from one area to another without either contact or mating with intervening populations. Although the genetic con-

sequences of both types of migration patterns, continuous and discontinuous, are the same—that is, homogenization of the populations involved
—discontinuous migration and admixture often produces new, genetically
distinctive *hybrid populations*. A hybrid population, which is the result of
the admixture of two or more recognizably very different gene pools, is
often referred to by such common terms "mixed race" or "mestizo."[1]

There are numerous unique hybrid populations in the world which are
the result of the discontinuous type of migration and admixture. The people
of Hawaii are a striking example of the results of the admixing of many
diverse gene pools. Immigrants of Caucasian, Chinese, Filipino, Japanese, Puerto Rican, and Korean ancestry have mated extensively with one
another and with the indigenous Hawaiians. Census figures show that in
1950, 18.9 percent of the Hawaiian population was classified as of "mixed
ancestry." In addition, the total amount of admixture in Hawaii is increasing each generation. For example, during the period 1948 to 1958 an
estimated 35 percent of all children born were of mixed racial origin, a
proportion twice as great as that of the parental generation (Morton,
Chung, and Mi, 1967). However, these figures undoubtedly underestimate
the actual case, based as they are on arbitrary census rules of racial
classification.

The countries of the western hemisphere also contain numerous distinctive hybrid populations. Immigrants of very diverse genetic origin from
Europe, Asia, and Africa crossed the barriers of the Pacific and Atlantic
Oceans to populate continents already occupied by numerous and varied
populations of American Indians. Here they all intermated to create such
populations as the *trihybrid* isolates of the United States. There are at
least thirty-five of these "triracial" isolates, ranging in size from less than
100 to 30,000 individuals, whose members are derived from Caucasian,
Negro, and Amerindian ancestry (Beale, 1957).[2] In other areas of the New
World similar trihybridization has been responsible for the present genetic composition of much larger populations, such as the peoples of northeastern Brazil, who are also mixtures of Caucasian, Negro, and Amerindian. In Brazil the admixture has gone about 97 percent of the way to total
racial panmixia (Krieger, *et al.*, 1965), a striking illustration of the homogenization of former widely diverse gene pools.

Of greater distribution and size than the trihybrid populations of the
New World are the *dihybrid* Negro populations, which are primarily the
result of admixture of Caucasians with the African Negroes who were

[1]The Spanish term "mestizo" is also used in a cultural sense to refer to a person or population which has assumed the Spanish language and culture of Latin America. However, in
another sense it may carry a biological connotation, referring to a person or population of
mixed Indian and European genetic ancestry.

[2]The genetic constitution of a number of those isolates has been studied, especially by the
physical anthropologist William Pollitzer (cf. Pollitzer, Menegaz-Bock, and Herion, 1966;
Pollitzer, *et al.*, 1967).

brought to the New World as slaves. Numerous workers have investigated the rates, amount, and dynamics of the gene flow from white populations into the gene pools of various New World Negro populations (see references at end of Reed article below). The articles in this chapter by T. Edward Reed and P. L. Workman and his associates, illustrate this type of admixture research in North American Negro populations.

Calculations of Admixture in American Negro Populations

Calculations to obtain the estimated amount of white admixture (M), that is, the proportion of genes derived from the white parental population, in the dihybrid Negro populations of the New World are done on the basis of the following formula (modified from Glass and Li, 1953):

$$M = \frac{q_h - q}{q_o - q}$$

For application of this formula it is necessary to know (1) the gene frequencies of the two parental populations (q_o and q), and (2) the gene frequency of the dihybrid population (q_h). The two basic assumptions which underlie the correct application of the admixture formula are: (1) that the hybrid population (q_h) is a true dihybrid in that two, and no more than two, parental populations have contributed to its gene pool,[3] and (2) that the gene frequencies used in the calculations are completely representative of the gene pools of the three populations involved (q_h, q, q_o).

Problems in Calculating Admixture

The first assumption holds that the North American Negro populations are true dihybrids, but this premise has not been conclusively demonstrated. Although some workers have discounted the likelihood of any significant gene flow from American Indian populations to the African slaves, their conclusion has been made strictly on the basis of the blood-group frequencies (cf. Glass, 1955). However, other studies have recorded various percentages of Negroes as claiming some American Indian ancestry; for example, in one sample of Mississippi-born Negroes, 69 percent claimed Indian ancestry (Meier, 1949). In view of such evidence, certain questions can be raised about the validity of discounting Indian genetic contribution to the Negro gene pools by gene-frequency analysis alone. Because the American Indian is much more similar in his blood-group frequencies to the European than to the African, the European and Indian contributions reinforce each other in shifting the American Negro gene frequencies away from those of the African. It is therefore ex-

[3]For calculating admixture in hybrid populations derived from more than two parental populations, the *least-squares* method is used, which considers simultaneously a number of loci and can be applied no matter how many populations are intermixing (Roberts and Hiorns, 1962).

tremely difficult to separate their respective contributions solely by examination of a hybrid gene pool. If one assumes only dihybridization, the mutual European-Indian reinforcement merely serves to make the estimates of the European contribution to the American Negro gene pool too high, in a degree proportional to the magnitude of the Indian contribution (Roberts, 1955). This problem of overestimating the Caucasian contribution is a difficult one to resolve. Accurate estimates of the original gene frequencies of the now extinct or highly admixed American Indian tribes of the South and East, who must have been the groups contributing to the American Negro gene pools, are impossible to obtain. To a certain extent, therefore, estimates purporting European admixture alone are provisional, especially where Indian ancestry is claimed.

The second assumption underlying the application of the admixture formula is that the gene frequencies used are representative, but it, too, is not fully substantiated in the American Negro admixture studies. As T. E. Reed points out in the article included here, four criteria must be met before the gene frequencies used can be considered totally representative of the populations necessarily assumed for the admixture calculations:

(1) The exact ethnic compositions of the two ancestral populations, African and Caucasian, are known.

(2) No change in gene frequency (for the gene in question) between ancestral and modern populations either of African Negroes or of American Caucasians has occurred.

(3) Interbreeding of the two ancestral populations is the only factor affecting gene frequency in U.S. Negroes—that is, there has been no selection, mutation, or genetic drift.

(4) Adequate samples (that is, samples that are unbiased, from correct populations, with small standard error) of the modern descendants of the ancestral African Negroes and U.S. Caucasians, and of modern U.S. Negroes, are available.

In practice these criteria are exceedingly difficult, if not impossible, to meet. Considering the work done on American Negro admixture, Reed states, "None of these criteria has been shown to be fully met in any study" (Reed, below). Yet, in all fairness, it should be noted that no investigator can possibly hope to resolve all the problems inherent in admixture studies of actual human populations.

For example, it can never be known how representative the gene frequencies drawn from the modern white and the West African Negro populations are of the original parental populations because it is not possible to identify the original populations with certainty. This problem is especially evident when one attempts to identify the original slave populations, because during the slave period a shifting from area to area occurred as different African Negro populations were drawn upon by the slave traders. Slave records provide some information about the sources. The most accurate records come from Charleston, South Carolina for the period 1733

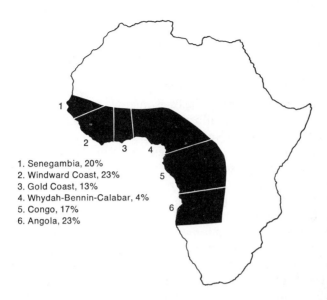

1. Senegambia, 20%
2. Windward Coast, 23%
3. Gold Coast, 13%
4. Whydah-Bennin-Calabar, 4%
5. Congo, 17%
6. Angola, 23%

FIG. 1. The six slave areas from which slaves were imported to Charleston from 1733 to 1807 (after Pollitzer, 1958).

to 1807. These records indicate that the major slave-trade sources may be grouped into six regions (Figure 1). On the basis of this evidence geneticists use modern West African tribal gene frequencies to represent the parental populations ancestral to the New World Negro populations. Yet there is no way of determining with precision the populations in West Africa from which the slave populations were drawn, and no way in which the modern West African gene frequencies can be weighted to take into account the different proportions of peoples who were transported to the United States (Roberts, 1955). The only solution is, as Reed suggests, to "draw inferences of varying degrees of vigor as suitable data become available."

With this goal in mind, Reed has reviewed the previously published estimates of M for American Negroes in an attempt to arrive at the most critical estimate possible within the limitations imposed by the data. He concludes that at present the best estimate of M is obtained by the Fy^a blood-group gene in that (1) good survey data are available on Fy^a gene frequencies in the African slaving areas; (2) Fy^a may be almost an ideal "marker" gene, existing at frequencies of about 0.43 in United States Caucasian populations as contrasted to frequencies of less than 0.02 in African populations from the slave area; and (3) strong selection does not seem to be operating at this locus. Using Fy^a, he calculates a best possible estimate of $M = 0.2195 \pm 0.0093$ for the American Negro population of Oakland, California. The reason given by Reed for using the Fy^a gene in his admixture calculations are unquestionably sound. However, because estimates derived from any locus are subject to sampling error, it is always preferable to use a number of loci for calculating admixture.

Variation in Negro Populations

An important point is that Reed's estimate is specific for a single American Negro population in California, and no inferences can be drawn from this estimate to any other Negro population. In fact, those estimates which are given for the over-all accumulated white admixture of the American Negro, are practically useless (for example, see Glass 1955). As would be expected, comparison of the gene frequencies from various Negro populations shows that there is no homogenous "New World Negro" group, and there is no biologically meaningful "American Negro" individual per se. There are only Negro *populations* and they vary widely in the composition of their gene pools as a result of the different environmental, cultural, and social forces to which they have been exposed. For example, the calculated values of accumulated admixture are significantly higher for Brazilian Negro populations than for those of any known United States Negro population. Within the United States alone the degree of hybridization of different Negro populations is quite variable. Especially noticeable is the greater white admixture exhibited by "nonsouthern" Negro populations as compared to "southern" populations (Table VI–2). The southern Negro population described in this chapter by Workman *et al.* shows significantly less admixture than that of Reed's California population. Workman and his associates give the best estimate of M as 0.104 in the Negro population from Claxton, Georgia. Its gene pool appears to contain approximately one-half the amount of white genes in the gene pool of Reed's California population.

Although gene-frequency analysis of the type used in these two articles shows that different Negro populations have undergone varying degrees of admixture, this type of study is only one, very limited means of approaching the whole problem of gene flow. Genetic analysis merely opens

TABLE VI-2. PERCENTAGE OF WHITE ADMIXTURE IN SOME
NEW WORLD NEGRO POPULATIONS

Negro Population	Estimated Admixture (approx. %)
Brazil (Source: Saldanha, 1957)	
São Paulo	55
Rio de Janiero	55
Bahia	35
United States (Source: Reed, below)	
Nonsouthern	19-26
Southern	4-11

the way for research directed to the historical and social forces which have controlled not only *entrance* of outside genes into a group, but also the *diffusion* of the genes within a group after their introduction.

Social Factors and Gene Flow

If two or more groups are in contact and the opportunity for mating exists, the first basic question involves the social factors that encourage or prohibit cross-mating; that is, the factors that encourage or prohibit the introduction of outside genes. These factors must be sought in the different contact histories of the parental populations involved. In Brazil, for example, where the Negro populations are more admixed than those in the United States, the social barriers to interracial matings have always been less rigid. The Portuguese were the predominant Caucasian population in Brazil until about 1872. In contrast to most other European groups, they participated relatively freely in cross-mating, and relative to their total numbers contributed disproportionately to the admixed gene pool.

While Portuguese males took part in miscegenation, reproduction of the African male slaves was checked. Slave reproduction was considered uneconomical as long as new slaves were available and cheap. Male slaves therefore often labored at some distance from the plantation house where the female slaves were employed, and the two sexes were commonly housed in separate slave quarters (Krieger, *et al.*, 1965).

In Chapter II some of the social and demographic factors were discussed which operate to keep modern Caucasian and Negro populations in the United States distinct and rates of admixture low. An extreme case of reduced gene flow is presented by the Gullah Negroes of Charleston, South Carolina. Comparison of the Gullah with other United States Negro populations shows that they have an unusually small amount of Caucasian ancestry ($M = 0.0366$) and that their gene frequencies are remarkably close to modern West African populations. The conclusion is that the Gullah have been quite genetically isolated, a situation paralleled by their social isolation, wherein they have retained a unique dialect and social customs which show certain similarities to those of West African tribes. It has been suggested that their genetic and social isolation may have resulted from the disproportionate ratio of the races in the coastal plantation area of South Carolina where the Negro slaves vastly outnumbered the whites. Apparently, whenever large numbers of Negroes are in close association with a smaller dominant white society the situation seems to operate against substantial Negro-white admixture (Pollitzer, 1958).

The mathematical formulas and statistical laws concerning expected rates of gene flow and the resulting gene frequencies and genotypic combinations do not, and generally cannot, take the nonbiological variables into account, important as they may be. For example, if the number of generations of contact between the parental generations is known, and if a constant gene flow through time is assumed, the average rate of gene flow per generation (m) can be mathematically calculated. The formula is

$$(1 - m)^k = \frac{q_n - q}{q_o - q},$$

where k is the number of generations of admixture.

Assuming ten generations of admixture, the modal value for gene flow per generation (m) in American Negro populations has been calculated to be between 0.02 and 0.025 (Roberts, 1955), somewhat less than the original figure of 0.036 given by Glass and Li (1953). But the rate of gene flow between the white and Negro populations has not been constant from the 1700's to the present, having been profoundly affected by such historical events as the Civil War, emancipation of the slaves, the Reconstruction Era, changing attitudes toward interracial marriage, and so on. Given the importance of the sociocultural variables in determining rates of gene flow, one is led to question the usefulness of an estimation of gene flow per generation which is based on purely mathematical grounds and unproven assumptions.

A second question regarding the dynamics of gene flow centers on the fate of the outside genes once they have been introduced into a population. Barriers to gene flow may exist not only between populations, but also within the population receiving the outside genes. For example, within the recipient population a certain degree of homogamy as dictated along the lines of educational and economic levels may parallel homogamy for physical traits. The combined homogamy drastically disturbs the equal chance for all gametes to meet, and acts to bring about observable stratification in the population so that certain physical and social traits are found in close association. In Kingston, Jamaica, for example, lighter skinned individuals constitute a greater proportion of the upper classes. It is obvious that in Jamaica diffusion of white genes has not been random throughout the total population, but has been determined to a large extent by nonbiological causes, as when the shade of one's spouse becomes a measure of social mobility.

In summary, the chances of any gamete in one population meeting any gamete in other cross-mating populations are not random, nor are the chances of any introduced gamete meeting any gamete within an admixed population random. Rather, the chances are governed by a number of intervening sociocultural variables.

Admixture and Natural Selection

The estimation of admixture is further complicated by the fact that natural selection may also have been simultaneously operative in changing the gene frequencies of an admixed population. When a population migrates, it may settle in a very different environment from that in which it evolved. With a change in the environment the selective forces operating on a population's gene pool may be significantly altered. The New World Negro populations are a case in point; the original immigrant slaves were brought to live in an environment very different from that of their native Africa. It would appear likely that both gene flow and selection may have been involved in determining the differences in gene frequencies observed between American Negro and West African populations.

As previously discussed in Chapter IV, given the lack of data with generational depth from human populations, it is difficult to investigate the actual effects of natural selection on human gene pools. One solution to the problem may lie in the analysis of admixed gene pools which evince the cumulative effects of selection and gene flow over several generations and may therefore enable the investigator to distinguish the effects of the selective forces on various genetic loci (Workman, 1968). The assumption underlying this approach is that if there is no differential gene flow with respect to the genetic loci involved, then the estimated admixture (M) should be identical for all loci and all alleles; if M varies between loci, the variation may be the result of selection operating differentially on the loci in question. The problem is, then, which genetic loci have been the most responsive to changes in the selective forces?

As discussed in Chapter IV, there is substantial evidence that in endemic-malarial areas selection is strongly operative on the hemoglobin, G-6-PD, and thalassemia loci. It could be predicated that eradication of endemic malaria would remove the selective advantage enjoyed by the heterozygote, and the abnormal gene frequency would decline in a gene pool under the relaxation of selective pressures. American Negro populations, therefore, should provide an excellent opportunity to examine the selection hypothesis. Here some gene frequencies would be expected to decrease rapidly under the combination of gene flow and the relaxation of selective pressure, while other frequencies would drop as a result of gene flow alone.

The study included here by Workman and his associates on the Claxton, Georgia Negro population is concerned with using admixture estimations in this way; that is, to detect those polymorphisms upon which natural selection is most likely to be operating. Their estimations of M for 15 different polymorphic loci were separable into two distinct groups (I and II). Group II, containing the alleles T (PTC tasting locus), Hp^1, G-6-PD, Hb^s, and Tf^{D1} (transferrin),[4] gave estimates of M (0.4 to 0.7) significantly higher than those of Group I M (0.1 to 0.2), which contained the common blood-group alleles. Gene frequencies from other New World

[4] A later publication by Workman (1968) indicates that the figures for TF^{D1} are not yet sufficiently reliable to be used to assess gene flow vs. selection.

Negro populations exhibit a similar trend (Motulsky, 1964; Workman, 1968). On the basis of the two different M estimates, the authors conclude that the Group II polymorphisms in the New World environment are unstable, possibly transient, polymorphisms. Certainly if one accepts the malarial hypothesis, the significant drop in the frequencies of the G-6-PD and Hb^s is predictable within an environment free of endemic malaria. It has also been suggested that the Hp^1 allele confers an advantage to individuals with abnormal hemoglobins, and that selection may operate simultaneously against both the Hp^1 allele and the abnormal hemoglobins in the New World (Workman, 1968). The evidence bearing on the possible selective forces operating on the PTC taste locus (T allele) is too scanty to permit any conclusions regarding the dynamics involved in its significant decrease in the Claxton Negro population.

There are limitations on the conclusions which can be drawn about selection on the basis of admixture estimations. For example, because the ancestral-parental-population gene frequencies are unknown, interpretations regarding the role that selection may have played in changing the gene frequencies of a hybrid population must be done cautiously. But as Workman (1968) states, it seems reasonable to suggest "that studies of gene flow can be successfully used to locate the alleles upon which selection has been operating in particular populations."

Bringing data from hybrid populations to bear on questions other than those directly related to gene flow per se illustrates the usefulness of admixture studies to analyze various components of human variability. For example, Morton and his associates, in their Hawaiian study, have called attention to the ways in which studies of admixed populations can assist in enabling an investigator to separate the genetic and environmental causes of specific phenotypic differences between populations. They were able to show in their study of admixed Hawaiians that the incidence of the congenital anomaly spina bifida is related to the environment and is independent of genetic racial affinities and unaffected by inbreeding or cross-mating. In addition, the analyses of cross-matings and admixed populations provide methods to answer questions about genetic factors such as degrees of dominance which are not amenable to investigation in small endogamous populations. As the breakdown of genetic isolates continues throughout the world at an ever accelerated rate, valuable data should be forthcoming from large admixed human populations. The data should provide new insights into the dimensions and dynamics of genetic differences between populations, insights which cannot be entirely provided by the traditional study of isolates.

American Negro Populations

See the following readings: "Caucasian Genes in American Negroes," by T. Edward Reed, p. 427; and "Selection, Gene Migration and Polymorphic Stability in a U.S. White and Negro Population," by P. L. Workman, B. S. Blumberg, and A. J. Cooper, p. 449.

REFERENCES

1. BEALE, C. L., 1957. American triracial isolates, their status and pertinence to genetic research. *Eug. Quart.*, 4: 187–196.
2. BIRDSELL, J. B., 1950. Some implications of the genetical concept of race in terms of spatial analysis. *Cold Spr. Har. Sympos. Quart. Biol.*, 15: 259–314.
3. CANDELA, P. B., 1942. The introduction of blood-group B into Europe. *Hum. Biol.*, 14: 413–443.
4. GLASS, B., 1954. Genetic changes in human populations, especially those due to gene flow and genetic drift. *Advances in Genetics*, 6: 95–139.
5. ———, 1955. On the unlikelihood of significant admixture of genes from the North American Indians in the present composition of the Negroes of the United States. *Amer. J. Hum. Genet.*, 5: 1–20.
6. ——— and C. C. Li, 1953. The dynamics of racial intermixture—an analysis based on the American Negro. *Amer. J. Hum. Genet.*, 7: 368–385.
7. HULSE, F. S., 1958. Exogamie et hétérosis. *Arch. Suisses d'Anthr. Gén.*, 22: 103–125.
8. KRIEGER, H., N. E. MORTON, M. P. MI, E. AZEVEDO, A. FREIRE-MAIA, AND N. YASUDA, 1965. Racial admixture in North-eastern Brazil. *Ann. Hum. Genet.*, 29: 113–125.
9. LERNER, I. M., 1954. *Genetic Homeostasis*. New York, Wiley.
10. MEIER, A., 1949. A study of the racial ancestry of the Mississippi College Negro. *Am. J. Phys. Anthro.*, 7: 227–240.
11. MORTON, N. E., C. S. CHUNG, AND M. P. MI, 1967. *Genetics of Interracial Crosses in Hawaii*. Monographs in Human Genetics, Vol. 3. New York, S. Karger.
12. MOTULSKY, A. G., 1964. Hereditary red cell traits and malaria. *Am. J. Trop. Med. Hyg.*, 13: 147–158.
13. POLLITZER, W. S., 1958. The Negroes of Charleston; a study of hemoglobin types, serology and morphology. *Amer. J. Phys. Anthro.*, 16: 241–263.
14. ———, R. M. MENEGAZ-BOCK, AND J. C. HERION, 1966. Factors in the microevolution of a triracial isolate. *Amer. J. Hum. Genet.*, 18:26–38.
15. ———, D. S. PHELPS, R. E. WAGGONER, AND W. C. LEYSHON, 1967. Catwaba Indians: Morphology, genetics and history. *Amer. J. Phys. Anthro.*, 26: 5–14.
16. ROBERTS, D. F., 1955. The dynamics of racial intermixture in the American Negro—Some anthropological considerations. *Amer. J. Hum. Genet.*, 7: 361–367.
17. ———, 1967. Incest, inbreeding and mental abilities. *Brit. Med. J.*, 4: 336–337.
18. ——— AND R. W. HIORNS, 1962. The dynamics of racial intermixture. *Amer. J. Hum. Genet.*, 14:261–277.
19. SALDANHA, P. H., 1957. Gene flow from white into Negro populations in Brazil. *Amer. J. Hum. Genet.*, 9: 299–309.
20. SCHREIDER, E., 1967. Body height and inbreeding in France. *Amer. J. Phys. Anthro.*, 26: 1–3.

21. SMITH, C. A. B., 1969. Local fluctuations in gene frequencies. *Ann. Hum. Genet.*, 32: 251–260.
22. SUTTER, J. AND J. M. GAUX, 1964. Decline of consanguineous marriages in France from 1926 to 1958. *Eugen. Quart.*, 11: 127–140.
23. TINDALE, N., 1953. Tribal and intertribal marriage among the Australian Aborigines. *Hum. Biol.*, 25: 169–190.
24. WORKMAN, P. L., 1968. Gene flow and the search for natural selection in man. *Hum. Biol.*, 40: 260–279.

Caucasian Genes in American Negroes

T. EDWARD REED

It is very difficult to describe the genetic history of a large, defined human population in a meaningful way. As a result there have been few opportunities, at the population level, to study the consequences of known genetic events in the recent past of modern populations. The Negro population of the United States, however, is one of the exceptions to these generalizations. The American individual to whom the term *Negro* is applied is almost always a biracial hybrid. Usually between 2 and 50 percent of his genes are derived from Caucasian ancestors, and these genes were very probably received after 1700. While it is obviously of social and cultural importance to understand Negro hybridity, it is less obvious that there are several pertinent genetic reasons for wishing to know about the magnitude and nature of Caucasian ancestry in Negroes. Recent data, both genetic and historical, now make possible a better understanding of American Negro genetic history than has been possible heretofore. Here I review and criticize the published data on this subject, present new data, and interpret the genetic significance of the evidence.

In order to put the genetic data in proper context, I must first give a little of the history of American slavery. The first slaves were brought to what is now the United States in 1619. Importation of slaves before 1700 was negligible, however, but after that date it proceeded at a high rate for most of the 18th century. Importation became illegal after 1808 but in fact continued at a low rate for several more decades (*1, 2*). The total number of slaves brought into the United States was probably somewhat less than 400,000 (*3*). Charleston, South Carolina, was the most important port of entry, receiving 30 to 40 percent of the total number (*4*). More than 98 percent of the slaves came from a very extensive area of West Africa and west-central Africa—from Senegal to Angola—and, in these areas, from both coastal and inland regions. Shipping lists of ships that brought slaves to the United States—and to the West Indies, often to be sent later to the United States— provide a fairly detailed picture of the geographic origins of the slaves and a less complete picture of their ethnic origins. Table 1 gives the approximate proportions of American slaves

T. Edward Reed, *a human geneticist, was born in Gadsden, Alabama in 1923. He received his B.A. at the University of California in 1948, and his Ph.D. at the University of London in 1952. At the present time he is Professor of Zoology and Anthropology and an Associate Professor of Pediatrics at the University of Toronto. His major field of interest and research is human-population genetics, particularly the study of natural selection as it relates to the human polymorphisms. He has done human-population field research in Caucasian populations in Michigan, California, Ontario, and Quebec; and Negro populations in California. His numerous publications deal with various aspects of human genetics. Some of his work of particular relevance to population genetics includes "Huntington's chorea in Michigan. 1. Demography and genetics," with J. H. Chandler, Amer. J. Hum. Genet., 10: 201–225 (1958); "Polymorphism and natural selection in blood groups," in B. S. Blumberg, Ed., Proceedings of the Conference on Genetic Polymorphisms and Geographic Variations in Disease (New York, 1961); and "A search for natural selection in six blood group systems and ABH secretion," with H. Gershowitz, A. Soni, and J. Napier, Amer. J. Hum. Genet., 16:161–179 (1964).*

brought from the eight major slaving areas of Africa. The contribution from East Africa is seen to be negligible, whereas the area from Senegal to western Nigeria contributed about half the total and the region from eastern Nigeria to Angola contributed the other half. An earlier tabula-

TABLE 1. AFRICAN ORIGINS OF SLAVES IMPORTED INTO THE NORTH AMERICAN MAINLAND [DATA OF CURTIN (37)]. DISTRIBUTION BY AREAS IS APPROXIMATE AND IS AN AVERAGE OF DATA FOR VIRGINIA (1710–1769), FOR SOUTH CAROLINA (1773–1807), AND FOR THE BRITISH SLAVE TRADE (1690–1807).

Coastal region of origin	Approximate present area	Peoples	Approximate proportion from region
Senegambia	Senegal and Gambia	Mainly Bambara and Malinke (from interior)	0.133
Sierra Leone	Sierra Leone	Sierra Leone, Guinea, Portugese Guinea peoples, plus Bambara and Malinke	.055
Windward Coast	Ivory Coast, Liberia	Various peoples of area	.114
Gold Coast	Ghana	About ¾ Akan people from southern part, the rest from northern part	.159
Bight of Benin	Togo, Dahomey, Nigeria west of Benin river	Peoples of Togo, southern Dahomey, and western Nigeria	.043
Bight of Biafra	Nigeria (east of Benin river) to 1°S (Gabon)	About ¾ Ibo, the rest Ibibio and people from Cameroon	.233
"Angola"	1°S to southwest Africa (Gabon, Congo, Angola)	Many peoples of the area, from the coast to far inland	.245
Mozambique and Madagascar			.016
Region unknown			.002

tion for entry at Charleston alone (5) is quite similar, except that the contribution from the Bight of Biafra is much less (0.021 as compared to 0.233) and that from "Angola" is appreciably greater (0.396 as compared to 0.245).

At some early point in American slavery, matings between slaves and Caucasians began to occur. Quantitative data are lacking, and we can say only that most of these matings occurred after 1700. Our concern here is the genetic consequences of the matings—the introduction of Caucasian genes into the genome (or total complement of genetic material) of the American Negro. We could, in theory, estimate the Caucasian contribution to American Negro ancestry in a very simple way *if* certain strict criteria were met. In practice it is not possible to show that all these criteria are met, but this fact has not stopped geneticists, including myself, from making estimates.

The usual estimation procedure is simple and direct. Consider some gene—say the allele A of the ABO blood group locus, whose frequency was q_a in the African ancestors of American Negroes and q_c in the Caucasian ancestors, while in modern American Negroes the frequency is q_n. If M is the present proportion of genes at this genetic locus (and, ideally, at every other locus too) which are derived from Caucasians, and if race mixture is the only process affecting q_n, then, by definition,

$$q_n = Mq_c + (1 - M)q_a \tag{1}$$

and therefore

$$M = (q_n - q_a)/(q_c - q_a) \tag{2}$$

This formula for M, or an algebraic equivalent, was used for all estimates of M given in Table 2 except one. [This one differed only in that three alleles were used simultaneously at one locus to obtain a maximum likelihood estimate for M; for each allele an equation of the type of Eq. 1 could be written, and used to estimate M (6)]. We see that if we know q_a, q_c, and q_n (for a defined area) without error and if there were no factors affecting q_n other than race crossing, estimation of M would be simple. Unfortunately, such is not the case.

Criteria for Critical Estimation of M

Critical evaluation of estimates of M requires complete specification of the needed criteria and judgment on the degree to which these criteria are met. These criteria are simple and obvious, but the demands they make have not always been appreciated. They are as follows.

1) The exact ethnic compositions of the two ancestral populations, African Negro and Caucasian, are known.

2) No change in gene frequency (for the gene in question) between ancestral and modern populations either of African Negroes or of American Caucasians has occurred.

3) Interbreeding of the two ancestral populations is the only factor affecting gene frequency in U.S. Negroes—that is, there has been no selection, mutation, or genetic drift.

4) Adequate samples (that is, samples that are unbiased, from correct populations, with small standard error) of the modern descendants of the ancestral African Negroes and U.S. Caucasians, and of modern U.S. Negroes, are available.

It should be said immediately that none of these criteria has been shown to be fully met in any study. In particular, point 1 is not met, because the detailed ethnic origins of slaves from the various slaving areas are unknown (4). Point 2 can never be met because ancestral gene frequencies are unknown and point 3, at best, can only be inferred from indirect evidence. Point 4 cannot be fully met for African Negroes, since the proportions of various ethnic contributions are only roughly known. The problem is simpler for U.S. Negroes and Caucasians, although marked heterogeneity in values of M between different Negro populations is now known to complicate the matter.

Somewhat more affirmative views on these criteria can also be given, however. *If* it can be shown that gene frequencies in neighboring modern tribes and in populations of adjacent former slaving areas do not differ appreciably, point 1 becomes less important. For example, this appears to be the situation for the ABO blood groups, the best-known genetic system throughout the slaving area. With regard to point 2, since the populations concerned usually were, and are, large, it is probable that this criterion is quite well satisfied. If point 1 is satisfied in the way suggested, point 4 may be met by using large, carefully collected samples. Unfortunately, it is less easy to overcome the problem posed by point 3. This is discussed below.

Review of Published Estimates of M

Table 2 is a tabulation of published estimates of M for American Negroes, beginning with the well-known estimate of 0.31 for Baltimore Negroes given by Glass and Li in 1953 (7). The estimates are grouped according to the authors' statements as to their validity or lack of validity (due to selection) as estimates of M. They are further classified as "southern" (estimates for Georgia, South Carolina, and Tennessee) or "non-southern." As has been noted elsewhere (6, 8, 9), among the presumed valid estimates, all "non-southern" estimates are greater than "southern" estimates. Also, the estimates presumed to indicate selection are usually appreciably higher than the estimates presumed to be valid. Among the "valid" estimates of M, that of Glass and Li (7) is by far the best known, and is often quoted as "the" estimate for the amount of Caucasian ancestry in "the" American Negro (see, for example, *10–14*). A revision of this estimate from 0.31 to 0.216 (*15*) appears to have escaped general notice.

The estimates of Table 2 must be considered in the light of the four criteria given above. As already noted, criterion 1 cannot be strictly satisfied for any estimate because the detailed ethnic origins of the slaves are unknown. The estimates for M in Table 2, however, do not even roughly meet criterion 1, since none of them is based on quantitative information on distribution of origins, such as is given in Table 1. Typically, data from only one or two regions of West Africa are taken to represent the whole slaving area. Ironically, for the best-known estimate, that of Glass and Li (7), Rh blood group data from East and South Africa, as well as from Ghana, were used to represent ancestral Rh blood group frequencies because better data were not then available. Glass, for his revised estimate (15), used only Rh data from Nigeria and Ghana. Of the 540 individuals from Nigeria studied (15), 105 were Ibos, who may be representative of ancestral inhabitants of the Bight of Biafra region, the area of origin of about 23 percent of American slaves (Table 1); the remaining 435 individuals from Nigeria may be representative of the slaves (4 percent) who came from Bight of Benin. The 274 individuals from Ghana studied (15) may be representative of the slaves (16 percent) from that region. The slaves (57 percent) from areas other than Nigeria and Ghana are unrepresented in Glass's revised estimate. These same Rh blood group data were used by later investigators in arriving at their own estimates (8, 9, 16, 17). These critical comments on the best-known estimate are made to illustrate the nature of the problem; similar comments could be made about each of the other estimates of Table 2.

With regard to criterion 4 (adequacy of samples), one can distinguish between (i) adequate representation (by the mean gene frequency used) of the entire slaving area and (ii) adequate sample size (as shown by a small standard error for M). If the gene used has a uniform frequency over the entire slaving area, any large sample from one part of the area could adequately represent the whole. The problem, of course, is to demonstrate uniformity. If, as one would expect, gene frequencies vary from region to region of the slaving area, appropriate samples over the whole area are needed if one is to obtain a properly weighted mean frequency. Neither of these approaches has been used in making any of the estimates. [I made an attempt to confirm the belief that the frequency of certain Gm alleles is near zero in African populations (6) but found that not enough surveys had been made.]

To make the problem more concrete, let us consider Glass's estimate of M (15) in the light of more recent Rh data. For the R^0 allele of the Rh locus, he used 0.5512 for the frequency in West Africa (on the basis of the data from Nigeria and Ghana). The frequencies in present-day U.S. Negroes and Caucasians were found to be 0.4381 and 0.0279, respectively, so that, from Eq. 2, we estimate M to be $(0.5512 - 0.4381)/(0.5512 - 0.0279)$, or 0.216. However, the frequency of R^0 in Liberia is 0.60 (18),

TABLE 2. PUBLISHED ESTIMATES OF THE PROPORTION (M), IN AMERICAN NEGROES, OF GENES THAT ARE OF CAUCASIAN ORIGIN. ALL ESTIMATES EXCEPT THOSE BASED ON GENES Fy^a, Gm^1, $Gm^{1,2}$, OR Gm^5 (AND PERHAPS AK^2) REQUIRE AN ESTIMATE OF AFRICAN GENE FREQUENCY APPRECIABLY DIFFERENT FROM ZERO. WITHIN REGIONS, LOCALITIES ARE LISTED IN CHRONOLOGICAL ORDER OF THE ESTIMATES. STANDARD ERRORS FOR M WERE NOT GIVEN (EXCEPT FOR REFERENCE 6).

Region* and locality	Gene(s)†	Sample size		M	Reference
		Negro	Caucasian		
		Estimates for M presumed by their authors to be valid			
Non-southern					
Baltimore	R^0	907	7,317	0.306	(7)
Baltimore	R^0	907	7,317	.216	(15)
Five areas	R^0, R^1, Jk^b, T, S	96 to 3,156	189 to 7,317	$\cong .20$	(16)
Cleveland and Baltimore	Gm^1, Gm^5	623	249	.310	(11)
Various	$R^0, R^1, R^2, r, M, S, Jk^b, k, Fy^b$			.232–.261	(17)
Chicago	$AK^{2\ddagger}$	1,063	1,315	.13	(14)
Washington, D.C., Baltimore, New York City	R^0, R^1, Fy^a			.20–.24	(8)
Oakland, Calif.	$Gm^1, Gm^{1,2}, Gm^5$	260	478	$.273 \pm .037$	(6)
Southern					
Evans and Bullock counties, Ga.	R^0, R^1	340	331	.104	(9)
Evans and Bullock counties, Ga.	$Gm^1, Gm^{1,2}$ G_C^1	189 231	295 292	.073 $\cong .10$	(12)
Charleston, S.C.	R^0, R^1, Fy^a	515		.04–.08	(8)
James Island, S.C. and Evans and Bullock counties, Ga.	R^0, R^1, Fy^a	394		.09–.12	(8)

432

Estimates of M *presumed by their authors to indicate selection*

	Hp^1	936	865(?)	$\cong .40$	(21)
Non-southern§					
Four areas, mainly Seattle					
Seattle	Hp^1	1,657	?	.478	(8)
Seattle	Gd^{A-}	658♂♂		.490	(8)
Southern					
Evans and Bullock counties, Ga.	T	285	314	.466	(9)
	Hp^1‖	167	145	.42–.70	
	Gd^{A-}	76♂♂		.34–.44	
	Hb^s	247		.46–.69	
Memphis	Tf^{D1}‖	133	107	.495	(8)
	Gd^{A-}	97♂♂		.175	

*An estimate of 0.34, from Hb^S data on 10,858 Negroes, is based on 11 sources in both the North and the South (38). It is therefore not placed in a regional category.   †Locus and alleles used are as follows. Blood groups: Rh (R^0, R^1, R^2, r), Kidd (Jk^b), M-N-S-s (M, S), Kell (k), Duffy (Fy^a, Fy^b); serum protein genes: Gm (Gm^1, $Gm^{1,2}$, Gm^5), haptoglobin (Hp^1), Gc (Gc^1), transferrin (Tf-D1); hemoglobin: HbS (Hb^S); red cell enzymes: adenylate kinase (AK^2), glucose-6-phosphate dehydrogenase deficiency (Gd^{A-}); phenylthiocarbamide tasting (T).   ‡Newly investigated gene. The African frequency of AK^2 is poorly known, but it is assumed to be zero. The 95-percent confidence interval for M is 0.03–0.23, according to my calculation.   §Seven non-southern estimates ranging from 0.270 to 0.685, obtained by Workman (8) (using Hp^1 or Gd^{A-}) on small samples (79 to 238 Negroes) are omitted here.   ‖ "Possibly" reflecting selection.

433

and in Bantu of the Congo (Leopoldville) it is also about 0.60 (*19*). If the true overall value for the slaving area were 0.60, the estimate for M would be 0.283.

With regard to the purely statistical accuracy of the estimates of M, as shown by standard errors, calculation of the standard errors for several pertinent estimates indicates that they may be much larger than the authors may have suspected (*20*). The standard error for Glass's estimate (*15*), for example, is 0.042, giving a 95-percent confidence interval of 0.133 to 0.299. The estimate in Table 2, of 0.13 for M for gene AK^2 (the lowest estimate for the non-southern region) has a standard error of 0.053, producing a 95-percent confidence interval of 0.025–0.234, overlapping Glass's estimate. This large error seems particularly surprising at first, in view of the large sample sizes, but it is explained by the very low AK^2 gene frequencies (< 5 percent). The standard errors of the other estimates appear to be of comparable size or larger (due to smaller sample sizes).

I have said enough to show the deficiencies of most of the estimates of Table 2 with regard to both African gene frequency and statistical accuracy. I should also comment on the classification of M estimates as "valid" (not affected by selection) or as indicating the effects of selection. Classification of an estimate in this way requires a "standard" M that is thought to be free from the effects of selection. Such a "standard" can then be used to determine whether an M estimated for some other gene demonstrates selection. The M estimates from Rh genes R^0 and R^1 have been assigned this role of "standard" by various investigators [Parker *et al.* (*21*) chose R^0 alone; Workman and his associates (*8, 9*) chose R^0 and R^1 in combination]. In addition, M estimates from frequencies of the Fy^a allele of the Duffy blood group locus (*8*) and the Gm^1 and Gm^5 alleles of the Gm serum group locus (*21*) have been considered as possible standards. Yet, as discussed above, it is not possible to prove directly that selection has not affected a particular gene frequency in American Negroes, and no evidence in support of the belief that it has not been offered. We can only draw inferences of varying degrees of rigor as suitable data become available. I attempt in the remainder of this article to draw and apply such inferences.

An Approach to a More Critical Estimate of M

To constitute a critical estimate in the light of the four criteria listed above, an estimate of M should substantially meet three of them—1, 3, and 4 (2 is, of course, untestable). This means that we should (i) have good survey data on gene frequency from most or all of the seven West African and west-central African slaving areas of Table 1; (ii) be able to calculate a mean African gene frequency properly weighted according to the origins shown in Table 1; (iii) have adequate data on Caucasians and U.S. Negroes; (iv) have samples large enough to give an acceptably small

TABLE 3. FREQUENCIES OF DUFFY BLOOD GROUP Fy(a+) IN
WEST AFRICAN AND CONGO (LEOPOLDVILLE) POPULATIONS.

Region	Sample size (N)	Proportion of Fy(a+)*	Reference
Liberia (many tribes)	661	0.00	(*18*)
Ivory Coast	163	.043†	(*18*)
Upper Volta	75	.00	(*18*)
Dahomey	20	.00	(*18*)
Ghana (Accra) and Nigeria (Lagos)	37	.00	(*39*)
Congo (Bantu)	501	.078‡	(*40*)

*Reacting positively with anti-Fy^a, indicating a genotype of $Fy^a Fy$ (most likely), or $Fy^a Fy^b$, or $Fy^a Fy^a$ (rare) (*39*).   †The true proportion is probably zero because the Ivory Coast positive reactions with anti-Fy^a are believed to be incorrect.   ‡The gene frequency for Fy^a is 0.040.

standard error for M; and, very importantly, (v) have some evidence that in U.S. Negroes the gene in question is not subject to strong selection. With regard to points (i) and (ii), an ideal situation is to have a gene which can be shown to be absent or rare in all parts of the slaving area but common in Caucasians. The problem of finding "the" African-ancestor gene frequency is then eliminated, and M is simply q_n/q_c. The Caucasian gene contribution is then directly determinable. It has been claimed that Gm alleles Gm^1, $Gm^{1,2}$, and Gm^5 are of this type (*22*); it is quite likely that they are, but not enough of the slaving area has been surveyed for Gm alleles for us to be sure (*6*).

The Fy^a gene may be almost an ideal "Caucasian gene" for estimating M. Available survey data for regions from Liberia to the Congo (Leopoldville), presented in Table 3, show that in this region (from which about 56 percent of the ancestral slaves came) the mean frequency of Fy^a is probably not over about 0.02. The mean frequency for all Africans of the slave area is probably less than 0.03. The frequency for U.S. Caucasians is about 0.43 (Table 4). Moreover, recent extensive studies in a population of California Negroes revealed no evidence for natural selection (evidence pertaining to fetal and infant growth and viability and to adult growth and fertility) associated with Duffy blood group phenotypes (*23*). Strong selection due to this locus seems excluded, so there is some protection against bias in the estimation of M. Table 4 presents available Fy^a frequency data for U.S. Negroes and for some U.S. Caucasians, and the resulting M estimates. The M estimates for the three non-southern regions studied do not differ significantly, so the estimate 0.2195 ± 0.0093 for California Negroes—the largest of the three samples —may tentatively be used as the best estimate of M for a non-southern

TABLE 4. ESTIMATES OF M DERIVED FROM Fy^a GENE FREQUENCIES FOR AMERICAN NEGROES FROM VARIOUS AREAS. THE FREQUENCY OF THIS GENE IN THE AFRICAN ANCESTORS OF AMERICAN NEGROES IS ASSUMED HERE TO BE ZERO; IF IT IS NOT ZERO, THESE ARE *maximum* ESTIMATES. N = NUMBER IN SAMPLE, q = Fy^a GENE FREQUENCY, S.E. = STANDARD ERROR OF THE ESTIMATE (ALL ESTIMATES BY T. E. REED).

Region and locality	Negroes		Caucasians		$M \pm$ S.E.*	Reference
	N	$q \pm$ S.E.	N	$q \pm$ S.E.		
Non-southern						
New York City	179	0.0809 ± 0.0147			0.189 ± 0.034	(39)†
Detroit	404	.1114 ± .0114			.260 ± .027	(41)
Oakland, Calif.	3146	.0941 ± .0038	5046	0.4286 ± 0.0058	.2195 ± .0093‡	(25)
Southern						
Charleston, S.C.	515	.0157 ± .0039			.0366 ± .0091	(5)
Evans and Bullock counties, Ga.	304	.0454 ± .0086	322	.422 ± .0224	.106 ± .020	(9)

*The q for Oakland Caucasians (who are of West European ancestry) was used in all estimates. $M = q_n/q_c$. †Two other New York City studies (42) are omitted because they involved selection for dark skin color. The data used here were grouped with both anti-Fy^a and anti-Fy^b studies. The observed distribution of four Duffy phenotypes differs from the Hardy-Weinberg expectation at the 0.025 level of significance. ‡If the frequency of Fy^a in the African ancestors were 0.02, this estimate would be 0.181.

436

area. The very small standard error of this estimate reflects both the discrimination power of this "Caucasian gene" and the large sample sizes for the Negro and Caucasian populations. The two estimates from the "Deep South" do differ significantly and should be kept separate. The smaller one, 0.0366 ± 0.0091 from Charleston, appears to justify the statement that these Gullah Negroes have an unusually small amount of Caucasian ancestry (5). It is clear that the data of Table 4 are especially useful in comparing M for different U.S. Negro populations, because the same gene, Fy^a, is used as the basis for all estimates. Any bias due to selection should operate quite similarly in the different Negro populations. The difference between "southern" and "non-southern" M values evident in Table 2 is also marked in Table 4 and must be regarded as real.

Thus Fy^a, for the reasons given, may be the best gene presently available for estimating M. When more African survey data are available, the "Caucasian" alleles Gm^1, $Gm^{1,2}$, and Gm^5 of the Gm locus, used jointly, may be as good. The AK^2 gene (Table 2) may be of some use if further African data establish a general zero frequency, but the low frequency, 0.047, of the AK^2 gene in Caucasians considerably reduces its discrimination power. The K gene of the Kell blood group system is sometimes thought of as a "Caucasian gene," but this is not strictly the case. This gene was present in 8 of 1202 Africans from the Liberia-Dahomey (18) and western Nigeria (24) region, at a mean frequency of 0.0033. The California Negroes of Table 4 (N = 3146) have a K gene frequency of about 0.0083, and the California Caucasians, a K gene frequency of about 0.046 (25). If we consider q_a to be zero, we obtain an estimate of 0.181 ± 0.026 for M for this population—clearly a maximum estimate and not reliable. This maximum does not differ significantly from the Fy^a estimate for this same population. The relatively large standard error here again reflects the low Caucasian gene frequency.

Although a zero q_a is generally preferable, there is one situation where a q_a value appreciably different from zero might yield a useful estimate of M. This could occur when there are sufficiently extensive and detailed data on African gene frequency to make it possible to calculate a mean African gene frequency, with weighting of regional gene frequencies according to the proportions of Table 1. At present, the ABO blood groups provide the only such usable genetic marker [the gene for hemoglobin S is known to be affected by selection, and much less information is available for other loci (26); for selection data on hemoglobin S, see (27)]. Table 5 gives the gene frequencies for genes A and B of the ABO system from relevant surveys in the seven major slaving areas of Table 1.

These extensive surveys reveal an overall uniformity in gene frequency, with the one exception of a somewhat low B frequency for the Bight of Biafra (Ibos). From these mean values for African frequencies of genes A and B and from extensive data on ABO-system distribution in California

TABLE 5. FREQUENCIES OF GENES A AND B OF THE ABO BLOOD-GROUP SYSTEM IN SURVEYS IN THE MAJOR SLAVING AREAS OF AFRICA (SEE TABLE 1); p = FREQUENCY OF A GENE, q = FREQUENCY OF B GENE.

Region	Peoples or population	Sample size (N)	$p \pm$ S.E.*	$q \pm$ S.E.*	Reference
Senegambia	Bambara, Malinke	2,120	0.159 ± .006	0.174 ± .006	(43)
Sierra Leone	Gbah-Mende	1,015	.159 ± .009	.151 ± .008	(44)
Liberia	> 18 tribes†	2,337	.143 ± .005	.148 ± .006	(18)
Gold Coast	Unspecified, from Accra	1,540	.130 ± .006	.122 ± .006	(45)
Bight of Benin	Yoruba of Lagos, Ibadan	1,003	.130 ± .008	.141 ± .008	(46)
Bight of Biafra	Ibo ("Eastern")	572	.161 ± .011	.089 ± .009	(47)
"Angola"	"Bantu"—8000 (mainly Bakongo) near Leopold-ville and 8000 from Angola	16,000	.152 ± .002	.138 ± .002	(48)
Mean frequencies‡ over the entire slaving area			.150	.131	(49)

*Maximum-likelihood estimate (49).   †Exclusive of Americo-Liberians.   ‡Calculated from values for p and q given in the body of the table, weighted by the proportions of Table 1 (after the removal of values for Mozambique, Madagascar, and "region unknown").

438

Negroes and Caucasians (*25*), a maximum likelihood estimate for *M* of 0.200 ± 0.044 was obtained (*28*). This estimate is not greatly affected by the accuracy of the proportions given in Table 1 or by the exactness of the values for individual regional gene frequencies (*29*). A good fit of the observed number of individuals in each of the eight race and blood-group classes with the corresponding number expected from the estimated parameters (gene frequencies and *M* values) tested by the chi-square method, indicates both that the estimation is reasonable and that there are no large selective differences between genes *A* and *B* in U.S. Negroes (*28*). This procedure therefore seems justified for the case of ABO blood groups. Practically, however, the large standard error for *M* indicates that, in spite of large samples, the estimate for this locus is too imprecise to be very useful.

Since there are now three different estimates of *M*, and since extensive data on other aspects of the problem, including selection, are available for this one large California population of Negroes, these estimates are presented in a single table, Table 6. We note that they do not differ significantly from each other; this is due at least in part to the relatively large standard errors for the Gm and ABO estimates. The marked superiority, for estimating *M*, of Fy^a over *A* and *B* for samples of equal size is evident (*30*), whereas, if the sample sizes were the same for Fy^a and the three Gm alleles, it would be found that these are equally efficient for estimating *M*. An extensive search for evidence of natural selection due to the presence of ABO blood groups in these Negroes, similar to the search reported above for the Duffy blood group, also failed to reveal any consistent selective effect (*23*). This finding, plus the good chi-square fit in the estimation of *M*, which implies that the *A* and *B* genes are not very different with respect to their selective values in U.S. Negroes, gives some assurance that the ABO estimate is not greatly disturbed by selection (*28*). No selection studies for Gm were made on these California Negroes, but extensive studies on a Brazilian population which was about 30 percent Negro, 11 percent Indian, and 59 percent Caucasian (*13*) revealed no

TABLE 6. ESTIMATES OF *M* CALCULATED FROM DATA ON GM SERUM GROUPS, DUFFY BLOOD GROUP, AND ABO BLOOD GROUP FROM NEGROES AND CAUCASIANS OF THE OAKLAND, CALIFORNIA, AREA. [DATA OF THE CHILD HEALTH AND DEVELOPMENT STUDIES (*6, 25*).]

Locus	Alleles used	Sample size (*N*)		*M*
		Negroes	Caucasians	
Gm	$Gm^1, Gm^{1,2}, Gm^5$	260	478	0.273 ± 0.037*
Duffy	Fy^a	3146	5046	.220 ± .009†
ABO	*A, B*	3146	5046	.200 ± .044†

*See (*6*).   †See text.

evidence of selective effect (*31*). Further evidence is provided by the good chi-square fit in the multi-allelic estimation obtained with the three Gm alleles (*6*). It seems reasonable to conclude that strong selective effects on these three estimates of M may be excluded. The existence of weaker effects, however, still sufficient to bias these estimates appreciably, cannot be ruled out. As more independent estimates on these and other genes become available, each having regard to the criteria listed above and including some safeguard against a strong bias due to selection and having a relatively small standard error (say, less than 0.02), it should become possible to obtain a "consensus" on the true value of M (for specified Negroes). Estimates biased upward or downward by selection will be separated from those little affected by selection, and so, in time, the former can be identified and rejected.

Use of M To Detect Selection

Several investigators (*8, 9, 21, 32*) have argued that selection for or against a gene may be clearly inferred from the M value that the gene produces. From the foregoing section it is clear that if (i) the true (unbiased) value of M (say, M_0) is known, (ii) the estimate in question (M_e) is calculated with regard to the criteria given above, and (iii) M_e differs significantly from M_0, then we may reasonably suspect that selection has caused the observed deviation. These conditions have not been met. In particular, we have no M_0. The M estimates obtained with R^0 (*8, 9, 21*), R^1 (*8, 9*), and Fy^a (*8*) were considered to be valid estimates unbiased by selection, but no objective evidence was offered to support these views. With one or more of these M estimates used as reference standards, it has been claimed that the deviant M estimates of the following genes demonstrate selection on these genes in U.S. Negroes: Hp^1, T, Gd^{A-}, Hb^S, and Tf^{D1} (see Table 2). These results can, at present, be considered only suggestive, but it must be admitted that the usually high M estimates obtained with Hp^1 and Gd^{A-} argue for an effect of selection (*27*).

A different approach was used to show that M estimates obtained with r, R^0, and R^1 alleles of the Rh locus ranked in this (decreasing) order of size for a Georgia population and also for two Brazilian populations (*32*). Accepted at face value, this is evidence of differences between M values from different Rh alleles. The investigators attribute these differences to selection. This same approach in these populations also indicates that M for the B allele is greater than $1.5M$ for the A allele (*32*). African Rh and ABO gene frequencies, weighted by slaving-area origins, were not used, however, although the African areas of origin of Brazilian Negroes are known (*2*). Again, these findings are interesting and suggestive but far from conclusive.

Workman (*8*), from inspection of A_1, A_2, and B allele frequencies in various West African, U.S. Negro, and U.S. Caucasian populations, concludes that there has been strong selection in U.S. Negroes against A_1 and for A_2. He identifies the various African data only as "West Africa," and does not use significance tests. Since Workman and also Hertzog and Johnson claim to find selection in the ABO system, it is pertinent here to recall that the M estimate obtained from ABO-system distributions that is discussed earlier in this article (an estimate based on *large* populations and good estimates for African gene frequency) did not suggest selective differences between the A and B alleles.

This critical review of claims for selection would be incomplete if I did not mention that there *is* an important theoretical reason to look for selection in hybrid populations such as the American Negro. As has been previously recognized (*6, 8, 32*), selection in U.S. Negroes over several generations can produce a cumulative effect in present-day individuals appreciably greater than the effect of a single generation of selection—the type of data usually available. There is thus a possibility of detecting, in hybrids, selection due to common polymorphisms which is too small [usually less than 50 to 10 percent of the mean (*23*)] to be detectable by ordinary one-generation studies. This possibility, together with the probability that some of the genes are selective [because it is most unlikely that a new genotype (the hybrid) in a new environment would be exactly neutral in selective value], makes the search for selection here especially worthwhile. Some of these selective genes may already have been identified.

Other Uses of M Estimates

In addition to the definite likelihood of their yielding valuable information on the action of natural selection in human populations, good estimates of the amount of Caucasian ancestry in U.S. Negro populations have at least two other "uses."

1) They provide objective information about the genetic heterogeneity among various populations of U.S. Negroes. Evidence of marked differences between southern and non-southern Negroes with respect to the amount of Caucasian ancestry, as shown in Tables 2 and 4, is the first clear result from this use of M estimates. As more good estimates from defined U.S. Negro populations become available, we may expect further heterogeneity to be revealed.

2) They provide an understanding of the distribution in American Negroes of those genetic traits, including diseases, that are due primarily to genes of Caucasian origin. There are few examples of such genes at present, but, aside from common genetic polymorphisms, like blood groups, few genes have been sufficiently studied to permit possible identification of racial differences in gene frequency. One probable example of such a genetic trait is phenylketonuria—a condition resulting from homozygosity for a rare autosomal recessive gene, producing a deficiency

of phenylalanine hydroxylase and resulting (if untreated) in severe mental defect. This occurs in about 1 in 10,000 births of persons of North European ancestry (33) but appears to be much rarer in U.S. Negroes (34). This rarity is understandable if the gene frequency in African Negroes is much lower than that in Caucasians (about 0.01). For example, if U.S. Negroes have, on the average, 20-percent Caucasian ancestry, the frequency of occurrence of phenylketonuria at birth in U.S. Negroes would be only 1/25th that in Caucasians, or roughly 1 in 250,000 —rare indeed.

An example of a disease which is not simply inherited but which may show a similar racial distribution is cirrhosis of the liver. A study in Baltimore Negro cirrhotics revealed, relative to Negro controls, a significant increase in Fy(a+b+) Duffy blood group phenotype and a decrease in Fy(a−b−) phenotype, whereas Caucasian cirrhotics showed no such difference from Caucasian controls (35). The simplest interpretation is that the disease is more frequent in Caucasians, and that Negroes with some degree of Caucasian ancestry, as shown by their Duffy blood group, are more likely to develop the disease than those lacking such ancestry (35). Other examples of traits whose frequency of occurrence in U.S. Negroes is affected by the amount of their Caucasian ancestry will surely be reported (36). Accurate information on M will be clinically useful here.

Summary

Published estimates of the proportion, in American Negroes, of genes which are of Caucasian origin are critically reviewed. The criteria for estimating this proportion (M) are discussed, and it is argued that all estimates published to date have either deficiencies pertaining to the African-gene-frequency data used or statistical inaccuracies, or both. Other sources of error may also exist.

Evidence is presented that the Fy^a gene of the Duffy blood group system may be the best gene now available for estimating M. Estimates based on Fy^a frequencies have been obtained for Negroes in three non-southern and two southern areas. The value of M is found to be appreciably greater in non-southern areas, the best estimate being 0.2195 ± 0.0093 (Oakland, California). This estimate is still subject to some uncertainty. The value of M in the South is appreciably less.

Natural selection can introduce a bias in the estimate of M. Claims that selection acting on certain genes in American Negroes have been demonstrated are reviewed, and it is concluded that they are not yet proved. The approach discussed here may be valuable in the future as a sensitive method for detecting the action of natural selection. In addition, knowledge of the amount of Caucasian ancestry may be of medical value in explaining the frequencies of occurrence of certain hereditary diseases in Negroes.

REFERENCES AND NOTES

1. J. H. FRANKLIN AND T. MARSHALL, in *World Book Encyclopedia* (Field Enterprises Educational Corporation, Chicago, 1966), vol. 14, p. 106.
2. P. D. CURTIN, personal communication (1969).
3. J. POTTER, in *Population in History: Essays in Historical Demography*, D. V. Glass and D. E. Eversley, Eds. (Univ. of Chicago Press, Chicago, 1965), p. 641.
4. P. D. CURTIN, personal communication (1968).
5. W. S. POLLITZER, *Amer. J. Phys. Anthropol.* **16**, 241 (1958).
6. T. E. REED, *Amer. J. Hum. Genet.* **21**, 71 (1969).
7. B. GLASS AND C. C. LI, *ibid.* **5**, 1 (1953).
8. P. L. WORKMAN. *Hum. Biol.* **40**, 260 (1968).
9. _____, B. S. BLUMBERG, A. J. COOPER, *Amer. J. Hum. Genet*, **15**, 429 (1963).
10. C. STERN, *Principles of Human Genetics* (Freeman, San Francisco, ed. **2**, 1960), p. 356.
11. A. G. STEINBERG, R. STAUFFER, S. H. BOYER, *Nature* **188**, 169 (1960).
12. B. S. BLUMBERG, P. L. WORKMAN, J. HIRSCHFELD, *ibid.* **202**, 561 (1964).
13. H. KRIEGER, N. E. MORTON, M. P. MI, E. AZEVÊDO, A. FREIRE-MAIA, N. YASUDA, *Ann. Hum. Genet*, **29**, 113 (1965).
14. J. E. BOWMAN, H. FRISCHER, F. AJMAR, P. E. CARSON, M. K. GOWER, *Nature* **214**, 1156 (1967).
15. B. GLASS, *Amer. J. Hum. Genet*, **7**, 368 (1955) (non-D^u-tested data for Africans, Negroes, and Caucasians; use of D^u-tested data for Africans and non-D^u-tested data for others gives $M = 0.281$).
16. D. F. ROBERTS, *ibid.*, p. 361. The estimate is "provisional"; 66 separate estimates were made, ranging from 0.0404 to 0.3341.
17. _____ AND R. W. HIORNS, *ibid.* **14**, 261 (1962). No sample sizes are specified.
18. F. B. LIVINGSTONE, H. GERSHOWITZ, J. V. NEEL, W. W. ZEULZER, M. D. SOLOMON, *Amer. J. Phys. Anthropol.* **18**, 161 (1960).
19. P. V. TOBIAS, in *The Biology of Human Adaptability*, P. T. Baker and J. S. Weiner, Eds. (Clarendon Press, Oxford, 1966), p. 161.
20. The following formula for the standard error (S.E.) of a ratio $R = y/x$ was used:

$$\text{S.E.} R = R \left[\frac{V_y}{y^2} + \frac{V_x}{x^2} - \frac{2C_{xy}}{xy} \right]^{1/2}$$

where the variance of y is V_y, that of x is V_x and the covariance between x and y is C_{xy} [see, for example, L. Kish, *Survey Sampling* (Wiley, New York, 1965), p. 207]. This formula is adequate for large or moderate-sized samples when it is unlikely that x is near zero. In terms of Eq. 2 for M,

$$\text{S.E.} M = M \left[\frac{V(q_a - q_n)}{(q_a - q_n)^2} + \frac{V(q_a - q_c)}{(q_a - q_c)^2} - \frac{2V(q_a)}{(q_a - q_n)(q_a - q_c)} \right]^{1/2}$$

where V represents the variance of the adjoining quantity in parentheses. The covariance between numerator and denominator of Eq. 2, due to the presence of q_a in both, is allowed for in this standard error.

21. W. C. PARKER AND A. G. BEARN, *Ann. Hum. Genet. London* **25**, 227 (1961). The total number of American, Canadian, and British individuals tested in the study reported is 865. A weighted estimate for the frequency of gene Hp^1 in U.S. Negroes, based on the data of Parker and Bearn, is 0.55; this value gives an M of 0.53.

22. A. G. STEINBERG, in *Symposium on Immunogenetics*, T. J. Greenwalt, Ed. (Lippincott, Philadelphia, 1967), pp. 75–98.

23. T. E. REED, *Amer. J. Hum. Genet.* **19**, 732 (1967); *ibid.* **20**, 119 (1968); *ibid.*, p. 129.

24. B. S. BLUMBERG, E. W. IKIN, A. E. MOURANT, *Amer. J. Phys. Anthropol.* **19**, 195 (1961).

25. T. E. REED, *Amer. J. Hum. Genet.* **20**, 142 (1968).

26. For gene distributions, see J. Hiernaux, *La Diversité Humaine en Afrique Subsaharienne* (Institut de Sociologie, Université Libre de Bruxelles, Brussels, 1968), figs. 2, 3, 7, 8, 12, 14.

27. There are good a priori reasons, entirely separate from M values, for expecting, in U.S. Negroes, a decrease in the frequency of the genes for sickle-cell hemoglobin, Hb^S, and for glucose-6-phosphate dehydrogenase deficiency, Gd^{A-}. (i) There is good evidence that in Africa the high frequency of the Hb^S gene is due to a selective advantage of heterozygotes for Hb^S in regions where malaria is endemic [see, for example, F. B. Livingstone, *Abnormal Hemoglobins in Human Populations* (Aldine, Chicago, 1967), pp. 105–107; A. C. Allison, in *Abnormal Haemoglobins in Africa*, J. H. P. Jonxis, Ed. (Davis, Philadelphia, 1965), pp. 369–371; D. L. Rucknagel and J. V. Neel, in *Progress in Medical Genetics*, A. G. Steinberg, Ed. (Grune & Stratton, New York, 1961), vol. 1, pp. 158–260]. There is strongly suggestive evidence that the Gd^{A-} gene in Africa is similarly kept at high frequencies due to selective advantage in malarious areas [see F. B. Livingstone, *Abnormal Hemoglobins in Human Populations* (Aldine, Chicago, 1967); A. G. Motulsky, in *Abnormal Haemoglobins in Africa*, J. H. P. Jonxis, Ed. (Davis, Philadelphia, 1965), pp. 181–185)]. (ii) Both genes are known to have selective disadvantages which can explain their rarity in nonmalarious areas. It is therefore to be expected that Negroes moved from their highly malarious homelands to the less malarious, and now nonmalarious, regions of North America would have lower frequencies of these two genes. This selective decrease would raise M estimates above the true value.

28. The computer program [see T. E. Reed and W. J. Schull, *Amer. J. Hum. Genet.* **20**, 579 (1968)] estimated M and Caucasian A and B gene frequencies, given the two African mean frequencies as constants and the two California populations determined by the gene frequencies to be estimated, subject to the constraints that, for *both* A and B, $q_n = Mq_c + (1 - M)q_a$. This equation is Eq. 1 applied to both alleles and is true when there is simple gene mixture without selection (see 6). Comparison of the observed numbers of the eight race and blood-group classes (2 races × 4 groups) with the corresponding numbers expected on the basis of parameter estimates gives a chi-square value of 5.910 for 3 d.f., $P > .10$.

29. When the negligible contribution from Mozambique, Madagascar, and "Unknown" is excluded, the proportions of Table 1, column 4, become (in order):

0.135, 0.056, 0.116, 0.162, 0.044, 0.237, and 0.249.   The corresponding proportions for South Carolina (1773–1807) are 0.197, 0.068, 0.164, 0.134, 0.016, 0.021, and 0.399 [data of Curtin (*4*)], yielding overall African mean values of 0.149 and 0.144 for p and q.   These two series differ appreciably with respect to the final two values, yet when the South Carolina series is used, the estimate of M is 0.256 ± 0.042, a difference of just over one standard error.   Also, q for the Bight of Biafra is the only markedly variant gene frequency among the frequencies for the seven regions, but replacing the p and q for this region by the p and q for the Bight of Benin or for "Angola" does not significantly change M (0.281 ± 0.040 or 0.251 ± 0.042, respectively, when corrected proportions of Table 1 are used).

30. The Fy^a estimate is based on $(0.044)^2/(0.0093)^2$, or 22 times as much statistical information as the ABO estimate.

31. N. E. MORTON, H. KRIEGER, M. P. MI, *Amer. J. Hum. Genet.* **18**, 153 (1966).

32. K. P. HERTZOG AND F. E. JOHNSON, *Hum. Biol.* **40**, 86 (1968).

33. V. A. McKUSICK, *Mendelian Inheritance in Man* (Johns Hopkins Press, Baltimore, ed. 2, 1968), p. 346.

34. For example, H. P. Katz and J. H. Menkes [*J. Pediat.* **65**, 71 (1964)] report the first definite case of phenylketonuria in a U.S. Negro. R. G. Graw and R. Koch [*Amer. J. Dis. Child.* **114**, 412 (1967)] report two "pale skinned" Negro brothers with phenylketonuria, bringing the total for U.S. Negroes at that time to five.

35. N. C. R. W. REID, P. W. BRUNT, W. B. BIAS, W. C. MADDREY, B. A. ALONSO, F. L. IBER, *Brit. Med. J.* **2**, 463 (1968).

36. Differences between Caucasians and Japanese with respect to gene frequency are instructive here.   Phenylketonuria appears to be much rarer (perhaps a tenth as frequent) among Japanese than among Caucasians [K. Tanaka, E. Matsunaga, Y. Hanada, T. Murata, K. Takehara, *Jap. J. Hum. Genet.* **6**, 65 (1961)].   Another single-gene trait, Huntington's chorea, also appears to be about ten times as frequent in Caucasians as in Japanese, according to T. E. Reed, and J. H. Chandler, *Amer. J. Hum. Genet.* **10**, 201 (1958). Examples of congenital abnormalities which are commoner in Caucasians than in Japanese, and vice versa, are given by J. V. Neel, *Amer. J. Hum. Genet.* **10**, 398 (1958).

37. P. D. CURTIN, *The Atlantic Slave Trade: A Census* (Univ. of Wisconsin Press, Madison, in press).

38. J. V. NEEL AND W. J. SCHULL, *Human Heredity* (Univ. of Chicago Press, Chicago, 1954), p. 255; J. V. Neel, personal communication (1969).   The estimate refers to "non-Negro" ancestry.

39. R. R. RACE AND R. SANGER, *Blood Groups in Man* (Blackwell, Oxford, ed. 5, 1968).

40. M. SHAPIRO AND J. M. VANDEPITTE, *Int. Congr. Blood Transfusion, 5th, Paris* (1955), p. 243; M. Shapiro, personal communication (1969).

41. H. GERSHOWITZ, unpublished data.

42. E. B. MILLER, R. E. ROSENFIELD, P. VOGEL, *Amer. J. Phys. Anthropol.* **9**, 115 (1951); R. Sanger, R. R. Race, J. Jack, *Brit. J. Haematol.* **1**, 370 (1955).

43. R. KOERBER, *C. R. Seances Soc. Biol.* **141**, 1013 (1947); R. Koerber and J. Linhard, *Bull. Soc. Anthropol. Paris* **2**, 158 (1951).   The frequencies for Bambara and Malinke do not differ significantly.

44. P. Julien, *Z. Rassenphysiol.* **9**, 146 (1937).

45. G. M. Edington, *West Afr. Med. J.* **5**, 71 (1956).

46. H. R. Muller, *Proc. Soc. Exp. Biol.* **24**, 437 (1927); J. P. Garlick, quoted in A. E. Mourant, A. C. Kopeć, K. Domaniewska-Sobczak, *The ABO Blood Groups* (Blackwell, Oxford, 1958), p. 173; J. P. Garlick and N. A. Barnicot, *Ann. Hum. Genet.* **21**, 420 (1957). The frequencies in these three surveys do not differ significantly from each other.

47. J. Hardy, *Roy. Anthropol. Inst.* **92**, 223 (1962). I have not used Hardy's data on "Onitsha Ibo" (sample size, 228) because I consider the subjects to be probably not of Ibo origin. I have not used data of J. N. M. Chalmers, E. W. Ikin, and A. E. Mourant [*Ann. Eugenics* **17**, 168 (1953)] on "southeastern" Nigeria (105 Ibo and 1 Tiv) because information on the type of Ibo was not given. Table 1 describes this region as "about 3/4 Ibo," and I could find no suitable data for the remaining quarter.

48. G. van Ros and R. Jourdain, *Ann. Soc. Belge Med. Trop.* **36**, 307 (1956); L. Mayor, *Bull. Clin. Statistics* **7**, No. 3, suppl. 126 (1954). The first study (for the Congo) gave values of 0.1556 ± 0.0030 for p and 0.1244 ± 0.0027 for q. The second (for Angola gave values of 0.1486 ± 0.0029 and 0.1514 ± 0.0030, respectively. An unweighted average of these values was used. The values for Angola do not differ significantly from the mean p and q values for seven central-coastal named tribes in Angola (total population, 1285) tabulated by Hiernaux (see *26*).

49. T. E. Reed and W. J. Schull, *Amer. J. Hum. Genet.* **20**, 579 (1968).

50. Preparation of this article was begun while I was engaged in work for the Child Health and Development Studies (Division of Biostatistics, School of Public Health, University of California, Berkeley, and the Kaiser Foundation Research Institute, Oakland, California), on leave from the University of Toronto, and was supported there by U.S. Public Health Service research grants HD 00718 and HD 00720 from the National Institutes of Health. The analysis was supported in part by a grant from the Medical Research Council of Canada. I thank Professor Philip D. Curtin for making unpublished data available and for commenting on the manuscript, Dr. Arthur E. Mourant and Mrs. K. Domaniewska-Sobczak for recent references to African blood group distributions, and Professors Curt Stern, Donald Rucknagel, and Peter Carstens for their comments. Dr. H. Gershowitz and Dr. M. Shapiro made available unpublished data on Duffy blood groups in Negroes.

[The following correspondence was generated by Reed's article and appeared in the "Letters" column of *Science* (December 12, 1969), p. 1353.]

T. Edward Reed ("Caucasian genes in American Negroes," 22 Aug., p. 762), has provided interesting evidence on the amount of non-African ancestry in the American Negro. Unfortunately, he never defines the population or populations included in the term. This is a serious omission, tending to prejudice the results of his work. In southern Louisiana, at least, many persons are called "white" who might very well be called "black" in the North. This would be a factor in making Southern Negroes seem less "mixed" than Northern ones. There is also the question of "passing" and related devices whereby African genes scatter through the "white" population. According to one estimate [R. P. Stuckert in *Physical Anthropology and Archaeology: Selected Readings*, P. Hammond, Ed. (Macmillan, New York, 1964), pp. 192–197], most Americans with African ancestry are "white," though, of course, their percentage of African ancestry must be small. We are dealing here with a sociological or "folk-scientific" classification. What is Reed's particular version of it? How did he arrive at his sample of Negroes? What possible statistical biases would be introduced by his procedures in defining and in sampling?

E. N. ANDERSON, JR.

Department of Anthropology,
University of California, Riverside 92502

Anderson's two main points require comment. A precise objective definition of "Negro" was not given because none was, or is, available. As implied in my first paragraph, the definition used in the various studies is generally the conventional one used by most persons in the particular area in question. I believe that this is the case for the five studies of Table 4, for example, but I can speak in detail only for the large Oakland, California, study. For this population, individuals (all married, age 17 years or over) were classified as Caucasian, Negro, or other race on the basis of the wife's statement (obtained by interview) about her own and her husband's ethnic background. If this background was said to be "mixed," the person was placed in a separate category (2 percent of the total) and was not used in my study. This classification as "Negro" or "Caucasian" did not differ in any obvious way from the conventional ethnic classification, used between persons, in this area.

Relative to the other biases I discussed, I think that the lack of a precise definition of "Negro" is a minor deficiency. I do not believe that this deficiency seriously affects the conclusions of my article. In particular, after discussions with colleagues who have lived in both the South and the North, I believe that the proportion of persons (relative to the total "Negro" population) who may be called "white" in one area of the United States but "black" in another area is not over a few percent. I would welcome objective information on this point.

The second point, African Negro genes in American "whites," does not bear importantly on my article but it does merit comment on its own. There is no doubt that the African genes introduced by "passing" have spread widely through the American "white" population. In several more generations it is likely that a majority of "whites" will have at least one gene of African origin. To a lesser extent, of course, genes from other non-Caucasian peoples will also be widely distributed. It is of interest to note, however, that in the California Caucasians I studied (who are representative of U.S. Caucasians), the proportion of genes which are of African origin is probably less than 1 percent. The genetic argument leading to this conclusion is the following: The R^0 allele (of the Rh blood group system) has a West African frequency of 0.55 to 0.60, a California Caucasian frequency (q_c, West European ancestry) of 0.0228 $\pm$ 0.0023, and an English frequency (q_e) of 0.0257 [with S.E. about 0.0037; $N = 2000$; from R. R. Race and R. Sanger, *Blood Groups of Man* (Blackwell, Oxford, ed. 5, 1968), p. 178]. The 95 percent confidence interval for $q_c - q_e$ is -0.0119 to $+0.0061$. If q_c is actually greater than q_e because of the contribution of African R^0 genes (and for no other reason), then it is about 95 percent probable that the proportion of "African" genes in the California Caucausians studied is *less* than $0.0061/(.55 - .0257) = 0.012$. At the present time "passing" may be important socially but it is unlikely that it is very important biologically.

T. Edward Reed

Departments of Zoology, Anthropology,
and Paediatrics, University of Toronto,
Toronto 5, Ontario, Canada

Selection, Gene Migration and Polymorphic Stability in a U. S. White and Negro Population

P. L. WORKMAN. B. S. BLUMBERG, and A. J. COOPER

In any population the frequencies of the alleles associated with a polymorphic locus are related to the total genetic constitution of the population and the environment in which it is situated; changes in either would result in corresponding changes in allelic frequencies. Since polymorphic traits provide an opportunity for rapid evolutionary changes (Ford, 1960), their distributions in appropriate populations can be used to study the directed (gene migration, selection, mutation) and non-directed (chance, drift) forces which have produced the recent evolutionary trends in man (Motulsky, 1960; Blumberg, 1961; Allison, 1962).

In American Negro populations, admixture with European Whites has altered the genetic constitution (Glass and Li, 1953; Stern, 1953; Glass, 1955; Roberts, 1955) and the change in their environment, by movement from Africa to North America, could have altered the adaptive values at the polymorphic loci. In this paper, frequencies of several polymorphic traits in Negroes and Whites living in the same Southern U. S. community will be compared with the frequencies of the same traits in contemporary West African Negroes in order to help evaluate the relative roles of selection, gene migration and drift in producing the present frequencies of the traits in the American Negroes. This comparison should also indicate which polymorphisms, if any, in the American Negro population are unstable; that is, traits with significantly different adaptive values in the two populations which have not reached stable frequencies, or traits which have adaptive disadvantage in the American Negroes, but are still present because of prior advantage in Africans (transient polymorphisms). The studies were conducted in Evans and Bullock Counties, Georgia. Claxton is the county seat of Evans, and the populations, for convenience, have been called the Claxton populations. A description of the populations studied and a discussion of the techniques used for identification of the phenotypes are given elsewhere in this issue. (Cooper, Blumberg, Workman and McDonough, 1963).

Reproduced by permission of the publisher and P. L. Workman from American Journal of Human Genetics, *15:71–84 (1963). Copyright, 1963, by Grune & Stratton, Inc. for the American Society of Human Genetics.*

Peter L. Workman, *a population geneticist, was born in Providence, Rhode Island. He received his B.S. in 1957 and his Ph.D. in 1962 from the University of California. Formerly a research biologist at NIDR, National Institutes of Health, in human-population studies, he is currently in the Departments of Pediatrics (Genetics) and Community Medicine at Mt. Sinai School of Medicine and is an Associate Professor of Anthropology, City College of New York. His major field of interest focuses on population studies of an interdisciplinary nature, combining anthropology, genetics, theoretical mathematics, and demography. In addition to his work on American Negro populations, he has more recently done genetic field studies in the American Southwest on the Papago Indians of Arizona and the Zuni of New Mexico. His publications include "Gene flow and the search for natural selection in man," Hum. Biol., 40:260–279 (1968); "The analysis of simple genetic polymorphisms," Hum. Biol., 41:97–114 (1969); and "Population studies on Southwestern Indian tribes. Part II. Local genetic differentiation in the Papago," with J. D. Niswander, Amer. J. Hum. Genet., 22:24–49 (1969).*

Analysis of the Data

The frequencies of the polymorphic traits studied in the Claxton populations are presented in Table 1, together with estimates of the frequencies of the same traits in other Negro and White populations and in West Africans. We have assumed that the remote ancestors of the American Negroes were mainly from West Africa (see, for example, Herskovits, 1941; Fage, 1959) and that the West African frequencies obtained from recent studies closely approximate those in the African populations from which the American Negroes have descended. While drift and mutation have probably had little effect on the African frequencies in the period since the Negroes came to North America, it is possible that changes in the selective values of particular traits in West Africa have caused corresponding changes in the gene frequencies in that period. Any conclusions drawn from the data must be considered with these restrictions in mind.

As seen in table 1, the frequencies of the traits in the Claxton White population are, in general, quite similar to those observed in other U. S White and English populations, the greatest differences being approximately 5 per cent. The frequencies in the Claxton Negro differ from other U. S. Negro populations by at most 4 per cent for the majority of the alleles considered. A comparison of the frequencies given in table 1 permits the assumption that genetic drift has had no appreciable effect upon the distribution of the polymorphic traits in either the Negro or the White population in Claxton. In addition, approximately 12 to 15 generations, or 350 years, have elapsed since the arrival of the first Negroes in North America (Glass and Li, 1953; Fage, 1959); and both the Negro and White populations in Evans County each contain more than 2,400 in-

Baruch S. Blumberg, *a physician and geneticist, was born in New York City in 1925. He received his B.S. in 1946 from Union College, his M.D. in 1951 from Columbia University, and his Ph,D. in 1957 from Oxford University. Presently he is the Associate Director for Clinical Research at the Institute for Cancer Research, and an Associate Professor of Medicine and Medical Genetics at the University of Pennsylvania. His major research interests are in medical genetics and in the anthropological approach to disease. In 1969 he received the Albion O. Bernstein, M.D. Award for his studies on the Australia antigen and heptatitis. He has done field work and published numerous papers on the genetics of such diverse populations as the Basques of Spain; the Yoruba and Fulani of Africa; various American Indian tribes of Alaska, Canada, and Ecuador; and groups in India, Japan, and the Pacific. He is editor of the book* Proceedings of the Conference on Genetic Polymorphisms and Geographic Variations in Disease *(New York, 1961).*

TABLE 1. FREQUENCIES OF ALLELES STUDIED IN THE CLAXTON POPULATIONS COMPARED TO FREQUENCIES FROM OTHER STUDIES

Allele or Segment	West African Negro	Claxton Negro	Other American Negro	Claxton White	Other U. S. and English White
R^0(cDe)	.594[4]	.535	.438[4]	.037	.026[6]–.028[4]
R^1(CDe)	.069[4]	.103	.158[4]	.426	.408[6]–.420[4]
R^2(cDE)	.086[4]	.108	.109[4]	.148	.141[6]–.150[4]
r(cde)	.211[4]	.230	.264[4]	.358	.384[4]–.389[4]
A	.148[4]	.158	.141–.188[8]	.246	.23–.29[4]
B	.151[4]	.129	.093–.147[8]	.050	.057–.09[8]
O	.704[4]	.713	.674–.733[8]	.704	.66–.70[8]
M	.476[4]	.485	.476–.532[7]	.508	.533–.547[7]
S	.134[4]	.155†	.160[6]–.186[4]	.281†	.327[4]–.377[4]
Fy^a	.0[6]	.046	.053[6]	.422	.414[6]–.434[7]
P	.780[4]	.757	*	.526	.542[6]
Jk^a	.783[4]	.743	.732[4]	.536	.514[6]–.523[7]
K	.009[6]	.005	.018[7]	.042	.046[6]–.066[7]
Lu^a	.036[6]	.044	*	.036	.039[6]
Js^a	*	.122	.103[1]	.002	.0[1]
Di^a	.0[13]	.03	.00[14]	.0	.0[14]
G6PD	.18–.21[11]	.118	.11[10]	.0	.0[11]
Hb^s	.08–.14[9]	.043	.02–.06[7]	.0	.0[7]
Hp^1	.60–.78[11]	.520	.531[3]–.539[2]	.41	.43[3]
Tf^{D1}	.035–.088[11]	.049	.055[12]	.01	.0
T	.795[5]	.670	.697[5]	.527	.455[5]

*No suitable estimate could be found.

†Estimated by S $= 1 - \sqrt{S(-)}$ for purposes of comparison with West African data.

[1]Giblett and Chase, 1954
[2]Giblett and Steinberg, 1960
[3]Sutton *et al.*, 1959
[4]Glass, 1955
[5]Glass and Li, 1953
[6]Race and Sanger, 1958
[7]Mourant, 1954

[8]Mourant *et al.*, 1958
[9]Allison, 1956
[10]Beutler, 1959
[11]Allison and Blumberg, 1962
[12]Parker and Bearn, 1961
[13]Gershowitz, 1959
[14]Layrisse, 1958

dividuals (Cooper *et al.*, 1963). From theoretical considerations, for this population size and time interval, it is unlikely that drift or mutation has has an appreciable effect on gene frequencies, (Kimura, 1956; Moran, 1962).

The unlikelihood of significant admixture between the American Negro and the American Indian population was discussed by Glass (1955). His conclusions are supported by our finding the Di^a allele, relatively common in American Indians (Layrisse, 1958), in only one of 188 Claxton Negroes.

If the assumptions discussed above are correct, then the frequency differences between the Claxton and West African Negroes can be ascribed almost totally to the effects of gene migration resulting from admixture between the American White and the American Negro population and to differences in the adaptive values of the traits in the West African and American Negro populations.

Selection and Migration

In order to evaluate the relative effects of selection and migration, estimates have been made of the total amount of gene migration from the American White into the Claxton Negro population using the method of Bernstein (1931) which assumes that the observed differences are due to migration alone. In the following calculations it is assumed that the frequencies of the traits in the Claxton Whites are representative of the frequencies in the White population which has contributed to the Negro gene pool. If q_W, q_N, and q_{Af} are the frequencies of an allele in the Claxton White, Claxton Negro and West African populations respectively, then the total amount of gene migration, m, is given by

$$m = \frac{|q_N - q_{Af}|}{|q_W - q_{Af}|}.$$

Such an estimate of gene migration, m, for a given locus or segment, is equivalent to an estimate of the admixture, or hybridization, which has occurred between two populations only if: (a) there is no assortative or preferential mating between the two populations with respect to the locus considered; (b) the gene migration is entirely from one population into the other; (c) individuals whose genotype is derived from both populations have no special bias with respect to fertility, social factors, geographic mobility, and other factors which would affect their contribution to the gene pool of the population. For example, for an organism in which the hybrids between two populations are not viable, no amount of admixture will result in gene migration. For the present study, since we shall assume only that there has been no preferential mating with respect to the traits under consideration, the relation between the estimates of gene migration and the actual amount of admixture can not be considered. In the absence of differences in the adaptive values of the traits in West Africa and in the United States the estimates of m, computed for each of the loci, should be equal. Then, an alteration of the adaptive values of the alleles resulting either from change in environment or from modification of the gene pool by admixture would result in differences between the m values calculated for the alleles. Small variation in the m values could be ascribed to sampling accidents, drift, or small inaccuracies in the estimates of the allelic frequencies in the West African population, as well as to small changes in the adaptive values. However, significantly different estimates of m must be the result of significant differences in the adaptive values of the alleles in the two populations.

Estimates of gene migration (m) have been calculated only for those alleles where reliable West African frequencies are obtainable and where the difference between the frequencies of the alleles in the Claxton White and West African populations, $|q_W - q_{Af}|$, is sufficiently large that the sampling error of the ratio is small. We have considered only those alleles for which $|q_W - q_{Af}|$ is at least .09. Table 2 gives the alleles considered and the corresponding values of $|q_W - q_{Af}|$ and m. The alleles or chromosome segments not suitable for this kind of analysis, for one or both of the above reasons, were O, M, R^2, K, Lu^a, Di^a, and Js^a.

Discussion

The most striking feature of the analysis is the apparent separation of the polymorphic traits into two distinct groups. In the larger group (Group I), including all the red blood cell antigens, the m values have a range from .094 (P) to .218 (B) and a mean value of .131. The other

TABLE 2. COMPUTED VALUES OF M AND $|q_W - q_{AF}|$

Allele or Segment	$|q_W - q_{Af}|$	m
Group I		
R^0	.562	.113
Fy^a	.422	.109
R^1	.357	.095
P	.266	.094
Jk^a	.247	.167
r	.147	.129
S	.147	.143
B	.101	.218
A	.098	.107
Group II		
T	.268	.466
Hp^1	.19–.38	.42–.70
G6PD	.18–.21	.34–.44
Hb^s	.08–.14	.46–.69
Tf^{D1}	.074	.495*

*The West African frequency for Tf^{D1} which was used (.088) was based on only two samples. See text for discussion.

group (Group II), which includes Hp^1 (haptoglobin), Hb^8 (sickle cell hemoglobin), G6PD, T (PTC-taste test) and possibly the Tf^{D1} (transferrin) alleles, has m values which range from .34 to .70, all considerably greater than those in the first group.

In order to determine which of the two groups contains alleles whose frequencies have been primarily altered by gene migration, we should consider estimates of m obtained from alleles whose frequencies would have remained approximately constant in the West African and American White populations over the past 300 years, and for which $| q_W - q_{Af} |$ is large. Since the frequencies of Rh alleles (which all fall in Group I) are considered to be quite stable over a period of several hundred years (see, for example, Mourant, 1954) we have assumed that the m values of the Group I alleles, in the Claxton Negroes, reflect primarily the effects of gene migration. The mean of the m values for R^0 ($| q_W - q_{Af} | = .562$) and R^1 ($| q_W - q_{Af} | = .357$), namely m = .104, can be considered the best estimate of m in the Claxton Negro population. The variation in the m values estimated for the Group I traits (*i.e.*, the red blood cell antigens) could result from either small differences in the adaptive values of the traits in West Africa and Claxton, sampling error or genetic drift. For the Group I traits the environmental selective forces appear to be similar in West Africa and in the Southern United States. This implies that they are maintained by selective forces which operate in both ecological settings. They are, however, balanced at different levels as shown by the gene frequencies for Africans and Whites in table 1, indicating probable differences between the gene pools of the given populations. Thus, barring unknown cyclic changes which could have occurred during the generations since the movement of the Africans to North America, or selective forces which have uniformly affected the m values for the Group I alleles, gene migration, resulting from admixture between the Claxton Negroes and the American Whites, appears to be the chief cause for the differences in the frequencies of these alleles in the West African and Claxton Negro populations.

For the Group II polymorphisms, gene migration alone cannot account for the m values which are all significantly larger than .104. Nor, as noted above, could either mutation or drift have significantly influenced the frequencies of these alleles. If the contemporary West African frequencies accurately reflect the population from which the Claxton Negroes are descended then these Group II traits must have significantly different adaptive values in West Africa and Claxton. Evidence from other studies supports this hypothesis. It has been suggested that heterozygotes for either Hb^s or G6PD have an adaptive advantage in a malarial environment (Allison, 1956; Motulsky, 1960). Selection against the heterozygotes for Hb^s or G6PD would lead to a rapid decrease in the frequency of the alleles. The T allele has been considered in relation to thyroid disorders

(Kitchin *et al.*, 1959). There is evidence from studies on Greek populations that the Hp^1 allele may be positively correlated with the thalassemia trait (Blumberg, 1963).

The Group II polymorphisms, T, G6PD, Hb^s, and possibly Hp^1 and Tf^{D1}, have values of m ranging from .34 to .70, indicating that these polymorphisms were, and probably still are, unstable. Since the G6PD and Hb^s alleles are almost completely absent in U. S. Whites, these traits may represent transient polymorphisms, present in the U. S. Negro because of a former adaptive advantage in the West Africans.

The present data do not provide any interpretation of the nature of the adaptive factors operating on the polymorphisms included in Group II. It should be stressed that the statistical analysis can only provide correlations between environmental conditions and allelic frequencies. Any valid interpretation of the differences in adaptive values must derive from medical or biochemical studies.

The statistical analysis of the data could be extended to a consideration of either the rate of gene migration per generation (Glass and Li, 1953; Saldanha, 1957) or the adaptive values for the traits. Such analysis would, however, entail assumptions such as constant rates of migration and fixed adaptive values, which are most unlikely, and the numbers produced would be of dubious worth.

The estimation of m for the Hp^1 gene is based on estimates of West African frequencies derived from populations in which many of the sera could not be typed because of absent or low haptoglobin levels (Allison, Blumberg and ap Rees, 1958). Although some ahaptoglobinemia is due to genetic causes (Giblett and Steinberg, 1960), much of it is probably due to the environment. Furthermore, in cases which have been recorded as ahaptoglobinemia at one time, but typable at another, the serum is often type 2-2 (Blumberg and Gentile, 1961). Hence the West African surveys may over-estimate the Hp^1 frequency. For example, if in the West African population sampled by Blumberg and Gentile (1961) half of the sera classified as type O were in fact type 2-2, the value of $|q_w - q_{Af}|$ would be too small to permit an accurate calculation of m. Furthermore, it is now known that there are at least three alleles commonly segregating at the Hp locus (Hp^{1F}, Hp^{1S}, Hp^2) (Smithies, Connell and Dixon, 1962) and the frequency of Hp^1 is actually the sum of Hp^{1F} and Hp^{1S} frequencies.

The calculation of m for the Tf^{D1} allele is based on only a small number of West African studies. Recently several slow moving transferrin variants determined by alleles other than Tf^{D1} have been reported and in some cases the transferrin phenotypes may have been misclassified in the earlier studies. Hence the Tf^{D1} frequencies reported may be high. Furthermore, it is not unlikely that the slow moving variants reported in non-Africans are determined by different alleles. The m value calculated for Tf^{D1} must be considered tentative.

Values of m greater than .20 have been reported by Glass (1955) and Roberts (1955) from comparisons between other American Negroes and Whites. Unfortunately, they compared the frequencies of polymorphisms in Negroes and Whites who did not live in the same community and used different populations to compare different alleles. Hence, from their studies it is impossible to analyze the variation in the m values which could reflect variation in the populations considered, different amounts of admixture, or in the influence of social factors as well as adaptive differences, drift, and so forth. The frequencies of the R^0 and R^1 alleles in the Claxton Negroes are closer to the West African frequencies than are those reported by Glass (1955); Pollitzer (1958), in a study of Negroes from Charleston, South Carolina, found frequencies almost equal to those in West Africa. This variation suggests that there may be significant differences in both the amount of admixture and the amount of gene migration in different U. S. Negro populations. That is, the high estimates of m obtained from studies on Negroes living in large Northern cities could reflect either different rates of admixture or a similar rate of admixture but differential rate of geographic movement of Negroes with a high proportion of white ancestry. In Pollitzer's (1958) study, both the gene frequencies and his anthropological studies on the relative isolation of the populations suggested a low rate of admixture.

The simultaneous analysis of the distribution of several polymorphic traits has served to isolate four (or five, including the Tf^{D1}) alleles whose frequencies are, or have been, significantly altered by selective pressures which are different in the West African and Claxton populations. Additional statistical and biological studies are required to determine the nature and amount of the adaptive differences of these alleles. Similar studies in other populations should reveal additional traits which are undergoing rapid evolutionary change. The same populations, and in fact the same blood samples, may be used to determine if newly-discovered polymorphisms are balanced (Group I) or unstable (Group II). Such studies are being undertaken with the serum Gm (gamma globulin), Gc (group specific) and beta lipoprotein groups.

The variation within the group least affected by adaptive differences (Group I) should be further analyzed not only in other American Negro populations, but in populations throughout the world. In this way, the loci most stable over many generations and in different populations could be determined and used for anthropological or historical studies in these populations.

Summary

The frequencies of more than 15 polymorphic traits were studied in an American Negro and White population living in the same rural Southern U. S. community and compared with the frequencies of the same traits in

West African Negroes and other American Negro and White populations. It is suggested that neither genetic drift nor mutation were likely causes of the variability observed. By estimation of the total amount of gene migration, m, from the Whites to the Negroes, (under the assumption of no selection) the polymorphic traits can be separated into two distinct groups. In Group I, the larger group, which contains the red blood cell antigens, the estimates of m (.1 to .2) are consistent with the hypothesis that migration alone can account for the differences in gene frequencies between the West African and the American Negro populations. The best estimate of m was found to be .104. In Group II, containing the G6PD, Hb^s, and T alleles (and possibly Hp^1 and Tf^{D1}) the significantly higher estimates of gene migration (.4 to .7) were concluded to result from both gene migration and different adaptive values of the traits in the West African and American environments.

REFERENCES

ALLISON, A. C. 1956. The sickle-cell and haemoglobin C genes in some African populations. *Ann. Hum. Genet.* 21: 67–89.

ALLISON, A. C. 1962. Natural selection in human populations. *Univ. Kansas Sci. Bull.* in press.

ALLISON, A. C., AND BLUMBERG, B. S. 1963. Polymorphisms in man. In preparation.

ALLISON, A. C., BLUMBERG, B. S., AND AP REES, W. 1958. Haptoglobin types in British, Spanish, Basque and Nigerian African populations. *Nature* (Lond.) 181: 824–5.

BERNSTEIN, P. 1931. Die geographische Verteilung der Blutgruppen und ihre anthropologische Bedeutung. *Comitato Italiano per lo Studio die Problemi della Populazione.* Rome: Institute Poligrafico deli Stato, pp. 227–243.

BEUTLER, E. 1959. The hemolytic effect of primaquine and related compounds; a review. *Blood* 14: 103.

BLUMBERG, B. S. 1961. Inherited susceptibility to disease. *Arch. Environ. Health* 3: 612–636.

BLUMBERG, B. S. 1963. Personal communication.

BLUMBERG, B. S., AND GENTILE, Z. 1961. Haptoglobins and transferrins of two tropical populations. *Nature* (Lond.) 189: 897–899.

BLUMBERG, B. S., KUVIN, S. F., ROBINSON, J. C., TEITELBAUM, J. M., AND CONTACOS, P. G. 1963. Alterations in haptoglobin levels. *J. A. M. A.* 184: 1021–1023.

COOPER, A. J., BLUMBERG, B. S., WORKMAN, P. L., AND MCDONOUGH, J. R. 1963. Biochemical polymorphic traits in a U. S. White and Negro population. *Amer. J. Hum. Genet.* 15: 420–428.

FAGE, J. C. 1959. *An Introduction to the History of West Africa.* Cambridge: Univ. Press.

FORD, E. B. 1960. Evolution in progress. In: *Evolution after Darwin, Vol. I. The Evolution of Life*, Sol Tax, ed. Chicago: Univ. of Chicago Press.

GERSHOWITZ, H. 1959. The Diego factor among Asiatic Indians, Apaches and West African Negroes; blood types of Asiatic Indians and Apaches. *Amer. J. Phys. Anthrop.* 17: 195–200.

GIBLETT, E. R., AND CHASE, J. 1959. Js^a, a new red-cell antigen found in Negroes; evidence for an eleventh blood group system. *Brit. J. Haemat.* 5: 319–326.

GIBLETT, E. R., AND STEINBERG, A. G. 1960. The inheritance of serum haptoglobin types in American Negroes: evidence for a third allele Hp^{2M}. *Amer. J. Hum. Genet.* 12: 160–169.

GLASS, B. 1955. On the unlikelihood of significant admixture of genes from the North American Indians in the present composition of the Negroes of the United States. *Amer. J. Hum. Gen.* 7: 368–385.

GLASS, B., AND LI, C. C. 1953. The dynamics of racial intermixture—an analysis based on the American Negro. *Amer. J. Hum. Genet.* 5: 1–20.

HERSKOVITS, M. J. 1941: *The Myth of the Negro Past.* New York: Harper.

KIMURA, M. 1956. Stochastic processes and distribution of gene frequencies under natural selection. *Sympos. Quant. Biol.* 20: 33–51.

KITCHIN, F. D., HOWELL-EVANS, W., CLARKE, C. A., McCONNELL, R. B., AND SHEPPARD, P. M. 1959. P.T.C. taste response and thyroid disease. *Brit. Med. J.* 1: 1069.

LAYRISSE, M. 1958. Anthropological considerations of the Diego (Di^a) antigen. *Amer. J. Phys. Anthrop.* 16: 173.

MORAN, P. A. P. 1962. *The Statistical Processes of Evolutionary Theory.* Oxford: Clarendon Press.

MOTULSKY, A. G. 1960. Metabolic polymorphisms. In: *The Processes of Ongoing Human Evolution*, G. W. Lasker, ed. Detroit: Wayne State Univ. Press.

MOURANT, A. E. 1954. *The Distribution of Human Blood Groups.* Springfield: C. C. Thomas.

MOURANT, A. E., ROPEC, A. C., AND DOMANIEWSKA-SOLECZAK, K. 1958. *The ABO Blood Groups.* Oxford: Blackwell Scientific Publications.

PARKER, W. C., AND BEARN, A. G. 1961. Haptoglobin and transferrin variation in humans and primates: two new transferrins in Chinese and Japanese populations. *Ann. Hum. Genet.* 25: 227–241.

POLLITZER, W. S. 1958. The Negroes of Charleston (S. C.); a study of hemoglobin types, serology and morphology. *Amer. J. Phys. Anthrop.* 16: 241–263.

RACE, R. R., AND SANGER, R. 1958. *Blood Groups in Man*, 3rd ed. London: Blackwell Scientific Publications.

ROBERTS, D. F. 1955. The dynamics of racial intermixture in the American Negro —some anthropological considerations. *Amer. J. Hum. Genet.* 7: 361–367.

SALDANHA, P. H. 1957. Gene flow from White into Negro Populations in Brazil. *Amer. J. Hum. Genet.* 9: 299–309.

SMITHIES, O., CONNELL, G. E., AND DIXON, G. H. 1962. Inheritance of haptoglobin subtypes. *Amer. J. Hum. Genet.* 14: 14–21.

STERN, C. 1953. Modal estimates of the frequency of White and near-White segregants in the American Negro. *Acta Genet.* (Basel) 4: 281–298.

SUTTON, H. E., NEEL, J. V., LIVINGSTONE, F. B., BENSON, G., KUNSTADTER, P., AND TROMBLY, L. E. 1959. The frequency of haptoglobin types in five populations. *Ann. Hum. Genet.* 23: 175–183.

CHAPTER **VII**

Genetics and
the Future of Man

Overview

In the following article, British zoologist John Maynard Smith discusses some of the possible biological and social implications of man's potential to alter his biological capacities. Man's control over his own genetic destiny is a possibility that may become reality in the not-too-distant future as molecular genetics, medical advancements, and improved cytological techniques bring the tools of control close at hand. The speculations of today may well be the realities of tomorrow.

Eugenics and Utopia

JOHN MAYNARD SMITH

There is no field of application of science to human affairs more calculated to arouse our prejudices than eugenics. I cannot hope to be free from these prejudices, but in this essay I will try to separate what we ought to do from what we can now do and from what we may in the future be able to do. These problems should be thought about because our ability to alter the future course of human evolution is likely to increase dramatically during the next hundred years.

There are three ways in which we may be able to alter man's biological capacities, which I shall call selectionist eugenics, transformationist eugenics, and biological engineering. Briefly, selectionist eugenics is the application to ourselves of the techniques which, since the Neolithic revolution, we have been applying in the breeding of our domestic animals and plants. In effect, we take measures to ensure that individuals with characteristics we like will contribute more to future generations than individuals with characteristics we dislike. These measures range from the simple to the sophisticated, from the slaughter of runts to the cold storage of spermatozoa for artificial insemination. The development of a science of population genetics has enabled us to estimate with greater accuracy the consequence of any particular interference with the breeding system and to choose between effective and ineffective methods of selection. But it has not altered the fact that this is an extremely slow and inefficient method of altering the genetic properties of a population and one whose speed can be increased only by increasing the intensity of selection; a bigger change is produced in the properties of the next generation if ninety-nine per cent of the males in the present one are selectively slaughtered or sterilized than if one per cent are so treated.

Recent advances in molecular genetics have raised the possibility of a different and far more effective method of genetic change which I shall call transformationist eugenics. At present, if we wish to eliminate an undesirable gene from a population, our only method of doing so is to reduce the breeding chances of those individuals carrying the gene; but now that we know something of the chemistry of heredity, it is possible to think of the direct alteration or transformation of particular genes. To-

Reproduced by permission of the publisher and John Maynard Smith from Daedalus, *Journal of the American Academy of Arts and Sciences, 94(2):487–505 (Spring 1965).*

John Maynard Smith, *a zoologist, was born in London in 1920. He received his B.A. in 1940 from Cambridge University and his B.S. in 1949 from University College, London. At the present time he is Dean of the School of Biological Sciences, University of Sussex. He has done research on the genetics, physiology, and behavior of Drosophila and studies in the field of animal movement. His interest in evolution is directed towards the practical problems of changing animal populations and the social consequences of the genetic differences in man. His books include* Theory of Evolution *(London, 1958) and* Mathematical Ideas in Biology *(Cambridge, 1969).*

day, this can be done only in micro-organisms, and then in only a very small proportion of cells exposed to the transforming agent. But it will be surprising if direct gene transformation does not become possible in man and higher animals during the next hundred years. If so, it will increase by many orders of magnitude the speed and economy with which the genetic properties of populations can be changed.

Finally, the continued development of surgical techniques, together with chemical methods of altering development and with tissue and organ culture, will make it possible to produce quite profound alterations in the biological properties of individuals without altering their genetic constitution. This I shall call biological engineering.

Before discussing in greater detail the technical and ethical considerations raised by these methods, there is one general point to be made: man is evolving anyway. That is to say, changes are taking place in the genetic properties of the human population whether we like it or not; and almost every political and social measure we take influences to some degree the nature and direction of these changes.[1] Probably the most important changes at the present time are due not to selective survival or fertility but to changes in the breeding structure consequent on increased population size and increased mobility. At one extreme is the reduction in the frequency of marriages between close relatives, which is likely at least in the short term (that is, for some hundreds of years) to have beneficial effects by reducing the frequency of diseases caused by recessive genes; at the other extreme is the increase in the frequency of inter-racial marriage, although social pressures have both minimized this change and rendered it almost impossible to evaluate its consequences accurately.

But in the literature of eugenics, more attention has been paid to selective effects, and in particular to the differential fertility of social classes and the consequences of improved medical care. The importance of the former subject has probably been exaggerated—the observed differences may not last long enough to have significant evolutionary consequences—but it is comforting in any case that measurements of I.Q. in 1932 and 1947 in Scotland showed a slight but significant rise on the second occasion although the negative correlation between the I.Q. score of children and the size of the family to which they belonged had led to the prediction that the mean I.Q. of the population must decline. Even if, as seems likely, the observed rise was the result of particular environmental rather than genetic factors, at least no measurable genetic decline has occurred.[2]

The effects of improved medical services should perhaps be taken more seriously because these are likely to be long-lasting. The argument is as follows. Improved medical and social care make it possible for people who in the past would have died to survive and have children. In so far as their defects were genetically determined, they are likely to be handed on to their children. Consequently, the frequency of genetically determined defects in the population is likely to increase; and an increasingly large proportion of the population will be engaged in keeping the rest alive. I think we have to accept the fact that there is some truth in this argument, but it is a little difficult to see what we should do about it. To ban the manufacture of glasses and the administration of insulin because these activities permit astigmatics and diabetics to breed seems inhumane. If we are going to administer insulin, there is no rational ground for refusing to undertake the more expensive and time-consuming job of feeding babies who suffer from a genetically determined form of mental defect known as phenylketonuria on a diet free of phenylalanine. And as our knowledge increases, the number of tasks of this kind which we shall feel obliged to undertake will increase. An extreme eugenist might suggest that although we should not ban the administration of insulin we should insist on the sterilization of those who require it. The difficulty with such a policy is that almost every human being possesses at least some characteristics, physical, moral, or intellectual, which we would prefer not to be transmitted to the next generation. The only humane answer at present appears to be that an increasing investment in medical and social care is a price we should be prepared to pay.

But there are two mitigating circumstances. First, as I shall discuss later, there may be a long-term way out of this dilemma. Second, not all the genetic consequences of improved medical services are dysgenic, and it may even be that most are not. To see why this is so, consider the case of malaria and sickle cell anemia. Briefly, there exists in man a gene S, which in homozygous* condition causes fatal anemia. Yet this gene occurs in high frequency in certain places, particularly in parts of Africa where malaria is or was until recently a common cause of death. The reason for this distribution is that individuals heterozygous for the gene S are resis-

---

*An individual who inherits the same gene, defective or otherwise, from both parents is said to be homozygous; an individual inheriting different genes from his two parents is said to be heterozygous.

tant to malaria. The frequency of S in Negroes of West African origin now resident in America is lower than those in West Africa, presumably because in the absence of malaria natural selection against homozygotes has not been balanced by selection in favor of heterozygotes; a similar decrease in frequency of S is likely to occur in Africa as malaria is brought under control. This is an evolutionary change consequent on improved medical care (that is, the eradication of malaria), but it is a desirable change since it has reduced the number of children dying of anemia.

The case of sickle cell anemia and malaria is rather a special one, but beneficial evolutionary consequences of improved medical care may be quite common. The greatest change so far produced by medicine (more particularly by preventive medicine) on the pattern of mortality is the reduction in the number of people dying of infectious disease. This has led to a reduction in the selection pressure in favor of disease-resistance and presumably to an increase in susceptibility. As long as we continue to control infectious disease by improved hygiene, inoculation, and so forth, this is probably a good thing. The reason for this is that in evolution, as in other fields, one seldom gets something for nothing. Genes which confer disease resistance are likely to have harmful effects in other ways; this is certainly true of the gene for sickle cell anemia and may be a general rule. If so, absence of selection in favor of disease resistance may be eugenic.

Now that death from infectious disease is rare in industrial countries, the main efforts in medical research are concentrated on diseases such as cancer and rheumatism, which usually affect older people. Cures for cancer (other than leukemia) would not have significant genetic consequences because, although they would prolong the life of many people, they would seldom add to the number of children born to such people. The main dysgenic effects of medical progress arise from cures of defects which are present at birth or appear before reproductive age, and in whose causation there is a genetic component.

So far, I have considered genetic changes which are happening or are likely to happen as unintended by-products of measures undertaken with other ends in view. I now turn to the possible effects of intentional eugenic measures, first considering measures of selective eugenics which either are already technically possible or are fairly certain to become possible in the near future. Such eugenic measures can be "negative," that is, concerned with the elimination or reduction in frequency of undesirable traits, or "positive," that is, concerned with improving the average performance, or the proportion of individuals capable of outstanding performance, of socially desirable tasks.

The probable effectiveness of negative eugenic measures depends first on whether the characteristic in question is genetically determined and, if so, whether it is caused by a dominant* or a recessive gene. For characters caused by dominant genes, negative eugenic measures could be effective but are usually pointless; for characters caused by recessive genes, they are ineffective.[3]

Consider first a character, Achondroplasia (dwarfism with short legs and arms), which is caused by a dominant gene. Since the gene is dominant, we can recognize all carriers of the gene. If we so wished, we could by sterilizing all achondroplastic dwarfs prevent any carriers from passing the gene on to the next generation. This would not completely eliminate the character from the population but would reduce its frequency to the number of cases arising by new mutation in each generation. But why should we wish to do this, since there is little to prevent an achondroplastic dwarf from leading a contented and useful life? If a dominant mutation is lethal or seriously disabling, selection will keep it at a frequency close to the mutation rate without our intervention; if it is not disabling, why should we interfere? Unhappily, there are exceptions to this easy excuse for inactivity. An example is Huntingdon's chorea, which is due to a dominant gene which does not manifest itself until middle life, after the affected person has had children. The condition differs from Achondroplasia in being fatal and very distressing for the sufferer. Additional distress arises if relatives of affected persons foresee, correctly, that they may develop the disease. In such a case, there are good grounds for discouraging any person with even one parent or sibling who has developed the disease from having children. I am satisfied that such people should be encouraged to undergo sterilization but doubt that sterilization should be compulsory; the case for compulsory sterilization will be stronger when we learn to recognize heterozygotes before the disease develops.[4]

Now consider a disease such as phenylketonuria, which is due to a gene which is recessive in the sense that only homozygotes suffer from the associated mental defect.[5] Approximately one person in 60,000 in Britain is homozygous for the gene; it follows that one person in 122 is a carrier.[6] Consequently, if we could not recognize the heterozygotes, then the sterilization of homozygotes (in fact, untreated homozygotes normally do not get married or have children) would remove only 1/245 of the mutant genes per generation. Eugenic measures would therefore be ineffective unless applied to heterozygotes, who can in this case be recognized biochemically although they are not mentally defective. But it seems likely that most people are carriers of at least one lethal or deleterious gene, although they cannot at present be recognized as such. It follows

---

*A dominant gene is one which produces an effect in single dose, being inherited from one parent only; a recessive gene has no effect unless inherited from both parents, or at least has no serious effect.

that as our ability to recognize heterozygotes increases, we could be led to sterilize almost the whole population on eugenic grounds, which is clearly absurd.

The ability to recognize heterozygotes for such conditions as phenylketonuria makes it possible in principle to eliminate the affected homozygotes by preventing marriage between heterozygotes. (The statement in the previous paragraph that almost everyone is heterozygous for something does not invalidate this conclusion: all that has to be avoided is marriage between two people heterozygous for the *same* gene; and this would rule out only a small fraction of possible marriages.) It is admittedly difficult to see how this can be achieved, but a start might be made by testing relatives of affected persons and partners in prospective marriages between cousins. There is also a sense in which such a measure would be dysgenic. By preventing the birth of affected individuals, it would remove any selection against the mutant gene; and this would lead to an increase in its frequency in the population.[7] To this extent, preventing marriages between heterozygotes could have dysgenic effects similar to the cure of a genetically determined disease.

In earlier discussions of eugenics, suggested measures of positive eugenics took the form of legislation designed to encourage particular classes of persons to have more children. Two examples of such suggestions are increased family allowances for university teachers and a tax on children, the logic behind the latter suggestion being that only the rich would be able to afford children and that wealth is at least an approximate measure of genetic worth.[8] But in recent years increasing attention has been paid to the possibility of artificial insemination. H. J. Muller[9] and, more recently, Julian Huxley[10] have suggested that we should try to persuade married women who have had one child by their husbands to have a second child by a donor of their choice. In view of the sources from which it emanates, if for no other reason, this suggestion merits careful examination.

First, how effective would such a measure be? I shall discuss the effects on a single metrical character; to be concrete I shall consider I.Q. score, since intelligence is the quality most usually prized by people in academic circles who propose eugenic measures. I shall make the following assumptions, which appear to be optimistic:

1) Among women, one per cent could be persuaded on eugenic grounds to have half their children by artificial insemination.
2) The husbands of such women would be a random sample of the population.
3) The mean I.Q. of the donors chosen would be one standard deviation above the population average. (Without intending to be either face-

tious or offensive, it is fair to ask what would be the relative popularities of Francis Crick and Ringo Starr.)

4) The realized heritability* of I.Q. scores is 0.5.

Given these assumptions, the mean I.Q. score of the next generation (allowing thirty years per generation) would be approximately 0.04 points higher than it would otherwise be. Compared with the rise of approximately two points observed in fifteen years in Scotland—which probably resulted from such things as the spread of radio and televison sets—I doubt whether such a rise would be worth the trouble.

Artificial insemination would be less effective in man than in domestic animals because a number of conditions can be satisfied in the latter case but not in the former. These conditions are:

1) It is possible to define the objective of selection—for example, growth rate or milk yield—and to accept deterioration in other characteristics—for example, mobility or intelligence.

2) It is possible to choose the male donors on the basis of this objective and to use progeny testing to ensure that the donors pass the appropriate characters on to their children.

3) It is possible to ensure that most females have most of their offspring by artificial insemination.

I assumed above that only one per cent of women could be persuaded to accept artificial insemination on eugenic grounds. Clearly, the effectiveness of the procedure would be increased if the proportion of women participating were greater. There have been societies in the past in which a large proportion of the women have been persuaded or coerced into a breeding system which had genetic consequences similar to the scheme suggested by Muller and Huxley. For example, among the Nambikuave Indians of central Brazil, a chieftain, nominated by his predecessor but dependent on popular consent, is the only member of the group to have a number of wives.[11] Although this practice does not seem to have been undertaken for genetic reasons, it cannot fail to have genetic consequences.

I do not believe that a larger proportion of the world's population will ever adopt such a system, using either artificial or natural insemination, but this belief may only reveal my prejudices. But it does seem possible that a small racial or religious group might adopt such a practice. If such a group could maintain a fair degree of genetic isolation from the rest of the population and if the great majority of women in the group bore at

---

*The realized heritability is defined as the progress under selection divided by the selection differential; that is to say, in the case of I.Q. score, it is the increase in the mean I.Q. score in a generation divided by the difference between the mean I.Q. of the selected parents and that of the population from which they were selected. The realized heritability normally lies between 0 and 1, and the value of 0.5 is fairly typical for a metrical character.

least one child by a donor of high I.Q. (the argument, of course, will apply to any character), then after a century the mean I.Q. of the group might have risen by one standard deviation, or fifteen points. In other words, a group might arise with an average intelligence similar to that of a group of students selected for a university. This seems hardly sufficient to justify the establishment of a new religion.

But what if artificial selection were continued not for a century but for a millennium? It is unlikely that the mean I.Q. would rise by ten standard deviations. Experience shows that if intense artificial selection for a single character is continued for a number of generations, the genetic response, although rapid at first, tends to slow down and even to stop. It is impossible to predict at what level this "plateau" will be reached.[12] But it seems quite likely that if a human community were to practice artificial selection for intelligence for a thousand years, there would be a rise of several standard deviations in the mean I.Q., and the community might contain several individuals with mental capacities greater than those of anyone alive today.

But as an estimate of what would happen if, for example, a number of groups such as the American Academy of Arts and Sciences were to campaign for artificial insemination on eugenic grounds, a rise of 0.04 points per generation seems optimistic. Nevertheless, it has been argued that artificial insemination is valuable in man because a small rise in mean score would produce a disproportionate increase in the number of people with exceptionally high scores. Thus, if it is assumed that I.Q. score is normally distributed and that a small change in mean I.Q. score does not alter the variance of the score (this need not be true, but it might very well be true),[13] then an increase of one point in mean I.Q. would be accompanied by an increase of twenty per cent in the proportion of people with I.Q.'s above 175. It is argued that although an increase of one point in the mean I.Q. might not be worth bothering about, an increase of twenty per cent in the number of geniuses is well worth striving for.

The argument is weakened by the fact that I.Q. score is not normally distributed; there are many more people with very high and very low scores than would be predicted on the assumption of normality.[14] Thus, if the distribution were normal, an increase of mean I.Q. of one point would lead to an increase in the proportion of people with a score greater than 175 from 3.3 per million to 4 per million. But since the distribution is not normal, the actual increase would be approximately from 77 per million to 85 per million.

But the main weakness of the argument lies in the assumption that an increase in the proportion of people with I.Q. scores above 175 would necessarily, or even probably, be associated with an increase in the number of people of outstanding ability as judged by their achievements. If this were so, it is difficult to explain why some quite small populations, for example Periclean Athens, should in a short time have produced such a number of people who, judged by their achievements, were of outstanding ability, whereas other larger populations, such as Greece during the Byzantine empire, should have produced hardly any. This is not to imply that outstanding achievements do not require unusual genetic endowments or that anyone could have written the *Principia* if he had had Newton's opportunities. What is suggested by a comparison of Greece in Classical and in Byzantine times is that any reasonably large population is likely to contain people genetically capable of outstanding achievements if social conditions are favorable. The same point is made perhaps more convincingly by referring to the frequent occurrence of simultaneous yet independent discoveries in science. It follows that a small increase in the proportion of people with I.Q. scores above 175 is unlikely to be important.

So far, I have accepted the four assumptions listed above as reasonable approximations. But one of the assumptions—that the husbands of women accepting artificial insemination would be a random sample—is manifestly false for two reasons. First, women accepting artificial insemination on eugenic grounds would hardly be a random sample; and, since mating in man, for intellectual and moral characteristics at least, is not random, their husbands would be likely to resemble them. Second, if the husbands agreed—and the results if they did not would hardly be desirable—they would presumably be above average in humility and unselfishness. It is at least possible to argue that these qualities are more desirable socially than the qualities for which the donors would be chosen. If so, the measure, in so far as it had any effects, would be likely to be dysgenic.

This raises the major difficulty of all suggested measures of positive eugenics, the problem of deciding what we want. It is fairly easy to recognize characteristics—blindness, mental defect, lameness—which we would wish to avoid in our own and in other peoples' children but much more difficult to define characters we wish to encourage, particularly when it is remembered that these characters may be mutually incompatible. Most experience with artificial selection in animals leads to the conclusion that selection in favor of a particular character—for instance, milk yield in cattle or the number of bristles in Drosophila—is effective in altering the selected characters in the desired direction; but the alteration is accompanied by changes in many other characters, changes whose nature cannot be predicted in detail but which are usually undesirable in that they lower fertility or the probability of survival, or impair performance in other ways.[15] This is only a restatement of the point made above when discussing disease resistance, that you rarely get something for nothing. It is a point usually forgotten in discussions of eugenics.

Two other points should be made concerning the problem of deciding what characteristics are desirable. First, it is probable that in man at least some desirable characteristics arise in genetic heterozygotes; if so, it is unreasonable to expect them to breed true. Second, it is far from clear that what we want is a genetically uniform population; indeed, societies seem much more likely to be workable if they contain individuals with a wide range of genetic capabilities.[16]

If our objective is to increase the proportion of genetically gifted people in the population, there is a method which is likely to become feasible in the fairly near future and which would be considerably more effective than artificial insemination. This is to make clonal "copies" of successful people. It has already proved possible to remove the nucleus from a fertilized frog egg and to replace it with the nucleus from one of the cells of a developing embryo; the egg then develops into a frog having the genetic characteristics of the embryo from which the nucleus was taken.[17] It will perhaps soon be possible to remove a fertilized or unfertilized human egg from the oviduct, remove the nucleus, and repace it with a nucleus from, let us say, a germ-line cell of some individual whose genotype we would like to reproduce. Implanted in a uterus, this egg would then develop the same genetically determined characteristics as those of the individual from which the nucleus was taken.

Leaving aside for the moment the desirability of such a "cloning" technique, let us turn to why it would be more effective than artificial insemination. In artificial insemination, only half the genes of the donor are transmitted. Therefore, their effects may be "diluted out" by the genes of the mother; and if the peculiar and desired characteristics of the donor depended on interactions between genes, these are likely to be lost. But in the cloning technique, an exact genetic replica, as in monovular twins, would be obtained.

How strong are the arguments for adopting this measure, supposing that it does become practicable? I do not want to be dogmatic on this point, but two arguments against it should be mentioned. First, the arguments outlined above for believing that human populations have an adequate supply of talented people to meet the problems of the time would, if accepted, show that there is little to be gained by adopting the cloning technique. Second, people "conceived" in this way could have severe and perhaps crippling psychological difficulties. Sons of famous fathers not infrequently suffer because too much is expected of them; much more might be expected of children known to be genetically identical to a famous "ancestor."

I now turn from selectionist to transformationist eugenics, from what we can do to what we may be able to do in the future. I want again to consider the case of sickle cell anemia, although there is a risk that this condition may come to play the same distorting role in evolutionary speculation today that the neck of the giraffe did in the last century. It is known that a person homozygous for the gene S differs from normal people because the hemoglobin in their red blood cells is insoluble and that this difference is due to the substitution of the amino acid valine for glutamic acid at a particular position in the β chain of their hemoglobin. It is reasonably certain that this abnormality is due to the presence of a single abnormal base in a DNA molecule in the chromosomes of blood-forming cells and that this, in turn, is due to the presence of a single abnormal base in a DNA molecule in the fertilized egg from which they developed (strictly speaking, there must have been four abnormal bases, since there were two homologous sets of chromosomes in the egg, each containing an abnormal base pair). When the details of the genetic code have been discovered, which is likely to be soon, it may be possible to specify which base has been substituted for which—for example, that adenine has replaced cytosine at a particular place.

People heterozygous for the gene S can be recognized, since their red blood cells contain about forty per cent of the insoluble hemoglobin and about sixty per cent of normal. A baby suffering from sickle cell anemia will be born only if two heterozygotes marry (except for new mutation or illegitimacy). As was pointed out earlier, the birth of anemic babies could be avoided by preventing the marriage of heterozygotes. It could also be prevented if it proved possible to transform a single base—say adenine to cytosine—in the sperm cells of the father, or in the oocytes of the mother, or in the fertilized egg. This would be an example of negative transformationist eugenics. It would have the immediate effect of preventing the birth of defective children without making it necessary to interfere with the choice of marriage partners and without having the dysgenic effect of causing a gradual increase in the frequency of deleterious genes.

Of the various methods of eugenics which have been or will be discussed in the essay, there seems little doubt that negative transformationist eugenics would be the most desirable. It would require the minimum interference with who marries or has children by whom; its effects would be confined to the limited and generally acceptable objective of preventing the birth of children with specific defects; and, far from having dysgenic effects, transformationist eugenics could provide a means of counteracting the long-term dysgenic effects of some types of medical care.

The drawback, of course, is that such methods are not at present practicable; and it is not yet possible to see how such transformation could be achieved. The major difficulty is the restricted nature of the transformation required. Thus a chemical procedure which transformed all or many of the adenine molecules in a nucleus into cytosine would certainly be fatal; instead, only one particular adenine molecule among the hundreds of thousands present must be transformed. Because of two properties of nucleic acid, namely, homologous pairing and recombination, the problem is not quite so hopeless as it sounds. The first of these two properties makes the following situation possible: if a normal "gene" (DNA molecule) for hemoglobin could be introduced into a cell carrying the mutant S gene, this normal molecule might pair base by base with the abnormal one. The second property raises the possibility that in some circumstances the normal molecule might replace the abnormal one in the chromosome. Something of this kind does occur in the phenomenon of bacterial transformation; unfortunately, it is now confined to bacterial cells, which are much readier to accept nucleic acid molecules than are animal cells, and even then it is possible to transform only a small proportion of the cells exposed to transforming nucleic acid. I find it impossible to say how much my conviction that transformation will become a practicable eugenic tool arises because the wish is father to the thought; but at least it seems rational for the next hundred years or so to attempt to cure or to make life possible for people with congenital diseases without worrying too much about the ultimate dysgenic effects.

Transformationist eugenics has its most obvious area of application in the negative field, in altering genes which give rise to obvious and gross deficiencies. It is possible to visualize positive application in animal breeding; if, for example, resistance to a particular disease, or ability to digest a particular food, could be shown to depend on the presence of a particular enzyme, then a gene determining that enzyme might be incorporated into the genotype of a domestic species. But the major application of transformationist genetics is likely to be in producing genetically changed micro-organisms designed to play particular roles in the manufacture of food and of other complicated chemical substances. It is more difficult to see positive applications to man. The production of individuals of outstanding intelligence will again be taken as an example, although even greater difficulties would arise if the characters chosen were, for instance, artistic ability or moral worth. The difficulty is simply stated: we do not know what changes in the egg's ability to synthesize specific proteins would lead to increased intelligence in the adult developing from the egg; therefore, even if we knew how to bring about specific gene transformations, we would not know what transformations to make. There is no reason to think that the problem is insoluble, but it would appear to be much further from solution than the problem of genetic transformation itself.

This brings me to the third technique available for the alteration of man's nature, that of biological engineering.[18] Here I have in mind the extension of existing medical techniques from the negative to the positive field. Today it is standard practice to attempt to cure many congenital defects by surgical or medical techniques, and there is no reason to doubt that treatment of congenital disease will be supplemented or replaced in the future by methods of treating the fetus so that the developmental process is altered and a normal child is born. But at this moment we do not use or contemplate using such techniques to produce outstanding individuals. For example, it would perhaps be technically possible through surgery to produce a man whose legs were so lengthened that he could run a mile in $3\frac{1}{2}$ minutes. But, sensibly enough, we prefer to let nature take its course and manufacture motor cars and airplanes if we want to move fast. But we do not hesitate to cure a lame child if we can.

It seems, then, that our present practice depends on a concept of normality, however difficult that may be to define. Since we are concerned here with a problem of what we ought to do rather than what is technically feasible, it is perhaps best to regard a characteristic as abnormal if it leads to a loss of function sufficient to cause its possessor to be unhappy. But we should ask also if there are circumstances in which we might wish to produce outstanding individuals. In the field of physical performance this seems unlikely, since it will always be easier to build a machine. There is, however, one exception: we cannot build machines to make us live longer. It is not at present possible to say whether we shall ever be able to produce a large increase in human life expectancy, even though we can already ensure that a larger proportion of people survive to old age. We do not at present know whether senescence is caused by a number of physiologically independent processes—in which case, even if we prevented one of these processes people would still die at much the same age of another—or whether there is one fundamental process of which the various superficial signs of senescence are merely symptoms. If the former assumption is correct, and the evidence suggests to me that it is, then a significant extension of the human life span is likely to prove very difficult.[19] It would also contribute disastrously to the present increase in world population. But should the world population problem prove soluble without war or famine, then an increase in human life span, if it could be associated with an appropriate decrease in human fertility, seems to me very desirable.

Olaf Stapledon, in his book *Last and First Men*, imagined the use of biological engineering to produce super-intelligences. Human neural tissue was permitted to grow and ramify through the corridors of a building and was supplied with sensory information and a motor output. In *Sirius* the same author imagined a dog whose intelligence, by surgical and other means, had been made equal to that of a man.

These feats are not at present technically possible, but there is no reason that it should not eventually be possible to bring about a dramatic increase in the size of certain parts of a human or animal brain by influencing development. It is, of course, by no means certain that such a simple procedure would lead to an equivalent increase in intelligence; it might equally well lead to idiocy. But there is one reason to suspect that an appropriate increase in size, together with other comparatively minor changes in structure, might lead to a large increase in intelligence. The evolution of modern man from non-tool-making ancestors has presumably been associated with and dependent on a large increase in intelligence, but has been completed in what is on an evolutionary scale a rather short time—at most a few million years. This suggests that the transformation in the brain which provided the required increase in intelligence may have been growth in size with relatively little increase in structural complexity—there was insufficient time for natural selection to do more. Of course, this process may have reached its limit, and further increase in intelligence may require a major reorganization of structure, which would be difficult to bring about by "engineering" methods.

On balance, it seems quite likely that within a hundred years or so it will be technically feasible to do the kinds of things imagined in Stapledon's books. But even if it is, it is not clear what the consequences would be. To ask oneself the consequence of building such an intelligence is a little like asking an Australopithecine what kind of questions Newton would ask himself and what answers he would give. One way of putting the problem is this: What questions could be asked or answered by a "super-intelligence" composed of neurons which could not be asked and answered by teams of investigators given time and the assistance of computers? It is quite possible that the answer to this question is "none." But I suspect that if our species survives, someone will try it and see.

The subjects discussed in this essay are diverse, so I will attempt to summarize my argument and draw some general conclusions. First, evolutionary changes are constantly occurring in the human species, and most legislative or social measures we take inevitably influence the nature of these changes. Some, but by no means all, of the genetic changes consequent upon improved medical and social services are dysgenic. At present, there is little that we can do to prevent these dysgenic effects, and the proper course for us to adopt is to regard them as part of the price we pay for being civilized. In any case these genetic changes are extremely slow in comparison with technical changes, and it is reasonable to hope that before they have become significant it may be possible to avert or reverse them by techniques of genetic transformation.

Deliberate measures to alter man's biological nature may be negative, designed to prevent or cure mental or physical defect, or positive, designed to produce individuals of unusually high performance in a desired area or to raise the mean level of performance. Techniques available can be classified as selectionist eugenics, transformationist eugenics, and biological engineering. Selectionist eugenics involves altering the relative number of offspring born to particular kinds of individuals or pairs. In most cases, these procedures are likely to be too ineffective to be worth bothering with. But it is worth making an effort to prevent individuals who carry deleterious dominant mutations which manifest themselves late in life from having children and to prevent the carriers of the same recessive lethal or deleterious gene from marrying one another, although the latter measure would have dysgenic effects in the long run. In the positive field, selectionist eugenics is again likely to be relatively ineffective. Probably the most effective procedure, and one which should become technically feasible in the fairly near future, would be some form of cloning.

Transformationist eugenics, involving direct alteration of specific genes in specific ways, is not at the moment possible, but may become so. If it does become possible, its use in negative eugenics would be desirable; but it is less clear what role it could play in positive eugenics.

Biological engineering in the negative field is simply another word for current medical practice. Problems, both technical and ethical, arise in the use of similar techniques to produce individuals of outstanding ability rather than to cure or prevent abnormality. Two major undertakings can be considered. One, a significant increase in the human life span, although dangerous unless the world population problem has been solved, will in the long run be desirable, but it is likely to prove very difficult and perhaps impossible. The other, the production of individuals generally resembling human beings but of outstanding intelligence, may prove relatively easy, although there is no guarantee that this is so; but even if it is technically feasible, it does not seem possible to predict what important results, if any, would ensue.

But these problems of transformationist eugenics, increase of longevity, and super-intelligence still lie in a future which is distant in historical terms even if it is immediate on an evolutionary time scale. Our immediate problem is what should be done with the means now available to us, and, more immediate still, what should geneticists and other biologists recommend be done.

I think the answer to this question is that we should not recommend that anything be done except the simple and limited negative measures suggested above. The reason for this is that I believe recommendations of positive eugenic measures can at the present only distract attention from more urgent and important questions. The most urgent message which biologists have to convey to the public is that if something is not done to arrest the present increase in world population, then that increase will be arrested by war, disease, and starvation. Eugenics can wait, birth control cannot.

REFERENCES

1. For a discussion of this point, see P. B. Medawar, *The Future of Man* (London: Methuen, 1959).
2. The evidence for a negative correlation between family size and I.Q. is summarized by L. S. Penrose, "Evidence of Heterosis in Man," *Proc. Roy. Soc.*, B. 144 (1955), p. 203. Penrose puts forward a genetic hypothesis which would account for this correlation and yet predict no change in I.Q. with time. His views have been criticized by P. B. Medawar, *op. cit.*, and by K. Mather, "Genetical Demography," *Proc. Roy. Soc.*, 159 (1963), p. 106. For a comparison of the 1932 and 1947 surveys in Scotland, see G. H. Thomson, *The Trend of Scottish Intelligence* (London: University of London Press, 1949).
3. The argument which follows is presented in J. B. S. Haldane, *Heredity and Politics* (London: Allen and Unwin, 1938).
4. Another genetically determined abnormality which, although not caused by a dominant gene, could be reduced in frequency by negative eugenic measures is translocation mongolism. Individuals of either sex heterozygous for a translocation involving chromosome 21 are themselves normal, but one third of their children will be mongolian idiots and one third will be "carriers"; only one third will be normal and likely to have normal children. Such people could be recognized if the chromosomes in a skin or blood sample were examined, and most of them would be found if all relatives of known mongols were examined. But the arguments for sterilization are perhaps less strong than in the case of Huntingdon's chorea because Mongolian idiots are commonly quite cheerful and contented. In any case, sterilization would not prevent the more common form of mongolism which is due to nondisjunction in the mother.
5. It has recently been found that a homozygous baby, if recognized at birth and subsequently kept on a diet free of phenylalanine, can develop normal intelligence. This does not affect the argument concerning the ineffectiveness of sterilization in the case of diseases caused by recessive genes, but it does illustrate the important point that genetically determined diseases may be curable.
6. This frequency omits cases known to have consanguineous parents. The frequency of carriers has been worked out from the Hardy-Weinberg ratio, assuming random mating.
7. It could be argued that this would not be dysgenic provided that marriage between heterozygotes was prevented. However, it is unlikely that we should succeed in preventing all such marriages. Also, the proportion of marriages contra-indicated on genetic grounds would increase. But it seems likely that long before these effects become serious some technique of negative transformationist eugenics will be available.

8. The idea that financial measures might be used for eugenic purposes was put forward by R. A. Fisher, *The Genetical Theory of Natural Selection* (London: Oxford University Press, 1930). The suggestion of a tax on children was made, perhaps not very seriously, by F. H. C. Crick in Gordon Wolstenholme (ed.), *Man and his Future* (London: Churchill, 1963), p. 276. The suggestion has the virtue of bringing out the necessary contradiction between financial measures suggested on eugenic grounds and those suggested by the humanitarian desire to protect children from the incompetence of their parents. On the same occasion, Crick made the more important point that the time has come to question our present assumption that people have a right to have children.

9. H. J. MULLER, *Out of the Night* (New York: Vanguard Press, 1935); and "Genetical Progress by Voluntarily Conducted Germinal Choice," in Gordon Wolstenholme (ed.), *op. cit.*

10. J. S. HUXLEY, *Eugenics in Evolutionary Perspective* (London: Eugenics Society, 1962).

11. C. LEVI-STRAUSS, *A World on the Wane* (London: Hutchinson, 1961).

12. For the occurrence of "plateaus" in selection experiments, see, for example, K. Mather and B. J. Harrison, "The Manifold Effects of Selection," *Heredity*, Vol. 3 (1949), p. 131; and I. M. Lerner, *Genetic Homeostasis* (New York: John Wiley & Sons, 1954).

13. A small change in mean will not alter the variance if the effects of a different gene on I.Q. are additive and if the frequencies of alleles for high and low intelligence are on the average equal.

14. C. BURT, "Is Intelligence Distributed Normally?" *Br. J. Statist. Psychol.*, Vol. 16 (1963), p. 175.

15. The occurrence of such correlated changes is not in doubt, although their explanation is still a matter of controversy; the subject is discussed in the references given under 12 above.

16. These two points are too important to be dismissed in a brief paragraph; my excuse is that they have been discussed more fully by T. Dobzhansky, *Mankind Evolving* (New Haven: Yale University Press, 1962).

17. T. J. KING and R. BRIGGS, "Serial transplantation of embryonic nuclei," *C.S.H. Symp. Quart. Biol.*, Vol. 27 (1956), p. 271.

18. Some possible developments of biological engineering, or "euphenics," were discussed by J. Lederberg, "Biological Future of Man," in Gordon Wolstenholme (ed.), *op. cit.*

19. For a discussion of this point, see G. C. Williams, "Pleiotrophy, natural selection and the evolution of senescence," *Evolution*, Vol. 11 (1957), p. 398; and J. Maynard Smith, "The Causes of Ageing," *Proc. Roy. Soc.*, B. 157 (1962), p. 115.

Glossary

Allele (allelomorph). The alternate form of a gene at a single gene locus.

Amino acid. The smaller constituent molecule of a protein. There are about twenty different amino acids which are combined in different numbers and arrangements, the average protein containing about 500 amino acids.

Anthropometry. The measurement of human dimensions to obtain the quantitative expression of body form.

Base. The nitrogenous component of a nucleotide. There are four bases of two general types—thymine and cytosine (pyrimidines), and adenine and guanine (purines).

Cephalic index. The index derived from the anthropometric measurement of the length and breadth of the head (B/L × 100).

Consanguineous mating. A mating between biologically related individuals, that is, "relatives."

Cytology. The scientific study of cells.

Demography. The statistical description and analysis of human populations, using the quantitative data of the numbers and distribution of people and their changes through time, the age and sex composition, the birth and death rates, and the mathematical interrelations among these quantities.

Diploid number. The chromosome number (2N) characteristic of the zygote of a particular species. In man the number is 46.

Dominant. The ability of a gene to express itself phenotypically in the presence of a different allele.

DNA (deoxyribonucleic acid). The molecule of a chromosome composed of a pair of DNA chains in a helix. It is composed of genes, probably linearly arranged, and through the intermediary of RNA (ribonucleic acid), it provides a template for the specification of the proteins being synthesized.

Enzyme. A catalyst; that is, a substance which, in small quantity, will make a chemical reaction proceed much more quickly than it would otherwise, yet without itself being permanently altered in the process. Every cellular enzyme is a protein.

Gamete. The mature sex or germ cell—the egg or ovum of the female and the sperm or spermatozoon of the male.

Gene. That interrupted portion of DNA which codes for a particular polypeptide chain of the protein. Depending on the number of polypeptide chains within a given protein, one gene may or may not code for one protein.

Genotype. The combination of genes that an individual possesses, as inferred from his phenotype.

Hemoglobin. The oxygen-carrying pigment of the human blood. It is a conjugated protein consisting of the protein globin and the associated iron compound heme.

Heterozygote. An individual who carries two different alleles at the gene locus or loci in question, for example *Hb-A/Hb-S*.

477

Homogamy. Positive assortative mating; that is, preference for a mate exhibiting similar characteristics—social, personality, physical, and the like.

Homozygote. An individual who carries the same alleles at the gene locus or loci on each member of the chromosome pair, for example Hb-S/Hb-S.

Hyperendemic, holoendemic disease. A disease present in a population throughout the year, infecting large numbers or all of susceptible people, respectively.

Inbreeding. The mating system composed of the total number of consanguineous matings in a population.

Independent segregation. The meiotic process by which each duplicated homologous chromosome is assigned to a separate germ cell. In this way chromosomes segregate themselves from each other, uninfluenced by the fact that they have resided together in the same nucleus.

Isogene. A line drawn through points (populations) on a map where the gene frequency under consideration is the same.

Karyotype. The chromosomal constitution of the nucleus of a cell.

Leptokurtic curve. A distribution curve which, relative to a normal curve with the same mean and variance, has more items near the mean and at the tails and fewer items in the intermediate regions.

Lethal equivalent. Either a single gene that is fully lethal in a homozygote, or a group of genes each of which, when homozygous, may cause death in a fraction of cases or, if separately homozygous in different individuals, would account for the average of one death.

Locus. The site of a gene on the chromosome.

Meiosis. Reduction division in the process of germ-cell formation so that the number of chromosomes is halved.

Mitosis. The process of replication and distribution of chromosomes to daughter cells.

Nucleotide. The monomer out of which a DNA chain is composed. Each nucleotide is a compound of phosphoric acid, the 5-carbon sugar deoxyribose, and one of the four nitrogenous bases (adenine, guanine, cytosine, uracil).

Phenotype. An individual's actual appearance which is the product of his genotype and the environment he has experienced.

Polygenic character. A phenotypic character determined by the cumulative effect of many genes. The variation of such characters in any population is quantitative rather than qualitative. These characters are studied by statistical measurement on some continuous scale.

Polypeptide. A compound containing two or more amino acids linked by a peptide bond.

PTC taste sensitivity. The ability, genetically determined, of some people ("tasters") to taste very dilute solutions of phenylthiocarbamide (PTC) and related substances. Others ("nontasters") are incapable of doing so.

Rad. The unit of radiation equal to the absorption of 100 ergs (an erg is a unit of energy) per gram of tissue.

Recessive. The inability of a gene to express itself phenotypically in the presence of a different allele.

Recombination. The process during sexual reproduction whereby chromosomes (and therefore genes) from different familial lines of descent are combined randomly in the zygote.

Somatic cell. A body cell as opposed to the germ cell, or gamete.

Subject Index

479

490

Index of Proper Names